ACS SYMPOSIUM SERIES **625**

Archaeological Chemistry

Organic, Inorganic, and Biochemical Analysis

Mary Virginia Orna, EDITOR

College of New Rochelle

Developed from a symposium sponsored
by the Division of the History of Chemistry,
the Division of Chemical Education, Inc.,
the Division of Analytical Chemistry,
and the ACS Committees on Education and on Science
at the 209th National Meeting
of the American Chemical Society,
Anaheim, California,
April 2–6, 1995

American Chemical Society, Washington, DC 1996

Library of Congress Cataloging-in-Publication Data

Archaeological chemistry: organic, inorganic, and biochemical analysis / Mary Virginia Orna, editor.

p.   cm.—(ACS symposium series, ISSN 0097–6156; 625)

"Developed from a symposium sponsored by the Division of the History of Chemistry, the Division of Chemical Education, Inc., and the Division of Analytical Chemistry, at the 209th National Meeting of the American Chemical Society, Anaheim, California, April 2–6, 1995."

Includes bibliographical references and indexes.

ISBN 0–8412–3395–0

1. Archaeological chemistry—Congresses.

I. Orna, Mary Virginia. II. American Chemical Society. Division of the History of Chemistry. III. American Chemical Society. Division of Chemical Education. IV. American Chemical Society. Division of Analytical Chemistry. V. American Chemical Society. Meeting (209th: 1995: Anaheim, Calif.) VI. Series.

CC79.C5A73   1996
930.1—dc20   96–4812
                                              CIP

Foreword

THE ACS SYMPOSIUM SERIES was first published in 1974 to provide a mechanism for publishing symposia quickly in book form. The purpose of this series is to publish comprehensive books developed from symposia, which are usually "snapshots in time" of the current research being done on a topic, plus some review material on the topic. For this reason, it is necessary that the papers be published as quickly as possible.

Before a symposium-based book is put under contract, the proposed table of contents is reviewed for appropriateness to the topic and for comprehensiveness of the collection. Some papers are excluded at this point, and others are added to round out the scope of the volume. In addition, a draft of each paper is peer-reviewed prior to final acceptance or rejection. This anonymous review process is supervised by the organizer(s) of the symposium, who become the editor(s) of the book. The authors then revise their papers according to the recommendations of both the reviewers and the editors, prepare camera-ready copy, and submit the final papers to the editors, who check that all necessary revisions have been made.

As a rule, only original research papers and original review papers are included in the volumes. Verbatim reproductions of previously published papers are not accepted.

Contents

INDEXES

Preface

TRACING CULTURAL EVOLUTION OVER CENTURIES and even millennia is the exciting task shared by archaeologists and archaeological chemists. Seldom financially rewarding, but of perennial interest, it has been the subject of nine major symposia at national meetings of the American Chemical Society. These symposia have attracted practicing archaeologists, chemists, biochemists, cultural anthropologists, and members of related disciplines from all over the world. The proceedings of five of these symposia have been collected in volumes published by the American Chemical Society (ACS), four in the Advances in Chemistry Series and the present volume in the Symposium Series.

New methods in analytical chemistry, particularly methods coupled to one another in tandem, have rendered biochemical samples almost as accessible on the ultratrace level as inorganic materials have been over the past few decades. Exciting new discoveries in the field, and the growth of the science fiction that often accompanies such discoveries, have contributed to a burgeoning interest in the biochemical aspects of archaeological chemistry. In this hemisphere, interest has centered on what chemical analysis can tell us about pre-Columbian civilizations. Although no one is claiming that "Jurassic Park—The Reality" is right around the corner, imaginations have been fired by the popularization of DNA reconstruction and replication. These developments have given rise to questions based on new possibilities in archaeological research.

Plans for the present symposium were initiated in 1992–93 to create a forum for presentation and discussion of these new possibilities in research. Special invitations were sent to some of the foremost workers in the field, and a call for papers was issued to every major venue of archaeological research worldwide. The resulting program contained papers from every major new area of archaeological research with an emphasis on the pre-Columbian and biochemical aspects. This symposium volume is a representative compilation of these papers. As such, it will be of interest to practicing archaeologists, archaeological chemists, anthropologists, historians of science, chemical educators, and all those interested in the story of how chemistry, not without some controversy, can help to trace the roots of humankind and the human environment through the millennia.

I sincerely thank each author for the time, effort, and cooperation it took to prepare this material for publication.

Mary Virginia Orna
Department of Chemistry
College of New Rochelle
New Rochelle, NY 10805

October 15, 1995

Dedication

This volume is dedicated to Lucy Roy Sibley, whose life and work exemplified that of a true scholar, mentor, teacher, and colleague. Her contributions to the field of archaeological textiles continue to guide us toward the future as we strive to understand the lifeways of the past.

Chapter 1

New Directions in Archaeological Chemistry

Mary Virginia Orna[1] and Joseph B. Lambert[2]

[1]Department of Chemistry, College of New Rochelle,
New Rochelle, NY 10805
[2]Department of Chemistry, Northwestern University,
2145 Sheridan Road, Evanston, IL 60208-3113

This recent symposium on archaeological chemistry emphasized studies in new areas of interest to archaeologists and archaeological chemists. Not only are the traditional areas of metal, glass, pottery, and stone treated, but also archaeological soils, fibers, dyes, bone, connective tissue, DNA, and organic residues.

Shedding light on the past by means of scientific examination received great impetus when major museums began to establish laboratories for that purpose on their premises. For example, the work of Alexander Scott in the 1920's gave rise to the world-renowned laboratories of the British Museum. While museums were mainly concerned with examination of their own holdings, many university laboratories in departments of archaeology, anthropology and chemistry found ample work by examination of materials from excavations worldwide. Recent research in these laboratories has concentrated on ancient metals, stone, pottery and glass, as evidenced from the reports of the biennial international archaeometry conferences and specialized journals in the field (1-3).

Archaeological Chemistry

The modern field of archaeological chemistry arose during the first thirty years after World War II as a result of the development of instrumental methods of inorganic analysis. which made it possible to develop new areas in archaeological chemistry. The methods of choice over the years have concentrated on elemental analysis, whether by atomic absorption or emission, X-ray fluorescence, plasma emission spectroscopy, X-ray powder diffraction, neutron activation analysis or mass spectrometry. These methods have lent themselves to the analysis of stone and lithic artifacts, ceramic materials, glass and metallic materials. Many of the methods of analysis, such as lead isotope provenance studies of marble or elemental provenance studies of obsidian, have passed through several generations of development. Some of the methods survived the tests of validity and of utility. Some have disappeared from the scene, either because the methods had fatal flaws or they did not really solve archaeological problems. Inorganic methods continue to be improved upon and added to, and a large body of literature has developed in this area.

0097-6156/96/0625-0001$12.00/0

The task of the archaeological chemist has become more complex than ever over the past decade. Once the domain of analytical chemists turned "amateur archaeologists," effective work in this area demands increasingly sophisticated equipment by way of advanced instrumentation, increased knowledge of statistical software packages for the assembling, processing and interpretation of coherent data-sets, increased interaction with members of related disciplines, awareness of the ever-burgeoning literature of archaeometry, archaeology and anthropology, and perhaps more important than ever, continual interaction and collaboration with members of related disciplines. Chemists working in this area must be aware of the fact that analytical data can be completely meaningless unless it is interpreted within the matrix surrounding the artifact or sample being investigated: professionally executed field work, well-documented excavation, proper sampling technique and meaningful scientific interpretation of the resulting data. Indeed, in the words of E. M. Jope (4), in order for any type of work in archaeological chemistry "to be effective, it must be collaborative. Integrations between excavators, investigators of excavated material, prehistorians and historians, and scientists of all sorts, is now growing, so that we are increasingly all one family seeking to present the past in a systematized and intelligible form."

The Present Symposium

While many of the goals of past volumes in this series (5, 6) coincide with the goals of this volume, namely, the mutual education of chemists, archaeologists and anthropologists in the use of new techniques on archaeological substrates and in the interpretation of data obtained therefrom, it is the choice of organic and biochemical substrates on the part of many of the investigators represented in this volume that makes this collection of papers unique. The solicitation of papers for this symposium contributed to a slant in this direction since only in very recent years has it been possible to analyze biological and organic materials with the significantly lower detection limits that enables these methods to lend themselves uniquely to solving some difficult archaeological problems.

Hence, the characteristic of the symposium represented by the papers collected in this volume was the virtual disappearance of many of the traditional materials normally reported on at archaeological and archaeometric meetings, namely, metal, pottery, stone and glass. For example, in *Archaeological Chemistry - III* (5), the proceedings of the Seventh Symposium on Archaeological Chemistry organized by the Division of the History of Chemistry's Subdivision of Archaeological Chemistry, fully two-thirds of the papers were devoted to inorganic materials, whereas only one-half of the papers in *Archaeological Chemistry - IV* (proceedings of the Eighth Symposium; 6) had inorganic materials as their subject. Of overwhelming interest to the contributors of the symposium that forms the basis of this volume were fibers, dyes, bone, collagen, archaeological soils, DNA analysis, and organic residues from a variety of artifacts including rock paintings and pottery. Two papers dealt with archaeological mineralized plant fiber. Three papers studied the identification of natural dyestuffs used for ancient textiles from Western and Eastern Asia. Ten papers dealt with archaeological bone and collagen, including dating, degradation determinations, analysis of lipid biomarkers, and quantitative measurements to reconstruct paleodiets. Six papers examined nucleotides in archaeological materials, including such technical difficulties as need for extraction, purification and isolation of high molecular weight fragments. The remainder of the papers reported on soil analysis, natural products, copper-based artifacts and pigments, and inorganic and organic residues on artifacts. Two papers each were presented on glass and metals; one paper each was presented on clay, flint, and obsidian. In summary, fully thirty-four of the forty-eight papers submitted for inclusion in this symposium dealt with organic substances or artifacts, a complete inversion of the ratio of organic to inorganic materials studied in comparison with the symposium of 1982 (5).

A major part of the symposium was devoted to papers dealing with gaining information about the peopling of the New World. One paper dealt with the possibility that the use of radiocarbon calibration alone will never resolve the debate over the date that the first humans entered North America. Another paper described the development of a chrono-cultural tool for determining contemporaneity of artifacts during the first contacts between Europeans and native North Americans.

The range of methodologies used to study the archaeological artifacts and remains was very broad. Simple wet chemical techniques were used in several cases. The most popular spectroscopic method was Fourier Transform Infrared Microspectroscopy (FT-IR). Instrumental Neutron Activation Analysis (INAA) was also utilized in several instances. However, it became quite clear that the instruments of choice are presently those which can be coupled with mass spectrometry. Many papers utilized gas chromatography coupled with mass spectrometry (GC/MS), high performance liquid chromatography coupled with MS (LC/MS), GC/MS-SIM, where SIM = selected ion monitoring, GC/MS/MS and inductively coupled plasma/MS (ICP/MS). One paper demonstrated for the first time that mass spectrometry can be used in the direct chemical examination of nucleotide bases in ancient materials.

An important feature of the symposium was an accompanying pedagogical symposium designed to provide information to practicing chemists on some of these new techniques, and at the same time to provide information on some of the more exciting substrates of archaeological chemistry such as the "Iceman" recently found in the Alps.

We can learn a great deal from the make-up of this symposium. We learn that molecular archaeology is making great strides by utilization of sophisticated instrumentation. We learn that simple wet chemical techniques are still viable; we also learn that some archaeological problems are not susceptible of solution unless a battery of highly sophisticated instruments and data-set analysis software is available. We learn that organic materials can yield important information despite the fact that degradation has taken place over centuries and millennia. Indeed, through special techniques reported in this symposium, it has been shown that the formidable challenge of reconstructing the original composition of organic material from its present degraded condition is a problem susceptible of resolution. Finally, we can note in many instances the use of sophisticated statistical software packages that can help in extracting meaningful scientific and archaeological information from analytical data.

Inorganic Materials

The rapid development of sophisticated analytical techniques for elemental analysis following World War II opened the way to archaeological chemical investigations of materials susceptible to elemental analysis, namely, ceramic materials, glass, stone and metals. For many years, these materials constituted the major set of substrates studied by archaeological chemists.

Neutron activation analysis and instrumental neutron activation analysis (NAA and INAA) became the method of choice for analysis of ceramics because of an array of factors: whole sample characterization, capability of high precision multi-element qualitative and quantitative analysis, and small sample size, to name a few (7). The greatly improved successor to emission spectroscopy (ES), Inductively Coupled Plasma (ICP) spectroscopy, with its greater sensitivity and wider range of analytes, overlaps with NAA with respect to only six elements with the required levels of precision (8). This means, of course, that both methods are useful in developing more broadly-based data-sets, but are virtually useless for comparison of data. Another powerful tool in the determination of the bulk chemical compositions of the clay bodies and glazes in ceramic materials is energy dispersive X-ray diffraction attached to a scanning electron microscope (9).

Many of the above-mentioned methods are also useful for the analysis of archaeological glass. However, electron-microprobe analysis (EMPA) and scanning electron microscopy (SEM) have been particularly useful because they are virtually nondestructive. Other methods such as ICPS and INAA are useful for analyzing unweathered glass, which is essential for valid comparisons between glass compositions. Glass analysis, in the past, has provided information regarding the sources of the raw materials used, the modification of glass colors, and the chemical characterization of glass products. Such information can provide the archaeologist with comprehensive data-sets susceptible to archaeological interpretation (10). Susan Frank has produced a comprehensive guide to archaeological glass in all of its aspects (11) and two excellent review articles by R. H. Brill (12) and D. Grose (13) summarize how modern chemical and physical analysis can reveal the sophisticated techniques of ancient Egyptian and Roman glassmakers. In chapter 2 of this volume, J. B. Lambert and co-workers utilize many of these techniques in analyzing 9th century Thai glass by ICPS in order to understand maritime trade patterns. In chapter 3 of this volume, R. G. V. Hancock, S. Aufreiter and J.-F. Moreau, using INAA as their analytical tool, compare the individual chemistries of three suites of turquoise glass trade beads with the previously established chemistries of well-dated beads from other archaeological sites.

Examination of stone and the lithic artifacts derived therefrom has usually taken the form of trace element analysis using many of the methods described above. Such analyses are very helpful in determining the sources of the raw materials, such as flint and marble, used to make artifacts (14, 15). Of even greater importance is examination of the factors that lead to erosion of stone and of chemical means that can be used to retard erosion. Threats to stone surfaces such as weathering, atmospheric pollution and attack by algae, fungi, mold and other microorganisms (16) is the subject of intense research at the present time. A fine review article by K. Lal Gauri summarizes some of the most promising approaches to this problem (17). Examination of stone also includes gemstones; NAA and other techniques have been used effectively to come to some understanding of a gem's religious, cultural and economic significance (18). A. R. Skinner and M. N. Rudolph, in chapter 4 of this volume, outline the possibility of using a relatively new dating technique, electron spin resonance (ESR), for determining the age of flint artifacts. They show that ESR has some advantages over the conventional technique, thermoluminescence (TL), but that some problems remain particularly in the area of sample selection. In chapter 5, F. R. Beardsley, G. G. Goles and W. S. Ayres summarize their INAA results on Easter Island obsidian in an attempt to trace the obsidian artifacts to their raw material source.

Analysis of metal artifacts by such methods as INAA and X-ray fluorescence (XRF) has been very valuable in determining trace metals embedded in a parent metal, giving valuable clues to metal provenance. Lead isotope ratios has also been valuable in provenance studies, and SEM has given insights into metal-working technology (19-21). In chapter 6 of this volume, J.-F. Moreau and R. G. V. Hancock, using INAA to determine eight elements in 500 copper based artifacts, show how these measurements allow them to assess the degree of homogeneity of these artifacts at the intra-site level as a means of determining which parts of an archaeological site may be contemporaneous. In chapter 7, A. A. Gordus and I. Shimada examine the gold-silver-copper ternary alloy contents of gold objects from an unlooted Peruvian gravesite. Using NAA and EMPA, they are able to show that surface depletion of copper and consequent enrichment of gold and silver were not deliberate, but the result of the repeated hammering-annealing required to produce the thin gold sheets used for the construction of the objects. G. F. Carter, in chapter 8, continues his study of the chemical composition of Roman coins using XRF as his method of choice. This analysis allowed undated coins previously estimated to have been struck between 15-16 A.D. and 22-23 A.D. to be confirmed. M. V. Orna, in chapter 9, shows how, in the Middle Ages, metallic copper was used to produce blue pigments that in some instances continue to defy characterization. Finally, in chapter 10, R. H. Tykot and S. M. M. Young summarize the archaeological applications of ICP/MS to stone and metal artifacts.

Archaeological Soils

Analysis of archaeological soils has historically centered around phosphate analysis. Phosphorus, in the form of phosphate, is a consistent indicator of human activity since its concentration increases through the life chain because of its relative chemical immobility. Phosphate analysis can therefore be used to detect the sites of human habitation even when all other traces of such habitation have disappeared (22). It can even be used, but with considerably more caveats, for the detection of bone that has virtually disappeared, leaving behind only a slight darkening of the soil known as a silhouette (23). A fairly recent critique of the use of soil phosphate analysis in archaeology (24) warns workers in this area to be aware of the fact that any archaeological site is dynamic and subject to change. Hence, phosphate could have been deposited either before or after settlement, or the general deposit of phosphate may not distinguish between human and higher animal use. In chapter 11 of this volume, H. Chaya describes a method for determining total phosphorus at an Aleutian Island site occupied by marine hunter-gatherers over a 1500 year period using molybdivanadophosphate color development rather than the more conventional molybdenum blue method. In chapter 12, another wet chemical technique is utilized by L. Barba, L. López, A. Ortíz, K. Link and L. Lazos in their analysis of residues on the lime plaster floors at the Templo Mayor of Tenochtitlan, and important Aztec archaeological site. From these analyses, they propose to be able to infer or reconstruct the ritual activities that took place at that site. In this case, these workers assume that the lime plaster floor, or the archaeological soil, was the baseline location for the deposition of other materials. Another departure from the conventional treatment of archaeological soils is the work of R. P. Evershed, P. H. Bethell, J. Ottaway and P. Reynolds in chapter 13. They employ GC/MS-SIM to provide a very sensitive and selective means of analyzing for characteristic steroidal marker compounds that can confirm sites of suspected cess-pits and latrines. In each of these cases of investigating an archaeological soil substrate, the analytes, the methodologies and the kinds of information sought were different. The advent of extremely sensitive and selective analytical tools, such as those used by Evershed and co-workers, will allow extraction of information from such substrates that was once impossible to obtain.

Organic Materials: Fibers and Dyes

Organic analysis has generally required larger amounts of materials for structure proofs by nuclear magnetic resonance (NMR), although mass, electronic and vibrational spectroscopies were successful in addressing many problems in the fields of organic dyes and foodstuffs. The development of solid state NMR methods in the 1970s finally enabled that technique to be applied to insoluble organic materials as well. Organic materials now may be analyzed with the same degree of success as enjoyed by inorganics two or more decades ago. The methods must pass through the same evolutionary path that tests their validity and utility. Among the organic materials of most interest to the participants in this symposium were fibers and dyes. Archaeological textiles may contain information regarding degradation, mineralization or other forms of alteration that can eventually lead to greater understanding of prehistoric environments and cultures. Elemental distribution (25) and isotope measurements (26) are often helpful in identifying and dating fibers. Fibers are rarely left in their raw natural state in use: they are normally dyed. The identification of these dyes can often be the beginning of a fascinating journey into history (27) and into the cultural and social significance of dyes (28).

In chapter 14 of this volume, R. D. Gillard and S. M. Hardman explore the mineralization of cellulosic and proteinaceous fibers through a laboratory simulation with oxygenated aqueous solutions. Using FT-IR microscopy they show that traces of organic material can survive long-term burial and even permit their identification under appropriate circumstances. H. L. Chen, D. W. Foreman and K. A. Jakes, in chapter 15, also study mineralization of fibers using XRD techniques to study the fibers' microstructure in order eventually to understand the mechanism of organic polymer degradation and replacement of the fiber by inorganic copper compounds. In chapter 16, K. A. Jakes and L. R. Sibley use IR spectroscopy in a different manner to study the cellulose, lignin and hemicellulose content that distinguishes types of fibers from one another. The next two chapters deal with the Shroud of Turin, a textile artifact that is associated by many with the suffering and death of Jesus Christ. A. Adler, in chapter 17, reviews the controversial status of the Shroud, particularly the nature of the image of a wounded human body that can be seen on it. In chapter 18, D. Kouznetsov, A. Ivanov and P. Veletsky highlight the inherent uncertainties of radiocarbon dating, particularly with respect to variations in conditions external to the artifact in question, indicating how this fact led them to question the accepted radiocarbon date of the Shroud of Turin. In their work, they devised a laboratory model to simulate the fire conditions to which the Shroud was subjected at Chambéry in 1532. Their results show that radiocarbon ages of experimental textile samples incubated under fire-simulating conditions are subject to notable error due to incorporation of significant amounts of ^{13}C and ^{14}C atoms from external combustion gases into the textile cellulose structure. They also take into account the known phenomenon of biological fractionation of carbon isotopes by living plants which can lead to enrichment of a textile by ^{13}C and ^{14}C isotopes during linen manufacture. Chapter 19 is a rebuttal to the Kouznetsov, et al. paper by a group at the University of Arizona headed by A. J. T. Jull. Jull and co-workers were members of the team that performed the radiocarbon dating on the Shroud of Turin in 1989 and found that the artifact had a carbon date of late-13th to mid-14th century, the time period when it first appeared in the historic record. Jull claims here, as he did at the original symposium, that the work of Kouznetsov, et al. is flawed since the Arizona team did not achieve similar results in their own fire-simulating experimentation. Editorially speaking, we must observe that the Arizona tests did not mimic the Kouznetsov tests since several of the experimental reactants were not present. These two papers are placed back-to-back in this volume to enable readers to compare them and draw their own conclusions. In chapter 20, the Kouznetsov team examines cellulose chains in archaeological textile remains, noting that they can contain a significant number of chemically modified β-D-glucose residues. Their work, using a capillary zone electrophoresis-mass spectrometric approach, shows a correlation between cellulose alkylation extent and calendar age of the textile samples tested. Their results suggest that if cellulose alkylation is the consequence of microbial activity, this phenomenon could be the basis of a new and efficient dating technique, at least among samples taken from a single site and subjected to a similar environment. In the final chapter in this section, chapter 21, Z. C. Koren discusses the application of high performance liquid chromatography (HPLC) to the identification of the natural dyes, anthraquinonoids, flavanoids and indigoids, found in a variety of archaeological sites as much as 3000 years old.

Biological Materials: Archaeological Bone, Connective Tissue, DNA, Radiocarbon Measurements

Biochemical analysis began with isotopic measurements on bone collagen to obtain dietary information. Ancient human diet also can be inferred from inorganic analysis of bone. For example, from a simple direct argument about strontium levels in flesh, strontium levels of bone can provide information about meat intake (*29*). Stable carbon isotope analysis of bone and connective tissue is also emerging as a powerful tool in diet reconstruction (*30*). For example, carbon isotope values can distinguish between C_3 and C_4 type plants in the terrestrial food web. In chapter 22 of this volume, P. T. McCutcheon discusses the uses of TL and thermogravimetric analysis (TGA) to date bone mineral. In chapter 23, J. H. Burton points out that while bone strontium faithfully reflects the dietary Sr/Ca ratio, other factors besides trophic-level significantly affect this ratio. He also discusses the cautions one should use in interpreting the Ba/Ca ratio in bone as well. D. M. Greenlee, in chapter 24, shows that the combined techniques of backscattered electron imaging and EMPA can be used to examine archaeological human bones from different post-depositional environments. The bones were shown to have different levels of structural preservation and highly variable elemental compositions relative to modern bone. Potential criteria for identifying the diagenetic processes involved and for recognizing diagenetically unaltered areas are also discussed. In chapter 25, N. J. van der Merwe, R. H. Tykot and N. Hammond point out that since the relative contributions of the protein, carbohydrate and lipid portions of the diet to bone collagen and bone apatite are still not fully understood, it is necessary to perform isotopic carbon analysis on both tissues for proper dietary reconstruction in all but the simplest food webs. They also assert that analysis of the flora and fauna available for human exploitation may be equally important. In chapter 26, A. M. Child discusses the effects of microbial decomposition on the rate of aspartic acid racemization in mineralized collagen by measuring the degree of racemization after a prolonged incubation period of sterile modern pig bones inoculated with bacteria and fungi.

The methods of molecular biology now may be applied to genetic remains of plant and animal organisms. These latter methods, based largely on the polymerase chain reaction (PCR), are in their infancy as applied to ancient organisms. Biological contributions will assume a prominent position in this field as methods become more sophisticated and smaller samples can be effectively examined. Studies on even very degraded fractions of mitochondrial DNA have allowed biochemists and molecular biologists to trace migrations and matings of various species over many centuries, thus shedding lights on population patterns and closest genetic links of present species with past species of living things (*31, 32*). In chapter 27 of this volume, M. W. Rowe and M. Hyman utilize PCR and phylogenetic analysis to aid in the identification of the organic binder and vehicle used in the 3000-4000 year old rock paintings (pictographs) painted on shelter walls in Seminole Canyon, Texas. The ultrasensitive method of PCR was essential for this analysis because of the seriously degraded small fraction of DNA that remained after so many centuries. PCR amplifies DNA and can produce millions of DNA copies from only a few enduring DNA fragments, thus enabling the replicated DNA to be sequenced in order to derive the requisite information. The sequences obtained reveal that the organic matter in the paintings was definitely of mammalian origin. In chapter 28, R. Vargas-Sanders and Z. Salazar isolated and characterized high molecular weight DNA fractions from bone remains of Mexican prehispanic populations. In chapter 29, M. W. Rowe and M. Hyman report on the development of a technique to remove organic carbon selectively from ancient pictograph paints without contamination from the mineral carbon in the rock substrate, mineral accretionary coatings or atmospheric carbon dioxide. Their technique is generally applicable to any pictographs which had organic matter added to the paints. C. M. Batt and A. M. Pollard, in chapter 30, demonstrate that questions concerning the earliest date that humans entered North America may not be answerable by radiocarbon dating alone. In chapter 31, D. L. Kirner and R. E. Taylor discuss techniques using accelerator mass spectrometry (AMS) whereby they can overcome the problem of background contamination in radiocarbon dating, thus opening the door to the use of microsamples.

Conclusion

While the papers contained in this volume do not cover all of the innovative work taking place at this moment in archaeological chemistry, they are a representative sample of such work and provide an overview for the interested archaeologist or chemist. In addition, a very substantial bibliography for each of the research areas discussed herein provides the reader with further material for study.

Literature Cited

1. *Archaeometry - Proceedings of the 25th International Symposium*; Maniatis, Y., Ed; Elsevier: Amsterdam, 1989.
2. *Archaeometry 90 - International Symposium on Archaeometry, 2-6 April, 1990, Heidelberg*; Pernicka, E.; Wagner, G. A., Eds.; Birkhauser Verlag: Basel, 1991.
3. Hughes, M. J. *Chemistry and Industry* **1992**, 7, 897-901.
4. Jope, E. M. In *Scientific Analysis in Archaeology and Its Interpretation*; Henderson, J., Ed.; Oxford University Press: Oxford, 1989; pp xiv-xv.
5. *Archaeological Chemistry - III*; Lambert, J. B., Ed.; Advances in Chemistry 205; American Chemical Society: Washington, DC, 1984.
6. *Archaeological Chemistry - IV*; Allen, R. O., Ed.; Advances in Chemistry 220; American Chemical Society: Washington, DC, 1989.
7. Neff, H. In *Chemical Characterization of Ceramic Pastes in Archaeology*; Neff, H., Ed.; Prehistory Press: Madison, WI, 1992; pp 1-10.
8. Harbottle, G.; Bishop, R. L. In *Chemical Characterization of Ceramic Pastes in Archaeology*; Neff, H., Ed.; Prehistory Press: Madison, WI, 1992; pp 27-29.
9. Tite, M. S.; Bimson, M. Archaeometry 1991, 33, 3-27.
10. Henderson, J. In *Scientific Analysis in Archaeology and Its Interpretation*; Henderson, J., Ed.; UCLA Institute of Archaeology, Archaeological Research Tools 5; UCLA Institute of Archaeology: Los Angeles, 1989; pp 30-62.
11. Frank, S. *Glass and Archaeology*; Academic Press: New York, 1982.
12. Brill, R. H. *Scientific American* **1963**, 209(5), 120-126.
13. Grose, D. *Archaeology* **1983**, 36(4), 38-45.
14. Craddock, P. T.; Cowell, M. R.; Leese, M. N.; Hughes, M. J. *Archaeometry* **1983**, 25, 135-163.
15. Germann, K.; Holzman, G.; Winkler, F. J. *Archaeometry* **1980**, 22, 99-106.
16. Urzí, C.; Krumbein, W. E. In *Durability and Change: The Science, Responsibility and Cost of Cultural Heritage*; Krumbein, W. E. et al, Eds.; John Wiley and Sons: Chichester, UK, 1994.
17. Lal Gauri, K. *Scientific American* **1978**, 238(6), 126-136.
18. Harbottle, G.; Wiegand, P. C. *Scientific American* **1992**, 266(2), 78-85.
19. Gale, N. H.; Stos-Gale, Z. A. In *Archaeological Chemistry - IV*; Allen, R. O., Ed.; Advances in Chemistry 220; American Chemical Society: Washington, DC, 1989; pp 159-198.

20. Manea-Krichten, M. C.; Heidebrecht, N.; Miller, G. E. In *Archaeological Chemistry - IV*; Allen, R. O., Ed.; Advances in Chemistry 220; American Chemical Society: Washington, DC, 1989; pp 199-211.
21. Carter, G. F.; Razi, H. In *Archaeological Chemistry - IV*; Allen, R. O., Ed.; Advances in Chemistry 220; American Chemical Society: Washington, DC, 1989; pp 213-230.
22. Eidt, R. C. *Science* **1977**, *197*, 1327-1333.
23. Keeley, H. C. M.; Hudson, G. E.; Evans, J. *J. Archaeological Science* **1977**, *4*, 19-24.
24. Bethell, P.; Máté, I. In *Scientific Analysis in Archaeology and Its Interpretation*; Henderson, J., Ed.; UCLA Institute of Archaeology, Archaeological Research Tools 5; UCLA Institute of Archaeology: Los Angeles, 1989; pp 1-29.
25. Jakes, K. A.; Angel, A. In *Archaeological Chemistry - IV*; Allen, R. O., Ed.; Advances in Chemistry 220; American Chemical Society: Washington, DC, 1989; pp 452-464.
26. Dinegar, R. H.; Schwalbe, L. A. In *Archaeological Chemistry - IV*; Allen, R. O., Ed.; Advances in Chemistry 220; American Chemical Society: Washington, DC, 1989; pp 409-417.
27. Saltzman, M. *American Scientist* **1992**, *80*, 474-481.
28. McGovern, P. E.; Michel, R. H. *Acc. Chem. Res.* **1990**, *23*, 152-158.
29. Lambert, J. B.; Simpson, S. V.; Szpunar, C. B.; Buikstra, J. E. *Acc. Chem. Res.* **1984**, *17*, 298-305.
30. Ambrose, S. H.; Norr, L. In *Prehistoric Human Bone: Archaeology at the Molecular Level*; Lambert, J. B; Grupe, G., Eds.; Springer-Verlag: New York, 1993; pp 1-38.
31. Edelson, E. *Mosaic* **1991**, *22(3)*, 56-63.
32. Menozzi, P.; Piazza, A.; Cavalli-Sforza, L. *Science* **1978**, *201*, 786-792.

RECEIVED October 9, 1995

Chapter 2

Analysis of Ninth Century Thai Glass

Joseph B. Lambert[1], Suzanne C. Johnson[1], Robert T. Parkhurst[1],
and Bennet Bronson[2]

[1]Department of Chemistry, Northwestern University,
2145 Sheridan Road, Evanston, IL 60208–3113
[2]Field Museum of Natural History, East Roosevelt Road at South Lake
Shore Drive, Chicago, IL 60605

Colored, blown glass fragments from two ninth century sites on opposite sides of the Kra Isthmus of Thailand have been analyzed for 17
elements by inductively coupled plasma methods. Most samples have
a mixed alkali matrix (medium proportions of K_2O), although a few
have the standard soda-lime matrix. Many samples also have very
high levels of CaO and Al_2O_3 that may have conveyed stability. The
blue, green, yellow, and violet colors come from significant amounts
of Co, Cu, Fe, and Mn. Attempts to cluster the entire ensemble of
samples resulted in groupings based only on color, and exclusion of
specific colorants was not effective. Cluster analysis of single-color
groups, however, led to separation between the two sites for the
majority of samples.

The Kra Isthmus is located at the narrowest portion of the Thai-Malay Peninsula that
ends at Singapore. The Field Museum of Natural History has recently completed
excavation of two sites on opposite coasts of the Kra Isthmus, Ko Kho Khao (K) on
the west coast facing the Andaman Sea and the Bay of Bengal and Laem Pho (L)
facing the Gulf of Siam and the China Sea. The isthmus could have provided an
overland nexus that would have avoided the dangerous Strait of Malacca. Neither
site is believed to have been inhabited before or after the ninth century AD and both
were found to contain large quantities of materials originating in China, India, and
West Asia. Objects found include glass beads, blown glass vessels, West Asian
earthenware, and Chinese porcelain and stoneware. The Chinese objects may be
dated stylistically to the period between 825 and 875 AD.

Southern Thailand was not a wealthy area at this time, yet remains of large
amounts of luxurious and presumably expensive goods were found. This observation
leads to the hypothesis that the inhabitants of the two sites were primarily traders
who imported goods mainly to ship onward to other locations. It is possible that
many other wares also were traded through these ports, such as spices, perfumes,

and textiles, but these more perishable materials would not have survived the moist, salty conditions of the sites.

The types of goods that did survive were almost identical at the two sites, so that it is possible that there was overland communication between them. The question then arises as to whether they traded with each other as well as with ships stopping at their respective ports, thereby providing a shortcut along the maritime silk route. This route linked the Far East, India, West Asia, the Mediterranean, and Africa between about 500 and 1700 AD and served to carry not only merchandise but also inventions and ideas. It lay almost entirely over water, so that the location of archaeological evidence is difficult. If the maritime silk route crossed land at any point, the potential for archaeological information is increased greatly. Way stations on the trans-peninsular route between K and L were searched for but not found.

Although the ceramics and glassware for trading were almost identical at the two sites, locally made pots were quite different in style. These more utilitarian artifacts suggest that the people of L were more closely connected with central Thailand, whereas the people of K were more closely connected with Malaysia as well as India. There was no evidence for local glass manufacture.

Thus the archaeological evidence suggests that K and L were occupied only during the ninth century, that their residents were of different origin or culture, and that they had some communication with each other. Chemical analysis of materials found at the sites may be able to provide additional information on these issues. In particular, numerous glass sherds were found at both sites, their shapes and appearances suggesting that they were from blown glass vessels. We have carried out a multielement examination of these materials by inductively coupled plasma (ICP) methods in order to compare the two Thai sites. In addition to characterizing the glass materials, we also have compared compositions between the sites. Differences could indicate that trade predominated from different sources and that the two sites did not have extensive communication with each other.

Raw materials for glass consist primarily of quartz sand or pebbles (SiO_2) and a mineral or plant ash source of alkali (Na, K) and alkaline earth (Ca, Mg). The addition of the Na or K salts lowers the melting point of silica, reduces the viscosity, and enhances the workability. The Ca or Mg salts provide stability and increase durability. Additional elements may be present either unintentionally (Ba, Fe, Al, Ti), or as colorants (Co, Cu, Fe), decolorants (Mn, Sb), or opacifiers (Sb).

Soda-lime glass typically has the following major elements (traditionally expressed as their oxides): SiO_2 (60-70%), Na_2O (12-18%), CaO (6-10%), and little or no K_2O. Potash glass contains K_2O (9-15%) and CaO (10-25%) but very little Na_2O. It is generally less durable than soda-lime glass and more difficult to work with. Mixed alkali glasses with up to 8% of both Na_2O and K_2O are more durable than either soda-lime or potash glass. Sayre and Smith (*1*) provided an early, comprehensive study of archaeological glass.

Materials and Methods

One hundred and sixty-five small, broken pieces of glass were collected from the two sites and were separated visually into 14 groups based on color. Some of these were not analyzed because they were multicolored or had surface decoration. Of the

146 pieces analyzed, 59 were from K and 87 from L (Table I). All the samples were found in the moist, salty sand of the sites. A few pieces had surface encrustation or showed weathering, but the majority appeared to be unaltered.

Table I. Thai Glass Groups Based on Color

Group	Color	No. of K samples	No. of L samples
1	Dark blue	6	3
2	Blue	4	3
3	Light blue	8	20
4	Greenish blue	2	1
5	Bluish green	8	9
6	Light green	7	18
7	Dark green	0	5
8	Greenish yellow	2	1
9	Yellow violet	6	1
10	Pale yellow	5	10
11	Very pale yellow	3	3
12	Light gold	3	0
13	Colorless	5	9
14	Pinkish violet	0	4

The method for elemental analysis was inductively coupled plasma atomic emission spectroscopy (ICP-AES). A Thermo Jarrell Ash Atomscan 25 Sequential ICP-AES instrument was used for all analyses. The high temperature of the plasma (9000-10,000 K) normally reduces or eliminates chemical and spectral interferences common to lower temperature flames such as those used in atomic absorption methods. The sequential instrument is programmed to move from the spectral line for one element to the next one, pausing long enough to collect a satisfactory signal. Thus many elements may be analyzed quickly and efficiently. ICP has been used previously to analyze British and Roman glass (2, 3).

Samples were removed for analysis with a Drexel hand drill and a diamond-coated drill bit. The surface layer was drilled off and discarded before material was collected for analysis. For a few, very thin samples, powdering was not possible, so small pieces were used. The materials were weighed into Nalgene low density polyethylene bottles, which had been presoaked with 5% hydrogen fluoride (HF) solution, and 10.0 mL of mixed acid (HF + 2 HCl) was added to each bottle with agitation (4). Ultrapure HCl and HF were purchased from Aldrich Chemical Co. Single element standards were purchased from Inorganic Ventures in 1000 ppm concentrations and were diluted with the stock HF/HCl mixture and deionized water until the desired concentration was reached. Standards and samples had the same acid concentrations. The instrument was standardized by a two-point line with a high standard and a blank, the standard being slightly higher than the concentration in the sample. Preliminary analyses were necessary to demonstrate complete dissolution and to determine approximate unknown concentrations for optimization of standards. Standards also were prepared by weight in Nalgene low density polyethylene bottles.

The quartz mixing chamber and the Tygon hoses in the ICP instrument were replaced with HF-resistant parts. Each sample was analyzed four times and the means taken. Optimized instrumental conditions may be found in Johnson (5). Precision of the instrument was tested by repeated analysis of a single sample, made from standards by dilution with HF/HCl in the same concentration range as the unknowns. The test for precision was repeated four times, and a water blank was run immediately without rinsing to test for sample overlap. These results are given in Table II for the 16 elements studied. Precision is excellent in all cases, and the blank readings exceeded the standard deviation only in the case of Pb. To test for accuracy, the National Institute of Standards and Technology (NIST) reference sample of soda-lime container glass (no. 621) was analyzed by the same procedure. Table III contains the observed and actual data for the seven elements common to

Table II. Test for Precision[a]

Element	Run 1	Run 2	Run 3	Run 4	Std. Dev.	Blank
Al	2.756	2.815	2.771	2.774	0.0219	0.0011
Ba	0.710	0.722	0.704	0.705	0.0072	0.0002
Ca	11.87	11.92	11.61	11.51	0.1721	0.0027
Co	0.510	0.516	0.487	0.499	0.0111	0.0084
Cr	3.025	3.046	3.008	2.990	0.0207	0.0004
Cu	0.530	0.530	0.517	0.509	0.0091	0.0001
Fe	1.998	2.039	1.989	2.012	0.0189	0.0026
K	2.686	2.669	2.548	2.457	0.0934	0.0038
Mg	12.57	12.85	12.67	12.54	0.1211	0.0012
Mn	1.280	1.306	1.266	1.269	0.0158	0.0024
Na	8.533	8.948	8.433	8.270	0.2504	0.0021
Ni	0.790	0.806	0.775	0.792	0.0110	0.0068
Pb	0.727	0.760	0.713	0.729	0.0171	0.0279
Ti	1.279	1.290	1.267	1.264	0.0103	0.0029
V	1.628	1.634	1.607	1.614	0.0108	0.0023
Zn	1.290	1.334	1.266	1.277	0.0258	0.0025

[a]Sample composed of standards diluted with aqueous HF/HCl. Concentrations are in parts per million (ppm).

Table III. Comparison of Standard Sample with Measured Values

Element	Reported[a,b]	Measured[b]
Al_2O_3	2.76 (0.03)[c]	5.78
BaO	0.12 (0.05)	0.15
CaO	10.71 (0.05)	11.47
Fe_2O_3	0.040 (0.003)	0.09[d]
K_2O	2.01 (0.03)	2.82
MgO	0.27 (0.03)	0.35
Na_2O	12.74 (0.05)	12.89
TiO	0.014 (0.003)	0.02[d]

[a]National Institute of Standards and Technology, reference sample No. 621.
[b]Concentrations in %.
[c]Standard deviation.
[d]Close to the detection limit.

this study and the report on the standard. Except for Al and Fe, the agreement is good, in particular for the key elements Na, Ca, K, and Mg (average deviation 20%). The differences for the major elements other than Al are still greater than desirable, presumably because of uncorrected matrix effects. The measured level for Al is high by a factor of two, a result that will have to be considered in the following discussion. Although the measured level for Fe also is high by a factor of two, the level of Fe in the standard is close to the detection limit and is much lower than in our materials, as also is the case for Ti. We expect that our analyses at the higher levels in the samples (1-3%) are much more accurate. Dilutions and standards used for the NIST sample were identical to those used for the Thai materials.

Results of Elemental Analysis

Inductively coupled plasma techniques were used for sequential analysis of 17 elements. The mean values for 146 samples are given in Table IV for 14 elements, as the results for Sn, Cr, and V invariably were below their detection limits. Thus Cr was not used in these samples as a colorant to produce yellow. The amounts for Ba and Pb also were extremely low and will not be discussed further. Individual compositions for all samples are given in Johnson (5).

Proportions of Alkali, Alkaline Earth, and Aluminum Oxides

The data in Table IV show that the glass from these sites has a relatively traditional soda-lime composition (about 17% Na_2O, 9% CaO) (6). The relatively high proportion of K_2O (about 4%) suggests a wood ash source of the alkali metals rather than

Table IV. Elemental Analysis of Thai Glass

Oxide	Mean wt.[a] (%) (standard deviation)		Mean wt.[b] (%)	Mean wt.[c] (%)
Al_2O_3	7.8	(6.1)	7.5	8.0
BaO	0.03	(0.02)	0.03	0.03
CaO	8.7	(3.1)	8.0	9.2
CoO	0.01	(0.04)	0.02	0.01
CuO	0.03	(0.05)	0.04	0.02
Fe_2O_3	1.7	(1.0)	1.7	1.6
K_2O	4.3	(2.4)	5.1	3.8
MgO	4.9	(2.1)	4.8	5.0
MnO	1.1	(0.9)	1.4	1.0
Na_2O	16.9	(8.5)	19.6	15.1
NiO	0.12	(0.17)	0.14	0.10
PbO	0.01	(0.02)	0.01	0.01
TiO	0.14	(0.08)	0.13	0.14
ZnO	0.07	(0.19)	0.11	0.04

[a]All samples.
[b]Only K samples.
[c]Only L samples.

a mineral such as natron. The wood ash source is substantiated by the high levels of Mg (about 5%), indicating, following Brill (7), that the wood ash had not been purified. The Mg and K levels, however, do not correlate ($r^2 = 0.00$). The Na and K levels do correlate highly ($r^2 = 0.60$), indicating a common source. The mix of Na with K may have conveyed additional stability to the glass. Hench (8) found that glass containing 12 mol% Na_2O and 3 mol% K_2O is twice as durable as glass containing 15 mol% Na_2O and no K_2O. The presence of Ca and Mg as stabilizers also reduces the aqueous solubility of soda ash and potash glasses. With about 8% CaO and 5% MgO, these Thai glasses have high concentrations of stabilizers. The high proportion of Al_2O_3 also might serve to stabilize the surface of the glass from dissolution, although the proportion listed in Table IV (about 8%) may be high by a factor of two, as indicated by comparison of our analyses with the NIST standard (Table III). Even at 4%, however, the proportion is high. In Hench's classification of glass surfaces (8, 9) high Al corresponds to a Type III-A surface, which is very durable and contains aluminum silicate or calcium phosphate. This combination of matrix and surface stabilizers suggests a very durable glass, as confirmed by the excellent preservation of the materials despite the wet, salty environment. The sum of the oxides of Na, K, Ca, Mg, and Al is 42.6%, leaving some 57.4% for silica, colorants, and decolorants. If the alumina content is high by a factor of two, then the figure of 61-62% for silica plus colorants is normal for mixed alkali glass.

Titanium has been suggested to be an impurity in the silica raw material (10). For the present glass materials, Ti correlates somewhat with Al ($r^2 = 0.33$), Fe (0.11), and Ca (0.30), all possible impurities in sand or other silica sources. Correlation of Al and Fe (0.19) is consistent with silica sources of these elements. Interestingly, Ca does not correlate with Mg (0.05), so that Ca may have come more from the silica than the alkali source. On the other hand, Al also somewhat correlates with K (0.16) and Na (0.11), so that alumina may have entered from both silica and alkali.

The observed overall composition is similar to that of 13 10th-11th century glasses from Kota Cina, Sumatra (63.5% SiO_2, 18.5% Na_2O, 3.6% CaO, 3.0% K_2O, 5.6% MgO, 4.7% Al_2O_3, and 1.2% Fe_2O_3 (11), although the Thai glasses have more CaO and Al_2O_3. The compositions, however, are distinct from lead and potash glasses manufactured in Asia (12). There also are similarities to the composition of nine Indian glasses analyzed by Brill (7) (68.3% SiO_2, 17.0% Na_2O, 2.9% CaO, 2.6% K_2O, 5.9% Al_2O_3, and 2.0% Fe_2O_3), although the Thai glasses have significantly higher levels of Ca, which was considered by Brill to be a key factor. Brill considered that the Sumatra glasses were close enough to the Indian composition to warrant a likely attribution to Indian manufacture (the Sumatra sites, like the Thai sites, showed no evidence for local manufacture). Although we cannot attribute the Thai glass exclusively to an Indian source because of the high levels of CaO, the high levels of Al_2O_3 are not characteristic of a Western Asian source. It is possible that there was a mixture of Indian and Western Asian sources.

Colorants

The blue colors of Groups 1 and 2 are caused, at least in part, by the presence of cobalt. The mean Co concentration for the dark blue Group 1 is 0.12% (calculated as CoO), and for the blue Group 2 is 0.06%. None of the light blue group 3 samples has more than 0.01% CoO, so that this element does not contribute to the color. Moreover, with one exception (6R at 0.04%), none of the remaining samples has CoO in excess of 0.02%. Copper also clearly contributes to the blue coloring, as Group 1 has a mean CuO concentration of 0.19% and Group 2 of 0.07. All other groups have mean CuO concentrations less than 0.02%, except for Groups 4 and 5, both at 0.026%. The correlation between cobalt and copper levels ($r^2 = 0.78$) is the highest by far for pairs of elements. The glassmakers clearly were using both elements (either from a single mineral source or from two sources) to obtain blue colors. It is interesting that Co and Cu covary with no other element except slightly with Fe ($r^2 = 0.21$ for Co and 0.18 for Cu), and that the intercept of the Cu/Co correlation is almost zero (0.01). This similarity of behavior is suggestive but not compelling for a single source of the blue colorants, a source that also contained Fe. The negligible levels of Pb indicate that Cu was not provided from bronze residues. The absence of a correlation between Cu and Zn ($r^2 = 0.04$) eliminates brass residues as a source of Cu.

Iron is present in significant concentration in all groups. At the high end are Groups 1 (2.70% calculated as Fe_2O_3), 2 (2.16%), 6 (2.37%), 7 (3.40%), and 8 (2.15%). Thus Fe contributes to the darker blues and greens. At the low end are

Groups 3 (light blue, 0.77%) and 13 (colorless, 0.77%). All other groups have Fe_2O_3 in the range 1.19-1.89%.

The second lowest level of Mn (calculated as MnO) is found in the light blue Group 3 (0.28%), so that this low intensity hue is achieved by reduced levels of both Fe and Mn. Interestingly, Group 7 (dark green) has the lowest level of MnO (0.17%). All other groups have at least 0.95% MnO. Particularly high levels are present in Groups 8 (greenish yellow, 2.85%), 9 (yellow violet, 2.30%), and 10 (pale yellow, 2.07%), so there is an association of the yellow variants with Mn. All other groups are in the range 0.94-1.80%. It is possible that Mn is serving as a decolorant in some cases. Thus in Group 8 the high level of Fe_2O_3 (2.15%) is juxtaposed with a very high level of MnO (2.85%), resulting in a low intensity greenish yellow hue. Similarly in Group 9, 1.48% Fe_2O_3 and 2.30% MnO result in a yellow violet color, and in Group 10, 1.59% Fe_2O_3 and 2.07% MnO result in a pale yellow color. Although calculated as MnO, Mn could exist in different oxidation states.

Nickel concentrations are extremely variable. Groups 4 and 5 have none (0.00%), and one sample (3A) has 1.13% (calculated as NiO). On the whole, Ni probably does not contribute significantly to coloring. Somewhat high levels are found in Groups 7 (0.23%), 8 (0.17%), 11 (0.23%), 13 (0.17%), and 14 (0.25%), i.e., both dark greens and lightly colored materials. Although NiO is known to cause a brown color, except in 3A the concentrations are too low for a direct effect.

Although Zn and Ti are associated with modern white pigments, both are present in significant amounts in these samples, but with no coloring pattern. Zn is concentrated only in Groups 1 (0.16%, calculated as ZnO) and 2 (0.42%), and in 11 samples of Group 3 (0.43%) (the remaining Group 3 samples all have 0.00% ZnO). The largest level (1.41%) is found in sample 2A. The remaining color groups have less than 0.1% ZnO. Ti is found in all groups in a somewhat uniform fashion, probably as an impurity in the raw silica, the lowest mean (Group 9) being 0.08% (calculated as TiO) and the highest (Group 8) 0.18%. Group 11 actually has a mean of 0.27%, but almost all the material is found in the single sample 11D (0.78%), the highest single value found. Thus we do not believe that these elements are contributing to coloring.

Statistical Comparison of Materials from the Two Sites

In order to determine whether there was communication or trade between the two sites, K and L, we conducted a statistical analysis of the data. If the composition of materials at the two sites proved to be identical, it is likely that the inhabitants of K and L communicated or traded with each other. Alternatively, but less likely, they traded with a common partner. On the other hand, significant differences in glass from the two sites could imply either that there was little trade between them or that burial conditions were different, so that their compositions diverged over time.

We used the SPSS/PC$^+$ software package for all statistical calculations. A normal distribution of data is preferred for these calculations. Measures of skewness (symmetry of the distribution) and kurtosis (relative proportions in the tail and in the main portion) indicated that almost all elements were not distributed normally. Such an observation is not at all surprising for the colorants. Thus the high levels of Co

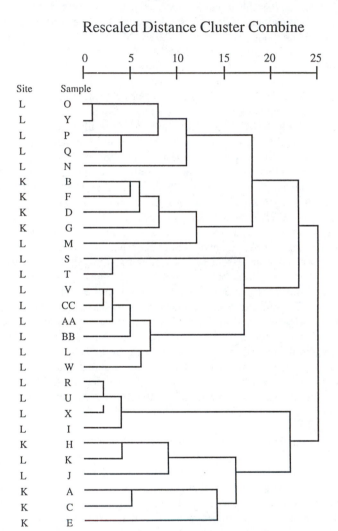

Figure 1. Results of clustering analysis for Group 3. The site K is Ko Kho Khao and the site L is Laem Pho. Samples are further specified alphabetically.

in the samples of Groups 1 and 2 results in a very large positive kurtosis. The alkali metals (Na and K) on the other hand are normal in both skew and kurtosis. Transformation of the data by simply taking the logarithm improved the situation appreciably, but use of the formula of Baxter (*13*) was even better. Thus all statistical calculations in this study used Baxter-transformed data. Cr, Pb, Sn, and V were excluded from consideration as they were at or below the detection limit in almost all samples, and Co was removed since its concentration clearly was related only to the single factor of the deep blue colors of Groups 1 and 2.

Cluster analysis of all samples using the remaining elements resulted in separation based almost entirely on color. To effect a separation based on site, it therefore is necessary initially to remove elements that determine color. With the groupings based entirely on color (Table I), we performed the ONEWAY test on all elements to determine which ones vary across these groups. Surprisingly, every single element exhibits variation by color at the 99% confidence level or higher. This study included Al, Ba, Ca, Cu, Fe, K, Mg, Mn, Na, Ni, Ti, and Zn. Each element either directly influences color or covaries with another element that does. Thus we cannot selectively exclude elements based on color in order to study site, as we would exclude all elements.

ONEWAY analysis based on site indicated that Al, Ca, K, Mg, Mn, and Zn are significantly different between the sites (see Table IV for elemental averages by site). Such differences, however, could be the result of accidental differences in color. We carried out a multivariate analysis of variance with both color and site as variables with the ANOVA procedure of SPSS/PC$^+$. We found that Ca, K, Na, Ti, and Zn covary with site and color. Thus, of the elements that differ between site, only Al, Mg, and Mn do not also vary with color. We were left with just three elements with which to determine differences between sites. We concluded that such an analysis would not be useful.

As an alternative, we sought differences between sites within each color grouping of Table I. We excluded Groups 4, 8, 11, 12, and 14, as they have too few samples, and Groups 7, 9, 12, and 13, as they come from essentially a single site. For the remaining six groups separately, we carried out the ONEWAY analysis by site and the CLUSTER procedure of SPSS/PC$^+$ based on agglomerative hierarchal clustering (sorting samples into clusters based on maximal internal similarities and external differences). The results of the cluster analysis are presented as dendrograms.

Group 1. The dark blue glasses (six from K, three from L) show no differences between sites for all elements but potassium (99%). Such an alkali metal, however, may be sensitive to diagenesis. The dendrogram (not shown; see ref. *5*) showed no separation of samples based on site.

Group 2. This small set of seven blue samples shows significant differences between sites for Ba and Ti at the 99% level and Ca and Mn at the 95% level. Cluster analysis resulted in two clusters, one composed of three samples from K (2G, 2H, 2I) and a second composed of three samples from L (2A, 2C, 2D) and one from K (2B). Thus within this small group there are differences by site.

Group 3. Ca, K, and Zn show differences between sites at the 99% level for these 28 light blue materials. The samples separate into five clusters (Figure 1). The top cluster consists of five L samples, the second of four K and one L samples, the third of eight L samples (this one may be considered as two clusters), the fourth

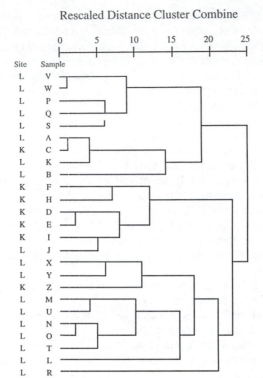

Figure 2. Results of clustering analysis for Group 6.

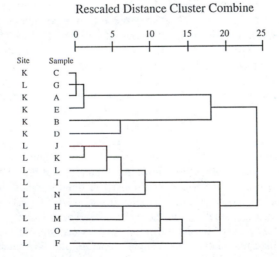

Figure 3. Results of clustering analysis for Group 10.

of four L samples, and the fifth of four K and two L samples. Thus the light blue glasses are separated well by site.

Group 5. Only Ni is found to differ between the sites for these 17 bluish green samples. As a result, the clustering procedure effects no separation by site.

Group 6. In this group of 25 light green samples, Ca, Fe, K, Mg, and Na all show significant differences between the sites. These are significant at the 99% level, except for Ca at 95% and Mg at 99.9%. This multiplicity of elements provides some separation by site in the cluster analysis (Figure 2). Five clusters and a single-element residue are evident. The first (from the top) and fifth clusters are composed entirely of L samples, and the third entirely but one of K samples. The second cluster has three out of four samples from L, and the fourth has two out of three from L. Thus clustering of this group by site is relatively effective.

Group 10. Ca, Cu, K, Na, and Ti show significant differences in these 15 pale yellow samples, the first three elements at the 99.9% level and the last two at the 95% level. Cluster analysis (Figure 3) shows an almost perfect separation between sites. The pattern is of two to four clusters. Five of the six in the top one or two clusters are from K, and all nine of the samples in the bottom one or two clusters are from L.

Thus four of the color groups (2, 3, 6, and 10), representing 75 of the samples, or about half, yielded an excellent separation by site according to composition. Two others (1 and 5), representing 26 samples, showed little or no separation by site. The remaining samples were excluded as described above. For a more comprehensive view, we combined Groups 2-6 (light blues and greens) into one large group and 8-13 (yellows) into another for cluster analysis. The results again were dominated by differences due to color rather than site.

Pairwise comparison between elements was examined for each site as well as for the combined set of materials. Site K had a much stronger correlation (0.90) between Co and Cu (and both of these with Fe) than did L (0.44), although the differences may simply represent poor statistics (L has only five pieces with significant Co). There are some interesting differences in the correlations with Ti. The Ti/Ca (0.45) and Ti/Fe (0.20) correlations are higher for K than for L (0.25, 0.09), and the Ti/Al correlation (0.14) is lower for K than for L (0.41). Although these differences are small, they may result from different proportions of Ca and Al in the silica and alkali sources, or differential loss of the elements from diagenesis.

Conclusions

Thai glass samples from the Kra Isthmus for the most part are of mixed alkali matrix, although some have the more traditional soda-lime composition. The blue, green, and yellow colors are explained in terms of normal effects of Co, Cu, Fe, and Mn. Because the effects of color dominated clustering attempts, we examined large color groupings separately in order to determine whether composition is related to site. For Groups 1 (dark blue) and 5 (bluish green), representing 26 samples, no compositional differences or clustering by site was observed. These samples are about equally distributed between K samples (14) and L samples (12).

For Groups 2 (blue), 3 (light blue), 6 (light green), and 10 (pale yellow), there are clear differences in elemental composition that result in discernible clustering of the samples based on site. Of the 75 samples in these four groups, 51 are from L and 24 from K. Although different elements are found to provide these distinctions, Ca differs between sites for all four of these groups, K differs in three of them, and Na and Ti in two of them. These differences may be attributed either to differences in diagenetic effects at the two sites or to differences between the original compositions at the two sites. The cluster analysis shows that there probably were many sources of the raw materials for the glass.

Acknowledgments

The authors thank the National Science Foundation for a grant to purchase the inductively coupled plasma instrument (DIR-9001644). We thank Mr. Paul A. Schittek for carrying out the interelemental correlation analysis.

Literature Cited

(1) Sayre, E. V.; Smith, R. W. *Science*, **1961**, *133*, 1824-1826.
(2) Heyworth, M. P.; Hunter, J. R.; Warren, S. E.; Walsh, N. *Archaeometry*; Maniatis, Y., Ed., Elsevier: Amsterdam, 1989, pp 661-670.
(3) Casoli, A.; Mirti, P. *Fresenius J. Anal. Chem.* **1992**, *344*, 104-108.
(4) Catterick, T.; Wall, C. D. *Talanta*, **1978**, *25*, 573-577.
(5) Johnson, S. C., **1993**, Ph.D. dissertation, Northwestern University, Evanston, IL.
(6) Goffer, Z. *Archaeological Chemistry: a Sourcebook on the Applications of Chemistry to Archaeology*; Wiley-Interscience: New York, NY, Chapter 9, 1980.
(7) Brill, R. H. In *Archaeometry of Glass*, Bhardwaj, H. C., Ed.; Indian Ceramic Society: Calcutta, 1987, section 1, pp 1-25.
(8) Hench, L. L. In *Proceedings of the XI International Congress on Glass*; Survey Papers Vol. II: Prague, 1977, pp 343-369.
(9) Hench, L. L.; Clark, D. E. *J. Non-Crys. Solids* **1978**, *28*, 83-105.
(10) Bayley, J.; Wilthew, P. In *Proceedings of the 24th International Archaeometry Symposium*; Olin, J. S., and Blackman, M. J., Eds.; Smithsonian Institution Press: Washington, DC, 1986; pp 55-62.
(11) McKinnon, E. E.; Brill, R. H. In *Archaeometry of Glass*, Bhardwaj, H. C., Ed.; Indian Ceramic Society: Calcutta, 1987; section 2, pp 1-14.
(12) Brill, R. H. In *Nara Symposium '91 Report: UNESCO Maritime Route of Silk Roads*; 1993, pp 7-79.
(13) Baxter, M. J. *Archaeometry* **1992**, *34*, 267-277.

RECEIVED October 9, 1995

Chapter 3

Chemical Chronology of Turquoise Blue Glass Trade Beads from the Lac-Saint-Jean Region of Québec

R. G. V. Hancock[1], S. Aufreiter[1], J.-F. Moreau[2], and I. Kenyon[3]

[1]Slowpoke Reactor Facility and Department of Chemical Engineering and Applied Chemistry, University of Toronto, Toronto, 200 College Street, Ontario M5S 1A4, Canada
[2]Laboratoire d'archéologie et Départment des sciences humaines, Université du Québec à Chicoutimi, Chicoutimi, Québec G7H 2B1, Canada
[3]Ontario Heritage Foundation, Toronto, Ontario M5C 1J3, Canada

Eighty turquoise blue glass trade beads from three archaeological sites in the Lac-Saint-Jean region of Québec have been chemically characterized by instrumental neutron activation analysis (INAA). Comparison of their individual chemistries with previously established chemistries of well dated beads from archaeological sites in Ontario, Nova Scotia and New York State allows us to estimate the tentative ages of the Québec beads, and hence to establish the time periods over which each of the three sites were in use.

In exchange for furs provided to European traders, the aboriginal peoples of North America obtained many kinds of trade goods including glass beads. On the basis of their morphology and archaeological contexts, certain glass bead types have been found to be an efficient tool for establishing both regional and continental cultural chronologies (1). Unfortunately, based on appearance alone, many common beads lack chronological specificity. We have established that chronological ordering could be achieved for the ubiquitous turquoise blue glass beads using their chemical compositions (2, 3). This is needed for sites that do not produce characteristic beads and for sites that have complex occupational histories.

The aim of this paper is to use instrumental neutron activation analysis (INAA) data to establish the chronologies of three sets of turquoise blue glass beads from the Lac-Saint-Jean area of Québec, relative to the chronology established for similarly colored beads from Ontario, New York and Nova Scotia. The three Québec sites were occupied for many centuries and so have occupational histories that are difficult to disentangle. As a result, it was necessary to analyze large numbers of samples using a technique that would provide non-destructive bulk elemental analysis.

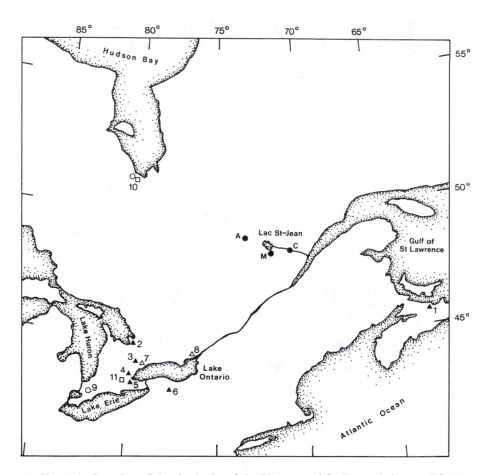

Figure 1. Location of sites in the Lac-Saint-Jean area of Québec and sites used for
comparison in Ontario, Nova Scotia and New York State. ● : Sites in Lac-Saint-Jean
region; A = Ashuapmuchuan, C = Chicoutimi, M = Metabetchouan; ▲ : Early
French regime (1580-1650) 1 = Pictou site (Nova Scotia), 2 = Molson, Ball, Auger,
Ossossane, and Train sites (Ontario), 3 = Kleinburg site (Ontario), 4 = Tregunno site
(Ontario), 5 = Burke and Sealey sites (Ontario), 6 = Adams, Cameron, Cornish, and
Warren sites (New York State); Δ : Late French regime (1660-1760) 7 = Bead Hill
site (Ontario), 8 = Fort Frontenac (Ontario); O : Early British regime (1760-1840) 9
= Bellamy site (Ontario), 10 = Moose Factory Level III (Ontario); □ : Victorian era
(1840-1900) 10 = Moose Factory Level I; 11 = Mohawk Village (Ontario).

Site Descriptions

The Lac-Saint-Jean area covers a hydrographic watershed of around 90,000 km^2, located to the northeast of Québec city. Its main feature consists of Lac-Saint-Jean, that empties into the St. Lawrence River via the 200 km long Saguenay River. The Saguenay River, and the rivers that feed Lac-Saint-Jean, provides a water access route north to Hudson Bay and west to the Saint-Maurice and Ottawa Rivers.

Among more than three hundred archaeological sites in the area, three (Ashuapmuchuan, Chicoutimi and Metabetchuan) have provided large (>200) glass bead collections (see Figure 1) (*4, 5*).

The Ashuapmuchuan site is located in the highlands, about 125 km northwest of Lac-Saint-Jean. It is a multicomponent site, whose first clearly defined component is Middle Woodland (5th to 10th century A.D.). A Late Woodland component (14th century) may extend to the early 17th century, with the arrival of European trade goods (*6-9*, and Moreau, J.-F.; Langevin, E. *Rapport de fouille, site DhFk-7, lac Ashuapmouc-houane (Lac-Saint-Jean), été 1990, Chicoutimi*, Université du Québec à Chicoutimi, Laboratoire d'archéologie, **1992**). Although European trade goods may have arrived at the site up to the 19th century, many of them, on typological and distributional grounds (*7-8* and Moreau, J.-F.; Langevin, E. *Rapport de fouille, site DhFk-7, lac Ashuapmouchouane (Lac-Saint-Jean), été 1990, Chicoutimi*, Université du Québec à Chicoutimi, Laboratoire d'archéologie, **1992**), seem to be earlier (*9*).

The Chicoutimi site is about 125 km northwest of the confluence of the Saguenay and St. Lawrence Rivers. It consists of two stratigraphically distinct components. The lower component, called *indian couche*, includes mainly aboriginal materials from the end of the Middle Woodland Period and the beginning of the Late Woodland Periods (11th century), through to European trade goods of the early 17th century (*10*). The second component consists mainly of European trade goods from a non-palisaded, 18th century trading post, with material from the late 17th to the 19th century (*11*).

The Metabetchuan site is located on the south shore of Lac-Saint-Jean, at the mouth of the Metabetchuan river. It may have been occupied as early as the Late Archaic Period, with Woodland Period, especially Late Woodland Period occupations confirmed by diagnostic pottery sherds (Laliberté, M.; Moreau, J.-F. *DcEx-1 - Un site traditionnel d'échange sur les berges du lac Saint-Jean*, Université du Québec à Chicoutimi, Laboratoire d'archéologie, **1988a**, and Laliberté, M.; Moreau, J.-F. *DcEx-1: les résultats de la campagne de fouille de 1987*, Université du Québec à Chicoutimi, Laboratoire d'archéologie, **1988b**). European contact components extend from the early 17th century to the 19th century.

Analytical Procedure

All eighty turquoise blue (copper-based) glass bead samples were assembled for non-destructive instrumental neutron activation analysis using the SLOWPOKE Reactor Facility at the University of Toronto (*2, 12*). Beads of mass 20 mg, stored in 1.2 mL polyethylene vials, were irradiated serially for five minutes at a neutron flux of 1.0×10^{12} neutrons cm^{-2} sec^{-1}. Two to five minutes after irradiation, their radioactivity was counted for five minutes using a hyper-pure germanium detector-based gamma-ray spectrometer.

Table I. Aluminum-, Antimony- and Cobalt-rich Glass Beads from
Ashuapmuchuan (A), Chicoutimi (C) and Metabetchuan (M)

	Al %	Ca %	Cl %	Co ppm	Cu %	Mn ppm	K %	Na %	Sn ppm	As ppm	Sb ppm
Aluminum-rich											
A51	2.13	2.7	≤0.16	≤25	0.43	75	≤1.6	12.9	≤1100	940	≤140
A54	2.24	2.8	≤0.14	≤38	0.44	84	≤1.7	14.2	≤1300	960	≤150
Antimony-rich											
A52	0.38	1.4	0.86	≤49	1.74	260	5.8	10.2	≤860	980	18600
C71	0.26	2.7	0.41	≤38	1.64	145	9.6	7.7	≤750	430	10400
Cobalt-rich											
A53	0.48	2.7	1.92	144	1.65	310	≤1.3	12.8	≤880	≤140	620
A55	0.59	4.8	2.15	172	1.13	170	≤1.5	13.2	≤800	360	≤100
C70	0.61	4.2	1.60	150	1.06	1940	≤1.5	12.7	2500	450	≤100
M56	0.67	5.2	2.15	200	0.80	230	≤1.6	13.3	≤1010	≤200	≤180
M58	0.59	4.8	2.15	179	1.06	610	≤1.6	13.2	≤1010	630	420
M59	0.68	5.0	2.44	185	0.85	230	≤1.5	13.8	≤870	300	≤110
M60	0.70	5.5	2.51	190	0.89	240	≤1.9	14.0	≤1000	≤190	≤170
M61	0.66	5.6	1.85	189	0.83	210	≤1.6	13.0	≤770	≤170	≤150
M62	0.65	5.1	2.24	202	0.84	250	≤1.4	13.4	≤890	290	≤110
M63	0.65	5.2	2.00	189	0.81	230	≤1.5	13.1	≤930	≤160	≤160
M65	0.58	4.0	1.96	180	1.16	320	≤1.2	12.2	≤780	460	≤90

Table II. Tin-rich Glass Bead Data from Ashuapmuchuan (A), Chicoutimi (C)

	Al %	Ca %	Cl %	Co ppm	Cu %	Mn ppm	K %	Na %	Sn ppm	As ppm	Sb ppm
Tin-rich											
A43	0.56	4.0	1.23	≤16	0.77	370	3.2	12.3	8600	≤90	≤100
A44	0.68	5.6	1.35	≤20	0.71	1500	≤1.5	13.3	10900	≤100	≤120
A45	0.60	4.8	1.14	≤20	0.61	1200	3.3	11.7	8400	≤90	≤100
A46	0.59	4.2	1.20	≤22	0.62	1300	3.0	11.7	9300	≤90	≤100
A47	0.59	4.4	0.95	≤14	0.60	1230	3.1	11.0	9400	≤80	≤90
A48	0.62	5.2	1.14	≤18	0.60	1300	3.2	11.1	9800	≤110	≤120
A49	0.60	4.8	1.15	121	0.57	1200	3.2	11.2	8900	≤100	≤110
C29	0.66	5.5	1.03	≤17	0.35	1660	3.9	11.0	7700	≤80	≤100
C30	0.74	5.7	1.18	≤23	0.41	1460	4.4	10.9	9500	≤100	≤110
C31	0.71	5.0	0.97	≤17	0.45	1570	4.4	9.3	11200	≤80	≤70
C32	0.64	5.4	1.11	≤20	0.52	1650	4.0	11.5	8700	≤80	≤90
C33	0.62	5.0	0.98	≤17	0.51	1500	3.4	10.1	11200	≤140	≤140
C34	0.68	5.7	1.04	≤17	0.50	1610	3.0	10.4	11300	≤110	≤110
C35	0.63	5.9	0.97	≤17	0.50	1580	3.6	10.5	11300	≤120	240
C36	0.63	5.9	0.85	≤15	0.32	1570	4.1	10.4	7500	≤140	≤140

This produced analytical concentration data for cobalt (Co), tin (Sn), copper (Cu), sodium (Na), aluminium (Al), manganese (Mn), chlorine (Cl) and calcium (Ca). The samples were recounted for five to thirty-three minutes the next day to measure the concentrations of the longer-lived radioisotopes of sodium (Na), arsenic (As), antimony (Sb) and potassium (K).

The Na measurements were used to link both counts. Elemental concentrations were calculated using the comparator method (*12*). Beads of larger mass were irradiated at proportionately lower neutron fluxes to generate enough radioactivity for reasonable chemical analyses.

The Al content was not corrected for the ^{28}Al produced by the ^{28}Si(n,p)^{28}Al nuclear reaction. This neglected correction may account for $\leq 0.12\%$ and, although important in absolute terms, is not relevant to the separation of the beads in this data set.

Results and Discussion

The analytical data for beads from the three sites in the Lac-Saint-Jean area are presented in Tables I to V. They have been sorted into groups based on their absolute and relative elemental contents. Each sample is identified by its provenance (A = Ashuapmuchuan, C = Chicoutimi, M = Metabetchuan). Some chemical groups are based primarily on high single element concentrations, including Al (2 samples; Table I), Sb (2 samples; Table I), Co (11 samples; Table I) and Sn (15 samples; Table II). The remainder of the beads were sorted using Ca, Cl, K and Na, since concentrations of these elements have been found to vary over time in turquoise blue beads (*2, 3*), giving rise to chemical groupings labelled:

Type I	Ca <4%	Cl/Na <0.14	K/Na <0.33	13 samples (Table III)
Type II	Ca >4%	Cl/Na >0.14	K/Na <0.33	16 samples (Table IV)
Type III	Ca >4%	Cl/Na >0.14	K/Na >0.33	21 samples (Table V)

Figure 2, a plot of Na versus Cl, illustrates some aspects of this separation. The aluminum-rich samples are clearly separated from the other beads, along with one of the antimony-rich samples. The tin-rich beads cluster to the left of the Type I beads, which in turn cluster to the left of the cobalt-rich beads. Below the latter two groups are the partially separated Type II and Type III groups.

A summary of the pre-bead-analysis archaeologically-derived chronological expectations is presented in Table VI. The larger beads from each site were expected to be from the 17th century, while the smaller beads were expected to span from the mid-17th century to the 19th century.

A simple plot of bead diameter versus length (Figure 3) shows that, while the small beads cluster together, the larger beads with Type I and tin-rich chemistries are clearly separable. Among the smaller beads, the cobalt-rich ones tend to be slightly more donut-shaped (i.e. short relative to their diameters) than Type II and Type III beads.

Possible Chronologies. Previous studies of 16th to the 20th century turquoise blue (copper-based) glass beads from Ontario and New York State (*2, 3*) showed enough differences in their elemental contents to allow the authors to establish a provisional chronology. This framework will be used below to assess the beads from the three Québec sites.

Table III. Type I Chemistry for Glass Beads from Ashuapmuchuan (A), Chicoutimi (C) and Metabetchuan (M)

	Al %	Ca %	Cl %	Co ppm	Cu %	Mn ppm	K %	Na %	Sn ppm	As ppm	Sb ppm
A37	0.53	2.5	1.27	≤18	0.81	270	3.3	12.4	≤780	≤170	≤170
A38	0.51	2.6	1.35	≤15	0.78	250	≤1.4	11.9	≤670	≤160	≤160
C01	0.70	4.3	1.59	≤20	0.76	170	≤1.5	12.5	≤790	470	≤180
C03	0.58	2.3	1.51	≤18	1.19	390	≤1.4	12.9	1500	≤170	≤170
C04	0.56	2.0	1.44	≤14	1.17	380	≤1.3	12.6	≤780	≤180	≤190
C05	0.53	3.1	1.48	≤15	1.05	165	2.7	12.1	≤600	≤160	≤170
C06	0.51	2.3	1.53	≤13	0.91	120	≤1.4	12.3	≤540	≤170	≤180
C07	0.56	2.3	1.44	≤18	1.40	370	≤1.4	12.7	≤730	≤180	≤180
C08	0.52	3.7	1.34	42	0.99	230	2.9	12.0	≤670	≤190	≤190
M39	0.66	3.0	1.54	35	0.77	190	3.6	12.0	≤540	200	≤110
M40	0.64	2.9	1.46	49	0.74	190	2.8	12.3	≤500	220	≤100
M41	0.59	3.9	1.44	≤14	0.66	150	≤1.4	12.4	≤650	≤90	≤100
M42	0.68	2.8	1.47	49	0.75	195	3.2	12.2	≤680	≤150	≤140

Table IV. Type II Chemistry for Glass Beads from Chicoutimi (C) and Metabetchuan (M)

	Al %	Ca %	Cl %	Co ppm	Cu %	Mn ppm	K %	Na %	Sn ppm	As ppm	Sb ppm
C12	0.72	5.7	1.81	≤25	0.66	380	3.0	10.0	≤820	≤130	570
C16	0.73	5.8	1.78	≤23	0.92	440	≤1.2	10.9	≤790	≤120	310
C19	0.80	6.0	1.68	≤18	0.89	1420	2.4	10.1	≤840	≤160	720
C20	0.57	5.5	1.68	≤17	1.20	1100	≤1.2	10.1	≤790	≤150	≤140
C22	0.54	4.4	1.84	66	1.18	650	2.7	10.1	1400	200	≤80
C24	0.98	6.0	1.56	≤29	1.11	600	2.9	9.1	≤980	160	280
C27	0.74	7.0	1.90	≤21	0.80	340	3.2	10.8	≤710	≤80	940
M57	0.68	5.9	2.17	89	1.16	340	≤1.3	11.3	≤870	550	≤150
M73	0.67	5.8	2.32	84	1.18	340	≤1.1	11.0	≤821	460	≤140
M74	0.68	6.4	1.93	88	1.16	310	≤1.1	10.9	≤790	550	≤120
M75	0.67	6.5	2.09	73	1.19	350	≤1.4	11.1	≤780	510	≤90
M76	0.68	6.7	2.08	86	1.25	280	≤1.1	10.8	≤800	460	≤901
M77	0.68	6.1	2.14	85	1.17	340	≤1.4	10.9	≤780	420	≤90
M78	0.64	6.2	1.83	80	1.16	300	≤1.3	10.6	≤680	450	≤100
M79	0.67	6.1	2.16	85	1.21	310	≤1.4	11.0	≤820	510	≤130
M80	0.66	6.5	2.19	74	1.25	300	≤1.2	11.1	≤780	490	≤90

**Table V. Type III Chemistry for Glass Beads from Ashuapmuchuan (A),
Chicoutimi (C) and Metabetchuan (M)**

	Al %	Ca %	Cl %	Co ppm	Cu %	Mn ppm	K %	Na %	Sn ppm	As ppm	Sb ppm
A50	0.66	5.7	1.50	≤18	0.56	1080	4.3	8.8	≤800	≤80	≤80
C02	0.67	2.4	1.87	≤31	1.19	180	≤3.0	11.2	≤1000	≤29	≤270
C09	0.84	5.8	1.77	≤31	1.06	1250	4.5	9.9	≤1000	≤130	720
C10	0.72	5.6	1.88	≤28	0.67	410	3.3	9.8	≤870	≤90	730
C11	0.79	6.4	2.04	53	0.87	350	3.5	9.9	≤860	≤130	580
C13	0.54	4.5	1.91	≤22	1.08	210	5.3	10.1	≤750	≤70	190
C14	0.52	3.9	1.75	35	1.29	190	4.3	9.9	≤690	≤160	≤170
C15	1.04	6.4	1.64	46	1.19	550	3.3	9.3	≤900	≤120	300
C17	0.54	4.6	1.87	≤21	1.11	210	4.1	10.0	≤720	≤80	190
C18	0.78	5.7	1.52	≤17	0.74	670	3.0	8.6	≤883	160	250
C21	0.54	4.1	1.88	≤24	1.11	190	4.7	10.3	≤820	≤160	≤150
C23	0.30	5.5	1.52	32	0.77	510	4.8	9.0	≤560	220	490
C25	1.01	5.8	1.52	≤18	1.18	570	3.6	9.0	≤960	190	260
C26	0.53	7.3	2.05	33	0.92	390	3.6	10.3	≤640	≤140	≤140
C28	0.78	5.5	1.67	≤28	0.71	690	4.4	8.7	≤930	210	260
C66	0.68	6.0	1.49	≤17	0.98	560	4.6	9.4	≤770	≤110	440
C67	0.56	6.1	1.59	42	1.02	550	3.8	9.3	≤650	≤130	460
C68	0.69	5.7	1.78	≤20	0.62	660	4.3	9.2	≤740	≤110	≤110
C69	0.68	5.9	1.59	≤17	0.58	1100	4.5	9.3	≤800	≤140	≤120
C72	0.71	6.3	1.52	≤23	1.19	490	4.7	9.3	≤1000	≤110	≤110
M64	0.74	6.0	1.93	50	0.76	320	3.3	9.4	≤800	≤100	500

**Table VI. Pre-analysis Archaeological Chronologies, Bead Dimensions,
Chemical Groups and Provisional Chemical Chronologies**

Samples	Archaeological Chronology*	Bead Size length x diameter	Chemical Groups	Chemical Chronology**
C01-C08	≈1625-1675	4.0x6.7-11.7x6.6 mm	I:01,03-08 III:02	≈1600-1750
C29-C36	≈1625-1675	5.8x3.6-9.4x4.2 mm	Sn:29-36	≈1640-1700
M39-M42	≈1625-1675	4.5x6.6-5.2x6.8 mm	I:39-42	≈1600-1650
A37-A38	≈1625-1675	4.6x6.1-4.6x6.3 mm	I:37,38	≈1600-1650
A43-A49	≈1625-1675	6.1x4.4-7.5x4.5 mm	Sn:43-49	≈1640-1700
C09-C28	≈1650-1725	1.4x2.4-3.4x3.1 mm	II/III:09-28	≈1700-1850
A52-A55	≈1650-1725	1.3x1.9-1.9x2.7 mm	Al:54; Sb:52 Co:53,55	≈1650-1950
M56-M65	≈1650-1725	1.2x2.1-1.7x3.5 mm	Co:56,58-63,65 II:57; III:64	≈1650-1750
C66-C72	≈1650-1725	1.7x2.7-2.8x3.3 mm	Co:70; Sb:71 III:66-69,72	≈1650-1900
M73-M80	≈1650-1725	2.3x2.9-2.7x3.0 mm	II(Co):73-80	≈1650-1750
A50-A51	≈1700-1800+	1.8x2.1-1.8x2.7 mm	III:50; Al:51	≈1750-1950

*   Based on shape, size and colour.
** Based on chemical similarities to known-age site beads (*2, 3*).

Summary:

A	≈1625-1800+	≈1600-1950
C	≈1625-1725	≈1600-1850
M	≈1650-1725	≈1650-1900

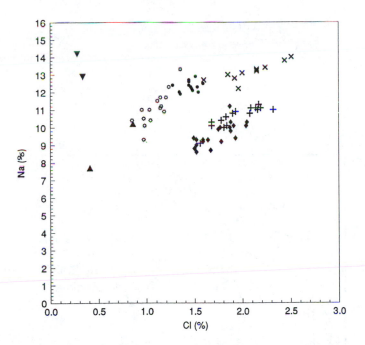

Figure 2. Scattergram of Na versus Cl. • : Type I; + : Type II; ♦ : Type III; ✿ : tin-rich; × : cobalt-rich; ▲ : antimony-rich; ▼ : aluminum-rich.

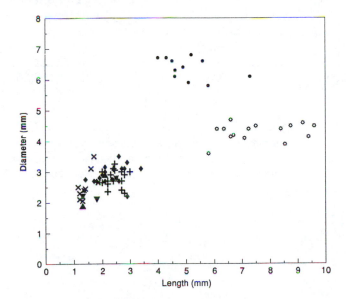

Figure 3. Scattergram of bead length versus bead diameter. • : Type I; + : Type II; ♦ : Type III; ✿ : tin-rich; × : cobalt-rich; ▲ : antimony-rich; ▼ : aluminum-rich.

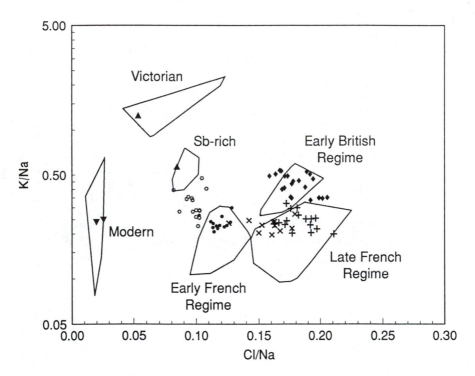

Figure 4. Scattergram of K/Na ratio versus Cl/Na ratio. • : Type I; + : Type II;
♦ : Type III; ✿ : tin-rich; × : cobalt-rich; ▲ : antimony-rich; ▼ : aluminum-rich.
The "convex hulls" were constructed from the comparative data for dated
Ontario and Nova Scotia sites, as described in the text. Note that the K/Na scale
is logarithmic.

Unlike the Québec sites, which were intermittently occupied for centuries, the comparative data derive from sites or portions of sites from Ontario and New York State that for the most part could be dated to periods of less than 60 years duration. On the basis of their archaeological and historical context, the comparative bead samples could be attributed to five periods (the latest beads were taken from ethnographic beadwork specimens) spanning a time of 350 years:

(1) Early French regime (1580-1650)
(2) Late French regime (1660-1760)
(3) Early British regime (1760-1840)
(4) Victorian era (1840-1900)
(5) Early twentieth century (1900-1930)

Changing sources of alkalis, and possible glass making recipes, over the study period, emerged in the glass bead chemistries, as indicated by varying concentrations of Na, K and Cl (*3*). Also, the use of As and Sb in later bead samples was found to be consistent with the historical record. For turquoise blue beads opacified with As or Sb, and for the early twentieth century, low Cl beads, there appears to be specific chemistries which might allow the dating of individual beads.

Figure 4, in which log(K/Na) is plotted against Cl/Na, serves two purposes. The first is to sort the Québec beads by their relative alkali metal and Cl content, while the second is to compare these data with the chemical groupings found earlier (*3*). The "convex hulls" or outlines of Figure 4 were developed from previously published data for the later periods (*3*) and from some unpublished data for the Ball, Auger and Pictou sites (see Figure 1) for the Early French regime. In constructing the convex hulls, a few outliers were removed.

In the Québec data, the two aluminum-rich samples sit within the zone allocated to modern 20th century turquoise blue beads, which are typically low in chlorine. One of the antimony-rich beads is located in the antimony-rich zone (*3*), while the other sits in the Victorian bead zone (*3*). The cobalt-rich beads scatter within or near the Late French regime zone (*3*). Most of the beads sampled from the Late French regime site of Bead Hill in Ontario also contained relatively high amounts of Co, unlike beads from earlier or later periods. The 15 tin-rich beads [only 3 tin-rich beads were found in the Ontario bead assemblages (*2*)] scatter between the antimony-rich zone and the Early French regime zone. The Type I beads cluster in the Early French regime zone.

The Type II beads scatter mainly within the Late French regime zone. The Type III beads scatter in and around the Early British regime zone, and seem to be separated from the Type II beads. This separation was not found in the simpler Na versus Cl plot (Figure 2).

Given that the zones in Figure 4 were established with relatively small numbers of beads (*3*), and so may not be all encompassing, the fit of the Québec data with the chemistries of turquoise blue beads from Ontario is rather remarkable. This gives rise to Table VII, which summarizes the results, and which from top to bottom, lists the glass bead chemical types in approximate chronological order. The dates given in the table are provisional.

Table VII shows that the majority (11) of the turquoise blue glass beads from the Ashuapmuchuan site are from the 17th century, with 4 more beads scattered through the next 3 centuries. Of note, it is the only site to produce two near-modern (aluminum-rich,

Table VII. Provisional Analysis of Québec Turquoise Blue Glass Trade Beads

Provisional Group	Chemistry	Sites and Sample Numbers			≈Period/Type Reference
		A	M	C	
Type I	Ca ≤4% Cl/Na ≤0.14 K/Na ≤0.33	2	4	7	≈1600-1650 (2) Periods II and III
Tin-rich	Sn >7,000 ppm	7	0	8	≈1640-1700 (2) #83,86; (3) #1
Cobalt-rich	Co >120 ppm	2	8	1	≈1650-1700 (3) #2-12
Type II	Ca >4% Cl/Na >0.14 K/Na ≤0.33	0	9	7	≈1700-1750 (3) #13-21
Type III	Ca >4% Cl/Na >0.14 K/Na >0.33	1	1	19	≈1750-1850 (3) #22-41
Antimony-rich	Sb >10,000 ppm	1	0	1	≈1800-1900? (3) #42-50,63-64
Arsenic-rich	As >30,000 ppm	0	0	0	≈1840-1900 (3) #51-58,62,79
Aluminum-rich modern	Al >2% Cl ≤0.3%	2	0	0	≈1900-1950 (3) #69-78,80

Types I, II and III are differentiated on the basis of Ca, K/Na and Cl/Na and are roughly equivalent to the Ontario material as follows:

Type I ≈ Glass Bead Periods II and III (2)
Type II ≈ Late French regime (Fort Frontenac, Kingston) (3)
Type III ≈ Early British regime (Moose Factory) (3)

By definition, Type I, II and III beads are relatively low in Sn, Co, Sb, As and Al.

A = Ashuapmuchuan   M = Metabetchuan   C = Chicoutimi chlorine-poor) beads. The Metabetchuan site provided 12 beads from the 17th century and 10 beads from the 18th century. It may be significant that no tin-rich beads were recovered from this site. Although the Chicoutimi site produced 16 turquoise blue beads from the 17th century, it is the only site of the three to give large numbers of turquoise blue beads (27) from the 18th and possibly early 19th centuries.

Conclusions

The bead chemistry chronologies match the archaeological expectations, based on other lines of archaeological evidence, of the periods of use of each site. This matching confirms that the relatively short-lived isotope producing elements that are analyzable by INAA are indeed appropriate for characterizing glass beads.

If the bead evidence is definitive for each site, all sites were in use in the 17th century; the Metabetchuan site was used into the early 18th century; and the Chicoutimi site was repeatedly occupied in the 18th to 19th centuries. Unlike the other sites, Ashuapmuchuan contains some near-modern beads.

Acknowledgments

Field work on the Metabetchuan and Ashuapmuchuan sites was made possible through various grants to Laboratoire d'archéologie de l'Université du Québec à Chicoutimi (UQAC) from the Programme d'actions spontanées and the Programme d'établissement de nouveaux cherchers of the Québec Fonds pour la création et l'aide à la recherche (FCAR) as well as a grant from the Social Sciences and Human Sciences Research Council of Canada. Chemical analysis of the glass beads was supported by part of a joint grant made available to UQAC by R. Oullet (Université Laval), principal investigator of a Programme équipes et séminaires grant from the Québec FCAR funds that includes J.-F. Moreau as a co-researcher. The analytical work was also made possible by an infrastructure grant from the Natural Sciences and Engineering Council of Canada to the SLOWPOKE Reactor Facility at the University of Toronto.

Literature Cited

1.   Kenyon, I.T. and Fitzgerald, W. R. *Man in the Northeast*, **1986**, 32, 1–34.
2.   Hancock, R. G. V.; Chafe, A.; Kenyon, I. *Archaeometry*, **1994**, 36(2), 253-266.
3.   Kenyon, I.; Hancock, R. G. V.; Aufreiter, S. *Archaeometry*, **1995,** (in press).
4.   Moreau, J.-F. *Recherches Amérindiennes au Québec*, **1994**, 24(1-2), 31-48.
5.   Moreau, J.-F. *Saguenayensia*, **1993**, 35(2), 21-28.
6.   Moreau, J.-F. In *L'Archéologie et la rencontre de deux mondes*; Fortin, M. Ed.; Musée de la Civilisation: Hull, Québec, **1992**; pp 103-131.
7.   Moreau, J.-F. In *Transferts culturels en Amérique et ailleurs (XVIe-XXe siècle)*; Turgeon, L.; Delâge, D.; Oullet, R., Eds.; *Les Presses de l'Université Laval*: Québec, **1995**, (in press).
8.   Moreau, J.-F. In *Mots, représentations. Enjeux dans les contacts interethniques et interculturels*, Fall, N. K.; Simeoni, D.; Vignaux, D., Eds.; Presses de l'Université d'Ottawa, coll. Actexpress: Ottawa, Ontario, **1994**, pp 343-374.

9.   Moreau, J.-F.; Langevin, E. *Recherches Amérindiennes au Québec,* **1992,** 22(4), 37-48.
10.   Chapdelaine, C. *Le site de Chicoutimi. Un campement préhistorique au pays des Kakouchaks.* Québec, ministère des Affaires culturelles, **1988,** Dossiers 61.
11.   Lapointe, C. *Le site de Chicoutimi. Un établissement commercial sur la route des fourrures du Saguenay-Lac-Saint-Jean,* Québec, ministère des Affaires culturelles, **1988,** Dossiers 62.
12.   Hancock, R. G. V. *J. Internat. Inst. Conserv., Canadian Group,* **1978,** 3(2), 21-27.

RECEIVED October 9, 1995

Chapter 4

Dating Flint Artifacts with Electron Spin Resonance: Problems and Prospects

Anne F. Skinner and Mark N. Rudolph

Department of Chemistry, Williams College, Williamstown, MA 01267

Electron spin resonance (ESR) dating has been applied to flint artifacts of recent origin (less than 10,000 years ago). ESR is a dating technique developed over the last two decades, using the accumulation of radiation damage as a chronological marker. Using this technique to study archaeological materials requires demonstrating that the ESR signal clock was reset to zero by thermal treatment during manufacture of the artifacts. Under probable heating conditions for these flints, it appears that resetting the signal is considerably more difficult than has been suggested by previous work. Comparison of the results with dates for the site obtained by C-14 yields apparent ages that for the most part are substantially older than the C-14 values, presumably because of insufficient heating in antiquity. However, some of the flints do fall within the generally accepted age range.

Whatever other information archaeologists require, almost always they need to know the chronological age of the material with which they are working. While many dating methods have been developed, all of them are limited in their applicability, either in terms of the time frame or in terms of usable material, or both. Thus any new method may be useful to fill some niche in the overall time and material matrix.

Electron Spin Resonance as a Dating Technique

Electron spin resonance (ESR) was proposed as a dating method as early as 1967 *(1)*, but its practical application began with the work of Ikeya *(2)* in 1975. Since then, there have been substantial contributions based on carbonate materials, bones, and quartz. Grün *(3)* has written a good general review of the subject. ESR, which detects the presence of unpaired electrons, is extremely sensitive to the electronic environment, and has been used for archaeological provenance studies *(4)* as well as for dating. For present purposes, the unpaired electrons of interest are those created as a result of radiation damage, which implies that ESR has many similarities as a dating technique to thermoluminescence (TL). It has some advantages over TL, primarily in the fact that the signal is not destroyed during measurement, and so it is easier to study a given sample under a variety of experimental conditions. However

0097–6156/96/0625–0037$12.00/0

ESR is at least in some cases less sensitive than TL; so far, for instance, it is not a practical method for dating pottery.

Any object is subject to radiation, but not all forms of radiation damage are sufficiently stable to serve as chronological markers. The general approach to establishing ESR as a dating tool has been to examine materials of interest to archaeologists and geologists to detect ESR signals related to radiation damage. If such a signal exists, it must meet the following criteria in order to be usable for dating studies.

(1) The signal must be detectable. ESR spectra are often complex, and the "dating" signal may be obscured by interference from other radical species. The sensitivity of detection puts a lower limit on the time frame for which the method is applicable. One of the primary aims of this work was to establish whether flint artifacts less than 10,000 year old were ESR-datable.

(2) The signal must be inducible by radiation, and must grow monotonically with the applied radiation dose. Every material will eventually reach saturation, when an increase in applied radiation no longer increases the signal, limiting the time range of ESR dating for that type of sample. However, saturation occurs at different levels in different materials, and so the age limit depends on the substance being studied.

(3) The signal must be stable. Although corrections can be made to the age if the mean life of the signal is known, Arrhenius-plot experiments usually require extrapolation over wide temperature ranges, and thus result in great uncertainty in the value of the mean life. In the case of flint, recent experiments (5) have suggested that laboratory heating yields mean life values that are so low that flint should not be dateable either by ESR or TL, and yet the determined age of artifacts is well within the limits of other evidence. Thus for flint we deduce that there exists a stable signal, but the exact limit of stability is not yet known.

(4) The signal must be zero at the beginning of the time period of interest. In many cases this is the moment at which the material is formed, e.g., the formation of a coral reef, or a stalagmite. Clearly, however, the age of formation of flint is not of interest to archaeologists, and the issue of zeroing the signal at the time of artifact production represents one of the major problems met in extending ESR dating from carbonate materials to flint.

Assuming that the material has been judged suitable for ESR dating, the actual analysis is virtually identical to that used for TL. Aliquots of the material are irradiated artificially and a growth curve of the ESR intensity is extrapolated to zero, yielding D_e ("equivalent dose" in Gy), or the amount of radiation (usually in gamma-equivalent form) needed to produce the observed natural signal. A sample of such an extrapolation is found in Figure 1. From D_e the age of the material in years can be found by dividing by the sum of the known radiation doses both external (D_{ext} in Gy/yr) and internal (D_{int} in Gy/yr) to the material, as shown in the relationship:

$$Age = D_e / (D_{ext} + D_{int}).$$

This of course requires that samples of the matrix in which an artifact was found be available for isotopic analysis. It also implies that the dose rate has been constant with time. For flint, this latter assumption is generally valid, since uranium and its daughter isotopes are in secular equilibrium.

The Archaeological Setting

Extensive stoneworking activities have been identified at a number of sites in north-central Florida. The known time periods documented for the area range from about

1000-10,000 years ago, placing these sites within the range of carbon-dating. However, artifacts are often recovered from contexts that contain no carbonaceous material, making the implementation of a second dating technique valuable.

Johnson Lake is a small pond in the extreme northwestern part of Marion County, Florida, between Gainesville and Ocala. On a bluff overlooking the south shore of the pond is a workshop area first excavated in 1959 *(6)*. The soil of the bluff is well-drained sand, quite dry for Florida. There is no flint outcrop on the bluff, but approximately one mile to the east is a quarry site, presumed to be the source of most of the numerous specimens recovered from the workshop site. However, there is considerable variability in the composition of some of the samples, raising the possibility that material may have been brought to Johnson Lake from other areas as well. Most samples show definite evidence of weathering.

The samples used in this study were collected by Dr. Barbara Purdy during the late 1980's. They were found at depths from 30-50 cm below the surface. The only criterion used in selecting them was that they appeared to have been heated, judging by color and surface appearance.

In addition to the flint samples, samples of the freshwater snail *viviparus georgianus* were collected from a shell midden at another Florida site expected to be roughly contemporaneous with Johnson Lake. Lake Monroe, in Volusia County, Florida, has been dated by radiocarbon to 4-6000 years Before Present (BP). The Johnson Lake site is estimated to have been occupied between 5-8000 years BP. Comparison of the ESR ages of the two materials should provide further cross-checking on the validity of the dating method.

Archaeological Use of Flint

Flint (or chert, there being no significant geochemical or petrographic differences between the two terms) is generally described as fine-grained silica. However, samples of flint may differ widely due both to the presence of various silica polymorphs and to trace elements in the non-siliceous matrix between grains. Since ESR is extremely sensitive to small changes in electronic environment, it is not surprising that a variety of ESR spectra can be obtained under the generic heading of "flint". However, there are several radiation-induced signals that are characteristic of siliceous material, and those are the ones being studied for use in dating.

Heating flint makes it easier to work into usable artifacts *(7-9)*. However, flint such as that from northern Florida must be heated gradually ("baked") or it will explode and leave no usable fragments. Therefore heating by early peoples would have required a technique such as the one described by Griffiths, et al. *(10)* of burying flints under campfires. This heating has the potential to anneal any radiation damage experienced by the flint since its formation, thereby setting the dating "clock" back to zero. Thus initial work in this area focused on methods of determining whether or not particular flint samples had been heated in antiquity (to be referred to as "archaeologically heated"). Physical appearance, such as luster, is one criterion. ESR can itself indicate the thermal history of the material. However, the extent of resetting of the ESR signal depends on both the temperature and the duration of heating, something much more difficult to establish.

ESR Spectra of Flint

The ESR spectra of flint are complex and can vary greatly depending on the source of the flint. However, there are two primary signals derived from the silica fraction of flint that have been studied as chronological markers. Signals are defined by their g-

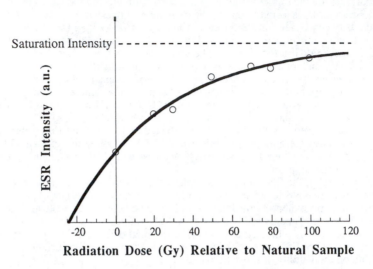

Figure 1. Example of growth curve for ESR signal in flint. D_e is given by the x-intercept. In this case the natural signal corresponds to approximately 23 Gy of gamma-equivalent radiation.

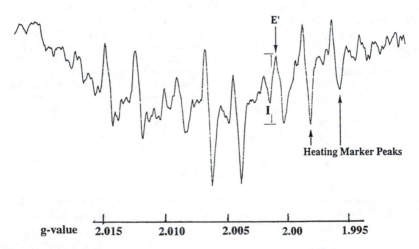

Figure 2. ESR spectrum of Florida flint before artificial irradiation, taken at room temperature, showing E' signal. The intensity (I) of the signal is taken as the E' peak height, as shown by the lines.

value, determined by studies on pure quartz. Figure 2 shows the room temperature spectrum of one of the flints used in this study. Visible on the high-field (right hand) side of the spectrum is a peak labeled E', attributable to an unpaired electron in an oxygen vacancy. The method of production of this defect is believed to be the

extraction of an oxygen ion, O^-, from an SiO_4 tetrahedron. E' defects can be annealed by heating at temperatures of approximately 350 °C, and if the sample is heated above 500 °C the ability of radiation to regenerate the signal is much reduced *(12)*. The number of E' defects in most flint appears to be small, and this signal saturates rapidly with artificial irradiation.

Also worth noting in this spectrum are the two small peaks immediately to the high-field side of the E' signal. These peaks appear to be absent from never-heated flint, but present in heated flint both from this site and from at least one other. Therefore their presence or absence in the natural spectrum can serve as a quick screening test for suitable samples.

The ESR spectrum can also indicate when the flint being studied is not microcrystalline quartz. Figure 3 shows a sample that is superficially indistinguishable from the one used to obtain Figure 2. However, it varied significantly in trace element composition, and this is reflected in the ESR spectrum. This sample proved completely unsuitable for dating.

Materials and Methods

The sample group described here consists of flint artifacts that were judged to have been archaeologically heated, plus some raw flint from the same region. On heating to 350 °C for 15 minutes, this flint changes color. However, prolonged heating is required to reduce the geological signal to a minimum. While it is likely that prolonged heating did occur (given the previously noted need to bake Ocala flint), the first observation is that color alone will not tell whether the sample is suitable for chronological study. The artifacts range in length from 5.5 cm to 11 cm, in maximum thickness from 10 mm to 15 mm, and in weight from 20 g to 109 g. All flints were sanded to remove the outermost surface, a step that eliminates the surface effects of alpha radiation from the environment. The larger flints were sampled in two ways: "edge" samples were taken from sections whose thickness was less than half the maximum value, and "center" samples were taken from thicker sections. Our intent was to see whether heating had been uniform; our assumption was that center samples would appear older because of greater residual signal. However, within the accuracy of measurement, no difference could be seen between the two types of sample.

Samples of both the archaeologically heated and natural flint were powdered in air, and the size fraction between 90-250 μm was washed with water and dilute hydrochloric acid to remove surface effects. The powdered sample was then divided and approximately half was heated as noted in the following section (for archaeologically-heated samples, results from this portion will be referred to as "reheated"). Irradiation of 8-10 aliquots each of heated and unheated powder was performed with a Cs-137 source at a rate of approximately 3 Gy/min. Post-irradiation treatment varied from none at all, to annealing at room temperature for two weeks, to annealing at 100 °C for one week. Annealing did not have a significant effect on the results.

The range of artificial irradiation has been shown to affect ESR results in other materials *(11)*. Large artificial doses may not truly mimic the effect of thousands of years of natural aging — experiments have shown that the signal of saturated signals in TL.   In theory these signals should yield the same value for D_e, and in fact one criterion for reporting these particular samples is that the D_e values for the two methods do agree within experimental uncertainty (+15%/-25%). Both analytical techniques can be criticized, however. The first relies heavily on the value determined for the reheated natural sample; the second relies heavily on the value determined for the natural archaeologically-heated sample. The second technique has an additional difficulty for these samples. The saturation of the E' signal in flint results in a sharply curved growth curve, as shown in Figure 1. Thus the interpolation of the natural signal onto the regenerated curve has more uncertainty than is reported when this technique is used in TL. The reported values of D_e in Table I are derived from extrapolation of the reduced growth curves. The R-values for the fit of these curves is ≥ 0.98.

Table I.   Results of ESR Analysis

Flint Sample	Fl4	Fl6	Fl 7	Fl8	Shell
D_e (Gy)	21.5	87.8	185	21.8	3.6
$U_{fl}{}^a$ (ppm)	0.71	0.68	0.97	2.35	0.1
$Th_{fl}{}^a$ (ppm)	0.219	0.136	0.208	0.147	---
$K_{fl}{}^a$ (ppm)	164	81	123	91	---
D_{int} (mGy/a)	600	482	698	1416	8
$D_{ext}{}^b$ (mGy/a)	745	745	745	745	468
Age^c ($x10^3$ years)	16.0	72	128	10.1	7.65

[a]U_{fl}, Th_{fl}, and K_{fl} refer to isotope concentrations within the sample.
[b]D_{ext} for flint was calculated from the following information: U_{sed} - 0.8 ppm; Th_{sed} - 1.5 ppm; K_{sed} - 200 ppm; cosmic dose 193 mGy/a.
D_{ext} for shell was calculated from the following information: U_{sed} - 0.7 ppm; Th_{sed} - 1.2 ppm; K_{sed} - 150 ppm; cosmic dose 210 mGy/a.
Water content of shell matrix: 12%
[c]Age uncertainties on ages are +15%/-25%.

    In converting D_e to age, a number of assumptions had to be made. The factor κ, which describes the relative efficiency of alpha radiation in inducing ESR-detectable damage with respect to the efficiency of beta and gamma radiation, was set at 0.1 in agreement with most other work (15). However, this value is derived from TL work on flint. There is no experimental evidence for this value for ESR measurements, and values from 0.05 to 0.2 can be found in the literature. Using a value of 0.05 would increase the reported age by approximately 10%; using a value of 0.2 would decrease the age by approximately 20%. The moisture content of the soil was assumed to be materials can appear to be increased by additional irradiation. Thus the upper limit used for analysis of these samples was 100 Gy.

ESR spectra were taken on a JEOL RE1X spectrometer. The E' spectra were recorded using a modulation amplitude of 0.5 millitesla (mT), a scan speed of 0.3 mT/min and at 0.1 mW of power. This signal is extremely sensitive to power, largely because it is not a single line. Artificial irradiation induces an additional signal at g=2.002. Annealing experiments suggest that this signal is short-lived, and it is only a minor component in the natural sample. At low power this signal can be distinguished from the "dating" peak at relatively low levels of artificial irradiation. Above approximately 200 Gy of artificial irradiation, measurement of the dating signal becomes problematic even at low power. However, since upper limit of additional dose was 100 Gy, this did not limit our ability to measure these samples.

Results

Results of Heating Experiments. Studies on Florida flint *(8)* indicate that improvement in flint working properties requires heating within a fairly narrow temperature range of 350-400 °C. Our heating experiments, therefore, were performed at 350 °C. Previous work *(12)* had suggested that heating flint to 350 °C for 30 minutes was sufficient to anneal completely the E' signal. The samples in that case were considerably older than the ones reported here. In contrast, we found that even after heating for 24 hours, there was a perceptible residual E' signal. In some cases this irreducibly minimum signal level was greater than 50% of the original signal. In itself this is not surprising. As noted, these are relatively young samples and therefore the signal due to regeneration since manufacture would be expected to be small. In older samples, where the archaeological signal is much larger (since it has been accumulating for a longer period of time), the residual signal is a small percentage of the original and might be overlooked considering the numerous other uncertainties of measurement.

The question then becomes: what happens when an archaeologically heated sample is reheated and subsequently irradiated with similar doses (i.e., total artificial dose approximately 100 Gy) to regenerate the growth curve? Does the residual signal show a different sensitivity from the original one, or does the regenerated signal depend on the size of the residual signal? Comparison of the growth curves of archaeologically heated and reheated artifact samples generally suggests that irradiation of reheated samples reproduces the shape of the archaeologically heated samples (see Figure 4).

Dating Results Using the E' Signal. Most of the artifacts in this sample set were not dateable. As noted, one was not siliceous flint. Others had not been heated at all, on the evidence of their natural signal and on the evidence of the rapid heating test suggested by Toyoda et al. *(13)*. Two samples had been heated, but when irradiated artificially they saturated almost immediately. The interpretation of this result is that the archaeological heating was insufficient to decrease the signal. Parenthetically, the "dating" of natural flint produced an "age" of approximately 500,000 years, but the uncertainty of this measurement is over 100% due to the saturation of the E' signal. We were most interested in four other samples where the results were more ambiguous. Two different growth curves were constructed for each of these samples. In one, the intensity of the residual signal was subtracted from all aliquots of the artifact signal, creating a reduced growth curve. This subtraction was based on the observation that the sensitivity was unchanged by further heating. In the other, a regeneration technique was used, analogous to the treatment of unbleachable

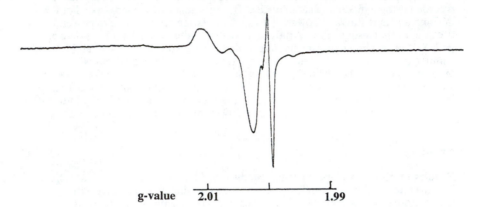

Figure 3. ESR spectrum of anomalous Florida flint, taken at room temperature.

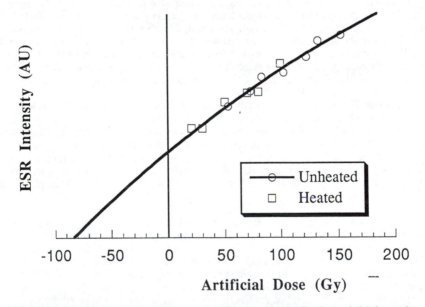

Figure 4. Comparison of growth curves of archaeologically-heated (Unheated) and reheated (Heated) flint. The dose values for the unheated sample have been shifted by an amount equivalent to D_e in order to show the superposition of the two curves.

the present-day value of 5%. Calculation of cosmic dose assumed a uniform burial depth of 40 cm.

Table I shows that flints from the same site do not necessarily have the same apparent age. The older ages are presumably due to heating that was insufficient to reset the E' signal from its geological value to its irreducibly minimum value. However, the youngest ages are not inconsistent with the other evidence of the age of the site, and are sufficiently promising to justify continued study of flints.

The last entry in Table I is the set of results for the freshwater snails from Monroe Lake. Non-marine snails are very sensitive dosimeters *(14)* and thus young samples can be dated. The date reported is in good agreement with radiocarbon results from the site (Purdy, B. A.; private communication, 1994), although at the upper limit of those results. While not directly relevant to the age of the Johnson Lake site, the results suggest that most assumptions about the external dosimetry in northern Florida are reasonable, but that a part of the reason for the flint ages being "too high" might lie in uncertainties in dosimetry.

Conclusions

Future work must concentrate in two areas. Use of the E' signal requires more information about the irreducible signal — whether the behavior of the Florida samples can be generalized to other flints, or whether it is related in some way to the microstructure of flint. In that case there might be other sites where reheating of the archaeological samples was not necessary to determine ages with a room-temperature technique.

Another aspect of the E' signal that needs further study is the dependence of the signal on microwave power. If a power level of 0.6 mW is used with these flints, D_e decreases substantially, bringing the calculated ages into very good agreement with other data. Ikeya *(15)* has noted that there seem to be good reasons not to use power levels above 0.1 mW for the E' signal, a conclusion supported by some preliminary work in this study. However, like so many other aspects of the study of flint, it may be that the optimum power level varies with the type of flint, and that more work might demonstrate a justification for higher power levels.

The second possible approach is to use a different marker. The signal shown in Figure 5 is attributable to the substitution of Al^{3+} for silicon. Clearly one cannot

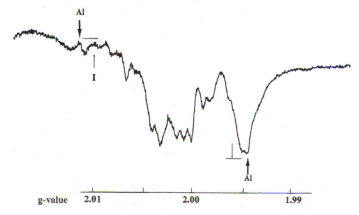

Figure 5. ESR spectrum of Florida flint, taken at 77K, showing Al signal. Note that the intensity (**I**) of the signal is measured from the <u>peak</u> of the high-g line to the <u>trough</u> of the low-g line.

rely on the presence of this signal, but in fact it is found in most flints. This signal must be measured at 77 K in order to localize the unpaired electron. It appears to be annealed more readily than the E' signal. Therefore an initial test for adequate heating in antiquity would be to determine if both signals give the same apparent age. However, in flint there appears to be an interfering signal *(12)* and signal subtraction was used to obtain Figure 5. This may introduce as much uncertainty in the Al^{3+} results as the irreducible signal does for the E' results.

Clearly this dating method is not ready for routine use on young samples. The uncertainties of measurement alone are too large for accurate determination in cases where 1000 years is a major time period. However, the present work has shown that small signals can be isolated and measured. And, although it is always desirable to find identical ages for samples from a given site, if a series of ages is found, this work indicates a theoretical basis for selecting the youngest as the most probable date.

Acknowledgments

Neutron activation analyses were performed at the Oregon State University Radiation Center with funding from the U.S. Department of Energy's University Reactor Sharing Program. The JEOL spectrometer was purchased with the assistance of NSF ILI grant number 915111. Sample irradiations were performed at the Wadsworth Laboratories of the New York State Department of Health with the kind assistance of Dr. Ulrich Rudofsky. Data analysis used the FIT-SIM program of Dr. Rainer Grün. The authors would also like to express their gratitude to Dr. Barbara Purdy, professor emerita at the University of Florida, not only for supplying the samples but for discussions on a wide range of archaeological topics.

Literature Cited

1.   Zeller, E. J.; Levy, P. W.; Mattern, P. L. *Proceedings of the Symposium on Radioactive Dating and Low Level Counting*; International Atomic Energy Agency: Vienna, 1967; pp 531-540.
2.   Ikeya, M. *Nature* **1975**, *255*, 48-50.
3.   Grün, R. *Quaternary International* **1989**, *1*, 65-109.
4.   Lloyd, R. V.; Smith, P. W.; Haskell, H. W. *Archaeometry* **1985**, *27*, 108-116.
5.   Porat, N. and Schwarcz, H. P. **1995**, *Appl. Rad. Isotopes*, in press.
6.   Bullen, R. P.; Dolan, E. M. *The Florida Anthropologist* **1959**, *12*, 77-94.
7.   Crabtree, D.; Butler, B. *Tebiwa* **7**, 7, 1-6.
8.   Purdy, B. A.; Brooks, H. K. *Science* **1971**, *173*, 322-325.
9.   Domanski, M.; Webb, J. A.; Boland, J. *Archaeometry* **1994**, *36* , 177-208.
10. Griffiths, D. R.; Bergman, C. A.; Clayton, C. J.; Ohnuma, K; Robins, G. V.; Seeley, N.J. In *The Human Uses of Flint and Chert*, Sieveking, G. G.; Newcomer, M. H., Eds.; Cambridge University Press: Cambridge, 1987; pp 43-52.
11. Barabas, M.; Walther, R.; Wieser, A.; Radkte, U.; Grün, R. *Appl. Rad. and Isotopes* **1993**, *44*, 119-130.
12. Porat, N; Schwarcz, H. P. *Nuclear Tracks Radiation Measurement* **1991**, *18*, 203-212.
13. Toyoda, S.; Ikeya, M. *Appl. Rad. and Isotopes* **1993**, *44*, 227-231.
14. Skinner, A. F.; Mirecki, J. *Appl. Rad. and Isotopes* **1993**, *44*, 139-144.
15. Ikeya, M. *New Applications of Electron Spin Resonance: Dating, Dosimetry and Microscopy*; World Scientific: Singapore, 1993; p 279.

RECEIVED October 9, 1995

Chapter 5

Provenance Studies on Easter Island Obsidian: An Archaeological Application

F. R. Beardsley[1,4], G. G. Goles[2], and W. S. Ayres[3]

[1]International Archaeological Research Institute Inc., Honolulu, HI 96826
[2]Department of Geological Sciences, 1272 University of Oregon,
Eugene, OR 97403–1272
[3]Department of Anthropology, University of Oregon, Eugene, OR 97403

This paper summarizes our work on geochemical characterization of the obsidian sources on Easter Island and its application to archaeological specimens. Obsidian was the primary and most durable tool material used by the Rapa Nui people during the prehistoric period. By tracing obsidian artifacts to their raw material source, a pattern of source utilization can be developed through the distribution of these "fingerprinted" artifacts, which in turn provides a foundation for the reconstruction of prehistoric patterns of raw material access, control, and trade. The source characterization focuses on (1) determining discrete geochemical properties of each source; (2) determining intra-source variation; (3) examining inter-source similarity. The island's four sources were systematically sampled, with selected samples from each source subjected to instrumental neutron activation analysis (INAA). Three trace elements—Sc, Zn, and Se—were found to provide the best source discriminants. Used in concert, the three elements provide a powerful tool for identifying each source. Artifacts from selected sites across the island were then subjected to INAA in an effort to examine source utilization. The focus of this phase of analysis was placed on distinguishing between the wholly black obsidians. Two of the four obsidian sources appear frequently in the artifact sample and display distributional patterns with potentially significant implications within the context of social, political, and economic systems.

Easter Island has captured the imagination of many western explorers and naturalists (*1-3*), writers and poets (*4, 5*), and anthropologists and archaeologists (*6-8*). It is among the more remote and environmentally impoverished islands in the South

[4]Corresponding address: P.O. Box 412, Wrightwood, CA 92397

0097–6156/96/0625–0047$12.00/0

Pacific, yet its long history of cultural isolation with little, if any, noticeable import of extra-island material culture, together with its geographic isolation and scant flora and fauna has produced an exceptionally rich and complex cultural tradition unequaled anywhere in the Pacific. The main attraction for those acquainted with the island is the ability of the founding population and its succeeding generations to adapt to and make use of this extreme subtropical environment, and to transfer the technology of the archaic East Polynesian tradition into the monumental architecture (*ahu*) and associated statuary (*moai*) that have become synonymous with the island. But these most visible and grand displays of the islanders' complex technology are only one component in the overall interpretation and reconstruction of the past. More crucial and fundamental elements in any reconstruction are the smaller, less spectacular industries—those ubiquitous components of the technological repertoire of the island, namely hand tools. Of particular interest are those implements and their production debris which were either in use or manufactured on a nearly daily basis and which display the indelible mark of the craftsman in form, tooling, and use-wear—those implements made from stone, bone, or other readily accessible and durable material.

For the archaeologist, the island is like an open air museum with many prehistoric sites left virtually untouched by 150 years of development, ranching, and agricultural activity. Yet this open air museum also has its drawbacks—namely, much of the traditional lore about the mechanics of everyday life and the social, political, and economic organization of the island as a whole disappeared with the capture and demise of a majority of the population during the last century. In piecing together the culture history of Easter Island, then, the archaeologist must sort through the scant oral histories and observations recorded (and sometimes embellished) by missionaries, explorers, natural historians, and other visitors to the island, draw on analogs from other Polynesian islands, and interpret the mute testimony of the material culture assemblage with the assistance of a variety of analytical methods.

The following pages present the results of the University of Oregon's (UO) Geoarchaeological Program's continuing research in reconstruction of the prehistoric era on Easter Island through provenance studies and the scientific analysis of various elements in the material culture assemblage—from their sources of origin, through their production and use, to their distribution across the island, and finally their discard or abandonment. In other words, our research spans the life of an artifact. This phase of the project is focused on obsidian, that glassy black stone which produces sharp edged implements, is easily worked into more complex shapes, and has a high luster with a virtually pure and dense coloration. Obsidian is a common and abundant material in artifact inventories across the island. It is found in virtually all archaeological sites, and occurs in a range of forms representative of the various stages of production and use, from the initial quarrying, to the manufacture and use of both formal and informal tools, to the final discard once their usefulness has been exhausted. As such, obsidian presents one of the best opportunities to examine the island's prehistoric social, political, and economic landscape and interaction patterns.

The cornerstone of a study of this nature is the determination of provenance—ideally of all or at least a large sample of the obsidian artifacts from archaeological sites across the island. The first step, naturally, is geochemical characterization of the island's obsidian sources, an essential operation in establishing the discrete fingerprint of each source. This, according to Leach (9), is just the initial stage of research in provenance studies; with respect to archaeological applications, it is merely the preliminary to the second (or final) stage of research—routine application to archaeological materials. The results of the first stage of our provenance research were published in 1991 (10); they are summarized below as a part of the overall discussion of provenance research on Easter Island and, in particular, our efforts to enter the second stage of provenance studies. The specific goals of our research are to document the prehistoric patterns of raw material use, control, and trade, and the broader social, political, and economic systems in which they are found; define regional patterns of access to the obsidian sources; and, determine the significance of such regional patterns for social group (*mata*) boundaries and exchange patterns.

Background

Easter Island is a small subtropical island in the eastern Pacific, in the extreme southeast corner of Polynesia. This lonely outpost of Eastern Polynesia is nearly 2,000 km from Pitcairn Island, 3,200 km from the Marquesas Islands, and some 3,700 km from the coast of South America. It is classified as a volcanic high island, with three volcanoes forming its main surface (Figure 1). Poike, the easternmost point of the island, is a composite cone about 3 million years old (*11*). Rano Kau, also a composite cone, forms the southwest corner of the island. It is younger than Poike, but no age has yet been determined (*11*). The youngest volcano on Easter Island is Terevaka, which forms the northern corner of the island. It is a complex fissure-volcano with a number of smaller eruptive centers on its flanks and although activity at Terevaka began roughly 300,000 years ago (*11*), the most recent eruptive episode took place about 2,000 years ago to form a parasitic cone on its slopes (*12*). The combined lava flows from Terevaka and its parasitic cones unite the three principal volcanoes to form the island surface (*11, 13, 14*).

Any description of Easter Island would be incomplete without mention of the three small precipitous islets off its southwest apex. Motu Kao Kao, Motu Iti, and Motu Nui figure prominently in the culture history of the island. According to Baker et al. (*11*), all three islets are likely the remains of parasitic centers on the now eroded and submerged flanks of Rano Kau.

All four of Easter Island's obsidian sources are concentrated in the southwest corner of the island, apparently on the slopes of Rano Kau though possibly on the superimposed slopes of Terevaka. Two of the flows are on the rim of Rano Kau or just below it—Rano Kau II on the rim and Rano Kau I (identified as Te Manavai by Bird [*15*] of the Australian Nuclear Science and Technology Organisation [ANSTO]) on the flanks of the volcano. The islet Motu Iti is the site of the third source; according to both Bird (ANSTO, personal communication, 1988) and Stevenson (Archaeological Services Consultants, Inc., personal communication, 1987), Motu Iti is nearly a solid block of obsidian, presumably a somewhat eroded dome. The fourth source is found on Maunga Orito, a parasitic vent on the northeastern slopes of Rano Kau or possibly the southern flanks of Terevaka (the specific geology of the area has yet to be defined). It is the largest source of obsidian on the island, covering an area of roughly 90 ha across the western peak and slopes of Orito. Ages of each of these flows have yet to be determined, but it is assumed (perhaps optimistically) that each was formed separately and at different times from the others, and as such have unique trace element contents that can be used to define discrete and identifiable fingerprints. According to Mahood (*16*), a volcanic center and the magma reservoir that sustains it may have a lifetime that extends a million years, during which time each successive eruptive unit taps the reservoir at a different stage in its chemical evolution. Thus, lavas of substantially different ages from the same volcanic center commonly have different compositions, especially with respect to trace elements.

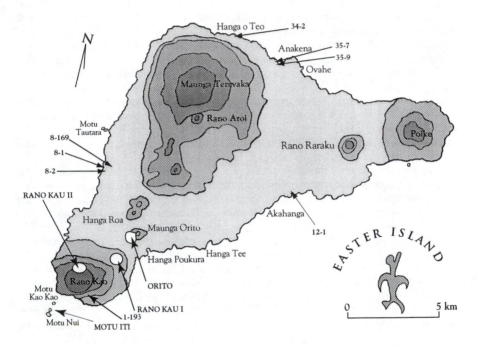

Figure 1. Easter Island, showing locations of obsidian sources and sampled archaeological sites.

Previous Research

Initial research on the island's obsidian flows was confined to geological investigations (*11-14, 17*), with the first research related to archaeological questions focused on obsidian hydration. Evans (*18*) was among the first to study obsidian from archaeological sites on the island, with the intent of developing a chronology for the sites by applying obsidian hydration dating methods to archaeological specimens using a general tropical rate (Tropical Rate B). In the process, he identified specific macroscopic features of the glass that could potentially affect the rate of hydration. Michels et al. (*19*) continued this obsidian hydration research and expanded it to include characterization of the obsidian (by X-ray flourescence [XRF] and atomic absorption) in an effort to determine the chemical component impact on hydration rates. They concluded that two rates were needed for the island—one for the Motu Iti source and one for the combined sources on Rano Kau and Orito. The latter sources, they found, were so chemically similar that only one rate was needed. Other hydration studies, such as those by Stevenson (*20*) and Stevenson et al. (*21*), concentrated on refining the rate of hydration for the island; these studies, however, have had some unforeseen consequences. In his ongoing research, Stevenson (Arch. Serv. Consult., Inc., personal communication, 1989) has determined that multiple rates are needed for archaeological specimens, each of which is dependent on the specimen's history of deposition. But the chemical characterizations of the obsidian sources to which these new rates apply rest on the earlier conclusions of Michels et al. (*19*)—that there are only two recognizable compositional populations for the obsidian flows on the island.

Provenance research related specifically to archaeological questions beyond the realm of chronology and obsidian hydration did not receive much attention until recently, with work undertaken independently in Australia by Bird (*15*) and in the United States by the UO's Geoarchaeology Program (*10*). Up to this time, it had generally been accepted that there were only three sources of obsidian on the island— Orito, Motu Iti, and one flow on Rano Kau (Te Manavai or Rano Kau I)—and that these represented two populations which are compositionally similar. Both Bird and we began our research with the recognition of four sources—two on Rano Kau (documented on soil and geological maps of the island, as well as in an archaeological atlas [*22*]) and two on Orito and Motu Iti.

Bird collected a reference set of obsidian in 1986 from each of the four sources and subjected them to PIXE-PIGME analysis (proton induced X-ray emission and gamma-ray emission), one of the few methods capable of distinguishing between sources, according to Leach and Warren (*23*). Bird identified three populations among the four flows. Two of the sources (Motu Iti and Rano Kau II), he found, were quite distinct on the basis of their chemical compositions, while the two remaining sources (Orito and Rano Kau I) could not be distinguished from one another but together exhibited chemical compositions distinct from both Motu Iti and Rano Kau II. He also noted that Easter Island obsidian can be distinguished from other known sources in the South Pacific, an important point from an archaeological perspective, especially for studies of inter-island contact and interaction. Contrary to these results, Leach and Warren (*23*) noted a similarity in the trace element composition between Mayor Island obsidian and that from Easter Island; their study, however, was within the context of provenance work on New Zealand obsidian sources, with no descriptions provided for the actual sources analyzed from Easter Island.

University of Oregon's Geoarchaeology Project

Our research, like Bird's, began with a systematic collection of obsidian from the sources. The three largest sources—Orito, Rano Kau I and II—were visited in 1987 by Beardsley, who mapped and sampled each one; the fourth source—Motu Iti—was not sampled at the time owing to rough seas which prevented access. However, samples from Motu Iti were obtained from both Bird and Stevenson. Four Department of Energy Reactor Use Sharing Grants from the Radiation Center at Oregon State University (OSU) provided the means to process 46 samples from all four sources and 40 archaeological specimens from eight archaeological sites across the island using instrumental neutron activation analysis (INAA).

The first stage of research, the source characterization, pursued three analytical tracks designed to determine: (1) the discrete geochemical properties of each source; (2) the range of chemical variation within the flow boundaries of each source; (3) the degree of similarity among the four sources. The results of this stage of work would influence the decision to either proceed to the second stage of research, application to archaeological materials, or seek an alternative method of analysis for the source samples.

The second stage of research was designed to provide at least a modicum of information on source utilization and distribution across the island, as well as contribute to an understanding of the political, social, and economic systems that influenced access to these sources and the probable control and trade of obsidian. Eight sites from across the island were selected for this phase of research—factors such as site location, proximity to the obsidian sources, and association with competing *mata* were considered in the site selections.

Sampling and Analytical Procedures. The procedures for sampling each obsidian source on Easter Island required a clear definition of the surface extent of the respective flows. For the two sources on Rano Kau, this meant a re-examination of the boundaries published in the *Atlas Arqueológico* (22). For Orito, it meant actually mapping the visible limits of the flow because no surface boundaries had been recorded prior to Beardsley's fieldwork. The boundaries of the Motu Iti source were not examined, but according to both Bird and Stevenson the islet itself is nearly all obsidian.

Topographic base maps from the *Atlas* were used during the mapping and sampling work at each obsidian source on Rano Kau and Orito. Samples of obsidian were collected from the surface at diverse points across each flow, along systematic transects used to establish the outcrop limits. The extent of the surface distribution within each source was recorded on the topographic maps, with the positions of all sampling locations triangulated and placed on the same base maps. Multiple samples of obsidian were randomly collected from each sampling location, which was defined as a circular area one metre in radius. In all, fifty locations were sampled across the three sources. Source boundaries for Rano Kau varied somewhat from those published in the *Atlas*, but the differences were insignificant. On Orito, however, the source appears to cover a substantial area of the western peak and flanks of this double peak formation.

It is worth noting that all three sources retain direct evidence of quarrying activity in the form of open pit mines. The pit mines are essentially shallow depressions, circular to oval in shape, excavated into the side of the slope. A slight berm or platform of debitage was formed around the outer edges of the pit as the mining and reduction debris was cast downslope (21). A number of these mines were encountered during the sampling work on both Rano Kau and Orito. At one mine on

Orito, for example, a large basalt hammer and anvil (used for the on-site reduction of the mined blocks or slabs of obsidian) were still in place, along with the debris from this reduction activity. The number of mines observed at each obsidian source, including mines on the steep slopes of the inner caldera wall at Rano Kau II, demonstrates that each flow was available and utilized during the prehistoric era. And, while obsidian continues to be used today in crafts predominantly manufactured for the tourist market, it is not mined; rather, obsidian is collected from the surface of the nearby source areas and archaeological sites.

A total of 46 samples was selected for analysis from the four source areas—9 samples from 7 sampling locations at Rano Kau I, 7 samples from 7 different sampling locations at Rano Kau II, 18 samples from 15 sampling locations at Orito, and 12 samples from 9 of the Motu Iti specimens obtained from Bird and Stevenson. Altogether, the samples represent a cross-section of the four obsidian sources. For the second stage of analysis, the archaeological application, a total of 40 pieces of debitage was selected from eight archaeological sites distributed across the island. From the west coast, 4 of the samples were from a small inland cave (Site 8-169), 5 samples were from the excavations at Ahu Ko te Riku (Site 8-1), and 1 sample was from Ahu Tahai (Site 8-2); at the southwest point of the island, 5 samples were collected from a house foundation (Site 1-193) on the slopes of Rano Kau, in the immediate vicinity of the obsidian quarries; from the south coast, 13 samples were collected from the stratified deposits in a small coastal cave (Site 12-1); and from the north coast, 5 samples were selected from a coastal cave (Site 34-2) on the northeastern slopes of Terevaka, 4 samples were taken from a small cave (Site 35-7) at Anakena, one of the more important areas in Easter Island culture history, and 3 samples were selected from an open site (Site 35-9) in the general vicinity of a small *ahu* at Anakena. These archaeological specimens represent a very small percentage of the obsidian found in sites across the island; this stage of research is just the beginning of what we hope to be more routine applications.

Prior to processing each sample from both the sources and site assemblages, visual examination of each was undertaken in an effort to determine if sortable categories existed. In general, four visual categories were identified, of which only one was source specific—a glossy black obsidian with white spherulitic inclusions found at Rano Kao I. The remaining three macroscopic categories were observed in the geological samples from all four sources. These included: a glossy, uniform black obsidian; a matte black obsidian that tends toward a charcoal to gray-green coloration (found principally at Orito but also at Motu Iti); and a black obsidian that ranges from dull to glossy and which contains irregular, elongated inclusions that give the material a rough appearance. Only artifacts which fell into these three latter macroscopic categories were selected for analysis; definite Rano Kao I artifacts which retained the white spherulitic inclusions were not chosen.

In preparation for the INAA work each sample was ground to a fine powder, of which about 0.5 g was encapsulated for irradiation; the exact amount of powder was measured to the nearest 0.1 mg. Samples from two locations—Orito CC and Motu Iti 1—were included in every irradiation set as a means of control and for comparative purposes; two samples from Rano Kau I were submitted for irradiation a second time owing to anomalous results from the first irradiation. As a further measure of control within our experiment, all samples were irradiated with and measured against a reference set of standard rock powders and liquid monitors, the geochemical properties of which have been published by the National Institute of Standards and Testing (NIST), the Radiation Center at OSU, and the Center for Volcanology at the UO (*24*).

Samples, together with standard rocks and monitors, were irradiated in the nuclear reactor at OSU. Once irradiated, the relative abundances of trace elements, including rare earths, within each sample, standard, and monitor were then calculated at various stages in the radioactive decay process—from a few days after irradiation (short counts), to a few weeks after irradiation (intermediate counts), to many weeks after irradiation (long counts). By observing the gamma emissions of induced radionuclides at different intervals after irradiation, we were better able to sort through energy interferences as well as to compare results obtained from each counting episode and from each laboratory. The Radiation Center at OSU calculated the results from the short counts and some of the intermediate and long counts; we, at the Center for Volcanology at UO, calculated results from the intermediate and long counts.

In both laboratories a gamma spectrometer was used to detect the gamma emissions. The spectrometer at the Center for Volcanology was equipped with a Ge(Li) detector. Output signals from the gamma emissions were fed into a preamplifier and then into the main amplifier. From the main amplifier, the output pulses were then fed into a multichannel analyzer. The raw data appear as a spectrum with a number of photopeaks, valleys, and plateaus spread across the 4096 channels within the multichannel analyzer. Each peak represents a series of constant energy events from the gamma emission of a specific radionuclide (some peaks are composite, but these instances are well known and corrections are routinely made for this kind of spectral interference). The Radiation Center uses their own computerized program developed specifically to select and analyze specific channels. At the Center for Volcanology, we select individual photopeaks based on observations of the spectral features displayed as histograms on a microcomputer interfaced with the multichannel analyzer.

The area within each photopeak was calculated, and that calculation then translated into a relative abundance for the corresponding element, commonly expressed in parts per million (ppm). All abundances within the samples were determined through comparisons with the known element abundances of the standard rocks and monitors that were irradiated with them.

Source Characterization: Stage 1.   The INAA results from the source characterization were published in the *Bulletin of the Indo-Pacific Prehistory Association* (*10*). Essentially, the similarity in the relative abundances of all but three trace elements and all of the rare earth elements was discouraging, with the four obsidian sources displaying little compositional diversity. Table I presents the aggregate results of all irradiation episodes; it is adapted from Tables I and II in Beardsley (*10*) and modified with the results of subsequent analyses. Only three elements—scandium, zinc, and selenium—exhibited differences unique to each source. In other words, all four sources could be geochemically identified and distinguished through the relative abundances of these three trace elements. Additionally, a general pattern of slight differences between relative abundances was recognized in a few of the other trace elements, but this pattern should not be considered significant and should certainly not be used as the principal discriminant between sources; it is useful primarily as confirmation of the recognized differences between sources.

To establish the geochemical identity of each source, the similarities and differences in element concentrations were reviewed in two stages. The first comparison was between samples from within the same source, intra-source variation; the second was between samples from the different sources, inter-source similarity. Each comparison was facilitated through graphic plots of element abundances and

Table I. Ranges of Mean Values for Trace Elements and Rare Earth Elements (ppm)[1]

Source/Quarry	Na2O %	FeO %	Co	Sc	Cr
Orito	5.49-6.02	2.63-2.91	0.17-0.46	0.46-0.57	6-19
RK I	5.61-5.86	2.70-3.42	0.40	0.45-0.62	7-11
RK II	5.15-5.52	2.20-2.59	0.31	0.20-0.30	8-14
MI	5.58-6.24	2.71-4.52	0.42-0.83	0.70-0.80	4-33

	Hf	Ta	Th	Rb	Cs
Orito	22.8-28.7	6.3-8.4	10.3-11.5	54-84	0.7-1.0
					(0.53-0.55)[2]
RK I	22.0-28.4	6.7-11.5	10.4-11.3	66-105	0.8-1.0
RK II	21.0-24.3	6.8-7.6	11.1-12.1	69-83	0.8-0.9
MI	19.0-38.1	6.4-9.1	9.8-11.1	49-92	0.5-1.0

	Sb	Zn	Se	
Orito	0.3-0.8	4.4-4.8	8.2-11.1	
RK I	0.3-0.5	6.5	bdl	
		(10.5-11.2)[2,3]		
RK II	0.3-0.4	4.1-4.8	bdl	
		(9.7-10.7)[2,3]		
MI	0.3-0.5	4.9-5.2	bdl	
		(10.3-10.4)[2,3]	(6.9-10.8)[2]	

	La	Ce	Nd	Sm
Orito	82-92	193-221	85-114	20.1-23.5
RK I	86-105	149-227	61-100	20.3-27.5
RK II	78-91	174-222	79-105	20.1-24.0
MI	91-90	177-215	85-104	19.7-23.5

	Eu	Tb	Yb	Lu
Orito	3.14-3.78	3.47-4.50	12.1-13.6	1.65-2.01
RK I	3.16-4.23	3.30-6.64	12.4-16.1	1.73-2.04
RK II	2.41-2.86	3.35-3.83	12.6-13.4	1.76-2.02
MI	3.41-3.85	3.25-4.21	11.9-14.3	1.65-1.89

[1]Revised from ref. 10; additional values added, ranges expanded; "bdl" stands for "below detectable limits".

[2]Variable readings from the Radiation Center at OSU.

[3]Normalization with Glass Mountain rhyolite.

Table II. Easter Island Artifacts: INAA Results of the Discriminant Trace Elements and Source Assignment

Sample Nr.	Sc	Se	Zn	Source
Set 1				
8-169-302	0.48±0.02	See note 1	4.6±0.15	Orito
8-169-348	0.49±0.02	below	4.5±0.15	Orito
8-169-365	0.46±0.01		4.4±0.15	Orito
8-169-313	0.48±0.02		4.5±0.15	Orito
1-193-216	0.55±0.02		4.7±0.15	Orito
1-193-222		values erratic; no determinations made		
1-193-232	0.50±0.02		4.7±0.15	Orito
1-193-223	0.49±0.02		4.6±0.15	Orito
1-193-207	0.48±0.02		4.4±0.14	Orito
34-2-1168	0.49±0.02		4.6±0.15	Orito
34-2-647	0.49±0.02		4.5±0.14	Orito
34-2-1130	0.52±0.02		4.6±0.15	Orito
34-2-528	0.50±0.02		4.5±0.15	Orito
34-2-829	0.52±0.03		4.5±0.15	Orito
35-7-798	0.49±0.02		4.5±0.14	Orito
35-7-694	0.48±0.01		4.5±0.14	Orito
35-7-929	0.49±0.02		4.5±0.14	Orito
35-7-549	0.71±0.02		4.9±0.15	Motu Iti
12-1-393	0.50±0.02		4.5±0.14	Orito
12-1-400	0.47±0.01		4.5±0.14	Orito
12-1-424	0.47±0.02		4.4±0.14	Orito
12-1-423	0.72±0.02		4.9±0.16	Motu Iti
12-1-409	0.49±0.02		4.6±0.15	Orito
21-1-473	0.48±0.02		4.5±0.14	Orito
Set 2				
8-1-13249	0.50±0.01	10.01±1.3	See note 3	Orito
8-1-13252	0.52±0.01	8.18±1.1	below	Orito
8-1-13254	0.45±0.03	9.36±1.1		Orito
8-1-13256	0.51±0.03	8.97±0.78		Orito
8-1-13260	0.75±0.04	9.24±1.3		Motu Iti
8-2-13263	0.48±0.03	9.42±1.6		Orito
12-1-13171	1.24±0.05[2]	12.6±2.0		Motu Iti
12-1-13177	0.54±0.04	8.83±1.1		Orito
12-1-13179	0.67±0.05	11.8±1.6		Motu Iti
12-1-13185	0.48±0.03	7.35±0.91		Orito
12-1-13198	0.71±0.04	10.91±0.85		Motu Iti
12-1-13202	0.48±0.03	6.94±0.73		Orito
12-1-13210	0.49±0.03	8.41±0.75		Orito
35-9-13269	0.54±0.03	8.13±0.78		Orito
35-9-13273	0.74±0.03	9.15±1.3		Motu Iti
35-9-13275	0.55±0.01	8.23±0.77		Orito

[1]Se values were not calculated for this set of artifacts.

[2]Sc values remained consistent through all counting episodes; resultant figure is consistent with Motu Iti values observed in multiple irradiations of the same sample.

[3]Zn values for Set 2 were calculated by the Radiation Center at OSU only; values were erratic and clearly inconsistent with the range of Zn values in Table I, and were therefore eliminated from this table.

tables of element ratios; 14 element ratios were reviewed as another comparative measure (*10*). Only those ratios with Sc provided any assistance in differentiating between sources. Graphic plots of the three critical elements proved more useful in displaying the differences between sources and similarities within a single source. Little variation was observed in the relative abundances of element concentrations between the sampling areas within a single source; in other words, each source appeared to be homogenous across the extent of its outcrop.

Between the four sources, the relative abundances of the three trace elements, Sc, Zn, and Se, were sufficient to detect differences, however slight, among the sources and to establish the foundation for the second stage of our research—the archaeological application. Of the three trace elements, Sc was used as the first discriminant between sources. It allowed us to identify three source populations— Motu Iti, Rano Kau II, and the combined Orito and Rano Kau I—almost immediately, with differences between source populations noted first in the short count episodes and confirmed through the consistent abundances in both the intermediate and long counts. These are the same source populations recognized by Bird in his analysis (*15*). From this point, we used Zn and Se to distinguish Orito and Rano Kau I. Either element can be used to differentiate the two sources, hence abundances of one element can be used to confirm the results based on abundances of the other. Unfortunately, both Zn and Se are difficult to determine and require intermediate and long counts. Zinc in particular is an exceptionally problematic element to determine because it occurs as part of a composite peak (a small bump on the low energy shoulder of a major Sc peak); long counts provide the most reliable measure because the half-life of the Zn isotope is so much longer than that of the major interference, the Sc isotope.

To supplement the Zn-Se differentiation between Orito and Rano Kau I, there is also a general recognizable trend in the relative abundances of some of the other trace elements in the compositional concentrations. This supplemental measure of discrimination, it should be noted, only reinforces the results of the Zn-Se distinction; it should not be used as the principal means to distinguish the two sources, especially as the relative abundances of the elements overlap. In general, the Orito source seems to have a smaller content of Rb on average than Rano Kau I; Cs, too, is on average less abundant in Orito, but both Ce and Cr tend to be relatively enriched. All other elements display virtually the same range in variation between the two sources.

An unexpected result of this first stage of research was the distinct differences noted in the results between the OSU and UO labs. The discrepancies between the two were, at least, consistent in that some of the reported contents of elements were systematically either higher or lower in one of the data sets relative to the other, including the source samples and standards (liquid monitors provide an additional reference set for UO only). These discrepancies are, however, problematic when attempting to establish the geochemical identity of a particular source. How does one reconcile the differences, and which laboratory results should be used in reporting element abundances? At least for our research, differences in the relative abundances of Sc were insignificant, but the relative abundances of Zn and Se exhibited distinct differences not only between the two labs but also between irradiation sets. For Se, distinction as a discriminant was established in the first two irratiation sets; for Zn, however, utility as a discriminant was recognized by the second irradiation set, but only confirmed in the third set with long counts. Unfortunately, the relative abundance of the Zn content in source samples was wildly variable between data sets produced at the OSU lab from intermediate counts (although the pattern distinguishing Orito from Rano Kau I remained consistent); by contrast, results produced in the UO lab from long counts were not only significantly lower than those from OSU but, and more appropriately, they reflected the visual character of the photopeak. The Zn values reported by OSU based on normalization with Glass Mountain rhyolite, at least, were relatively consistent with those from UO. As disconcerting as this whole exercise may be, it illustrates the variability inherent in results from different laboratories. Caution, then, should be the watchword when interpreting and comparing laboratory results.

Overall, however, the first stage of our research—source characterization— was successful. We found that we could determine diagnostic geochemical properties unique to each of the four obsidian sources; that intra-source variation was not measurable—instead, a single source proved to be homogenous in its chemical composition across the extent of its outcrop; and that inter-source variability was present and could be recognized through the relative abundances of these three trace elements. The next step, application of the results of the source characterization to archaeological specimens from sites around the island as a means of tracing the artifacts to their raw material source and reconstructing the prehistoric cultural landscape, is described below.

Archaeological Application: Stage 2. Eight sites widely distributed around the island were selected as the first test of the results of our source characterization. Originally the sites were excavated as part of an extensive excavation program conducted by Ayres in 1973 (25) and represent a wide range of site types—from an open site near a small *ahu*, to a house foundation, to ceremonial sites (*ahu*), to residential caves—within a time frame extending from about the 8th century A.D. to the 18th century. One site (Site 1-193), a house foundation that dates roughly to the 17th century, is located on the southern slopes of Rano Kau just below the rim of the caldera. It is located in the immediate vicinity of all four obsidian sources, the closest being Rano Kau II. Rano Kau I and Orito are somewhat more distant, but still reasonably close. Even Motu Iti is close in terms of distance, but it requires a much more arduous journey that entails scaling the cliffs of Rano Kau, crossing rough shark infested waters, and traversing yet another set of cliffs on the islet. For the remainder of the sites selected in this study, the closest obsidian source is Orito.

The three sites on the west coast are roughly half way between Rano Kau and Terevaka, in the Tahai region of the island and within the territory of the Miru *mata*. One of the sites is a small inland cave (Site 8-169), the other two are ceremonial sites or *ahu*, both of which are located on the coast. Ahu Ko te Riku (Site 8-1) dates from about the 12th century to about the 17th century; Ahu Tahai (Site 8-2) dates from perhaps as early as the 8th century to about the 17th century.

One small, well stratified coastal cave (Site 12-1) is located in the territory of the Ngaure *mata* at Hanga Runga Va'e, roughly at the mid-point of the south coast. Use of the cave appears to date between the 15th and 18th centuries. The site number was changed to 12-210 during a later survey of the area, and recorded in the archaeological atlas by this later number; however, as the recovered materials from Ayres' excavations retain the original site number, Site 12-1, it will be used here to avoid confusion.

The three remaining sites are located on the north coast. Site 34-2 is a coastal cave known as Ana Papa te Kena, on the northeastern slopes of Terevaka. It is within the territory of the Raa (Miru) *mata* and dates to at least the 15th century, but according to local reports continues to be used on occasion. Site 35-7 is a small coastal cave in Anakena, one of the most important areas in Easter Island culture history as it was said to be the landing place of the founding population and remained the home of the island's paramount chiefs (*ariki*), who were members of the Miru *mata*. Use of the cave dates from the 14th century to at least the 17th century. Site

35-9, the last site in this stage of our research, is an open site within the immediate vicinity of Ahu Ihu Arero in Anakena, and dates roughly between the 15th and 18th centuries. Anakena is within the territory of the Miru *mata*.

All specimens included in the analysis fall into one of the two black categories— either matte black or glossy black, neither of which can be attributed to any one source. Obsidian which could be clearly traced to Rano Kau I, with its white spherulitic inclusions, was not included in the irradiation samples. Of the total collection of debitage examined in the initial sampling process, RKI obsidian constitutes roughly less than 10 percent; the remainder of the collection is populated with black obsidians, which presents a formidable task in the determination of their source of origin.

The results from the INAA work indicated that only two sources were represented among the artifact samples—Orito and Motu Iti (see Table II for INAA results). Orito, as the largest obsidian source on the island, would be expected to dominate the obsidian component of artifact assemblages. But in view of the difficulties associated with acquisition of Motu Iti obsidian, one would expect it to represent a very small percentage of an assemblage or, perhaps, be confined only to high status sites such as *ahu*, sites in Anakena, or sites within the territory of the Miru *mata*, because of its prestige value.

All but one of the samples from Site 1-193, in the midst of all four obsidian sources, were assigned to Orito. The single exception was a specimen for which anomalous results were obtained; as such, it could not be reliably assigned to any one source. For the three west coast sites, four specimens from the inland cave (Site 8-169) were assigned to Orito, while artifact samples from the two *ahu* (Sites 8-1 and 8-2) contained both Orito and Motu Iti obsidian. The single south coast site (Site 12-1), from which 13 samples were selected, contained both Motu Iti and Orito obsidian. At Ana Papa te Kena (Site 34-2) on the north coast, all five artifacts were defined as Orito obsidian. And at the two sites in Anakena (Sites 35-7 and 35-9), the specimens represented both Motu Iti and Orito obsidian.

Orito obsidian was the main source represented among the total number of samples selected for this stage of research, with Motu Iti obsidian representing less than 20 percent of the sample collection. The predominance of Orito obsidian was expected; however, the far ranging occurrence of the Motu Iti obsidian in both geographic distribution and site type was not. Its appearance within the Miru territory was expected, as the Miru *mata* was the highest ranking clan of the island and was also the source of the *ariki*, the island's line of paramount chiefs. As such, items of high value or prestige (e.g., items that represent difficulties in access, such as the Motu Iti obsidian, complexities in design and manufacture, and the like) would be expected to occur within the boundaries of the highest ranking *mata*. Its occurrence, however, within the boundaries of the Ngaure *mata* (Site 12-1) on the south coast suggests that the Motu Iti obsidian was either more readily available than expected or that it was perhaps under some distributional restrictions. Was obtainment of the Motu Iti obsidian direct or indirect for *mata* members outside the Miru *mata*? Or was there some social, political, or economic connection between the Miru *mata* and the occupants or users of Site 12-1 that is not readily discernible with the geochemical data alone? Of course, a more accurate picture of source use could be drawn if a larger sample of the artifact assemblage were submitted for analysis. According to Leach (9), whole assemblages should be analyzed from each site in order to insure all sources at a site are represented. McKillop and Jackson (26) add that in provenance work, sample size is clearly an important factor as it may introduce biases into the data.

The results of this initial stage of analysis demonstrates that Orito, at least, was commonly exploited during the prehistoric era. But beyond this finding, our results are at best preliminary regarding the implications for the social, political, and economic organization of the island. Who had access to the Orito source, for example? Was it commonly exploited by all inhabitants of the island or was access limited to a few, with acquisition of the source material part of a more complex pattern of exchange? The ubiquity of the Orito obsidian in the archaeological sites sampled suggests that most had access to the source, whether direct or indirect. The same questions could be posed for the Motu Iti source; however, its more limited appearance in the sample collection suggests that source access was limited. Further, its occurrence within the Miru lands hints at some restrictions on access; perhaps it was reserved for the high clan use. Yet its presence in the Ngaure *mata* raises questions about the relationship between the Miru and Ngaure—do the Motu Iti specimens represent items of exchange? Or were there more complex social, political or economic alliances between the two *mata* which are as yet obscured?

Naturally, a study of this nature—provenance research—raises more questions than it answers, and leads to one conclusion, in order to address more thoroughly questions on the cultural landscape of Easter Island during the prehistoric era, more provenance work needs to be completed on the archaeological collections, with a far greater number of specimens subjected to analysis. In addition, the provenance work should be accompanied by a more complete lithic analysis which examines the full complement of the lithic production industry—the size and amount of debitage, the number of complete and partial tools and possibly ornaments in a collection, the amount of cortex present, the presence of cores and blocks of unworked material, and the frequency of reworked implements. As a supplement to the provenance work, lithic analysis can provide insights into the value of the material—for example, are smaller pieces being reworked more frequently at sites more distant from the sources? There are, of course, many more questions along these same lines; suffice it to say, there is certainly much more work that needs to be completed with, perhaps, other components of the portable material culture assemblage added to the analysis.

Conclusions

Until our study, only three source populations for the four obsidian sources had been identified on Easter Island. Our work illustrates that four sources can be distinguished from one another through INAA, and that these results can be readily applied to archaeological specimens. This process, however, is time consuming, demanding, and expensive. And, as we have used it, it is also a destructive technique that requires the sample studied to be ground to a fine powder. While this is not of great concern when using debitage in the analysis, it does present a dilemma when considering application to rare artifacts such as the obsidian pieces in statue eyes. This limitation, however, can be circumvented in future studies, at the cost of additional effort.

Our work has successfully established the foundation for provenance studies on the island, although not without first overcoming an unexpected analytical problem, namely differences in the data generated from the two laboratories used in this study. The basis for these differences can be traced to the processes employed at each lab— OSU, for example, calculated element abundances from energy events (photopeaks) selected by their computer program at specific channels; at the Center for Volcanology at UO, on the other hand, element abundances were calculated from photopeaks selected from a visual display of spectral data. The latter process is more time consuming, but it ensures that the entire photopeak is included in a calculation and that the necessary adjustments are made in calculating abundances from composite peaks; the former process, computer selected peaks/channels, is poorly adapted for dealing with slight shifts of peak positions and composite peaks, introducing a potential risk of calculating abundances from either partial or compound peaks, defined by the range of channels targeted for a specific energy event by a computer program. In a positive light, the differences in laboratory results were, for the most part, relatively minor (with the exception of Zn), with consistently differing element abundances reported. As a cautionary note and by-product of our research, we have demonstrated that no two laboratories produce the same information and one should be prepared to normalize and reconcile the resultant data sets.

Within the range of criteria developed by Leach (*9, 23*) as the basis for routine provenance work on archaeological collections, the results of our research are at best uneven. These criteria include: (1) the method should be capable of reliably distinguishing between all potential sources available during the prehistoric era; (2) the technique should be non-destructive so that the cultural property can be analyzed at a later date; (3) the method should be rapid, inexpensive, and not too complex; (4) the method should be available for general use by archaeologists for large runs of samples numbering in the thousands.

Our analytical technique of choice, INAA, has proven to be a powerful method for differentiating the four obsidian sources on Easter Island (criterion 1), but it is, as yet, a destructive method that is both time consuming in regard to preparation time (and processing time if one chooses to do the work oneself) and expensive, and as such would be prohibitive for large sample sets. Bird at the Australia Nuclear Science and Technology Organisation is at present trying to refine the PIXE-PIGME technique to recognize the four source populations distinguished by INAA. If he is successful, Leach's four provenance criteria could be met. There are, too, other analytical methods (e.g., electron microprobe) which remain to be explored in their applicability toward Easter Island obsidian source identification. But until such time as Bird is able to refine his methods or other techniques are tested, INAA remains the most useful technique for distinguishing all four obsidian sources on the island and determining the raw material sources of archaeological specimens.

Acknowledgments

Our research was supported by a series of grants from the Department of Energy Reactor Use Sharing program (1987-88, 1988-89, 1989-90, 1990-91). Special thanks must be extended to Roger Bird and Chris Stevenson, who supplied obsidian samples from Motu Iti; Eric Sonnenthal, who developed the INAA reduction program used by the Center for Volcanology at UO for this project; Art Johnson, former director of the Radiation Center at OSU; and especially Bob Walker, who, before his untimely death, administered the DOE grants and conducted the analytical work at OSU.

Literature Cited

1. Thomson, W. J. In *Annual Report of the Board of Regents of the Smithsonian Institution*; U.S. Government Printing Office: Washington D. C., 1889; pp 447-552.
2. Routledge, K. *The Mystery of Easter Island*; Sifton, Praed and Co. Ltd.: London, 1919.
3. Heyerdahl, T. *Aku-Aku: The Secret of Easter Island*; Ballantine Books: New York, 1958.

4. Dos Pasos, J. *Easter Island: Island of Enigmas*; Doubleday: Garden City, NY, 1971.
5. Neruda, P. *The Separate Rose*; O'Daly, W., Trans.; Copper Canyon Press: Port Townsend, WA, 1985.
6. Metraux, A. *Ethnology of Easter Island*; Bernice P. Bishop Museum Bulletin 160; The Bishop Museum Press: Honolulu, HI, 1940.
7. Heyerdahl, T.; Ferdon, E., Jr., Eds.; *Reports of the Norwegian Archaeological Expedition to Easter Island and the East Pacific: Archaeology of Easter Island*; Monographs of the School of American Research and the Museum of New Mexico, Number 24, Part 1; Forum Publishing House: Stockholm, 1961.
8. Mulloy, W.; Figueroa G.-H., G. *Chile: The Archaeological Heritage of Easter Island*; Report for the United National Educational, Scientific, and Cultural Organization; 1966.
9. Leach, B. F. *New Zealand Archaeological Association Newsletter* **1977**, *20(1)*, 6-17.
10. Beardsley, F. R.; Ayres, W. S.; Goles, G. G. *Bulletin of the Indo-Pacific Prehistory Assn.* **1991**, *11*, 179-187.
11. Baker, P. E.; Buckley, F.; Holland, J. G. *Contributions to Mineralogy and Petrology* **1974**, *44*, 85-100.
12. Hanan, B. B.; Schilling, J.-G. *Journal of Geophysical Research* **1989**, *94*, 7432-7448.
13. Bandy, M. C. *Bulletin of the Geological Society of America* **1937**, *48*, 1589-1610.
14. Chubb, L. J. *Geology of Galapagos, Cocos, and Easter Islands*; Bernice P. Bishop Museum Bulletin 110; The Bishop Museum Press: Honolulu, HI, 1933.
15. Bird, J. R. In *Archaeometry: Australasian Studies 1988*; Prescott, J. R., Ed.; University of Adelaide, Dept. of Physics and Mathematical Physics: Adelaide, 1988; pp 115-120.
16. Mahood, G. A. In *Obsidian Dates IV: A Compendium of the Obsidian Hydration Determinations Made at the UCLA Obsidian Hydration Laboratory*; Meighan, C. W.; Scalise, J. L., Eds.; Monograph XXIX, Institute of Archaeology, University of California Press: Los Angeles, CA, 1988; pp 105-112.
17. Bonatti, E.; et al. *Journal of Geophysical Research* **1977**, *82*, 2457-2480.
18. Evans, C. In *Reports of the Norwegian Archaeological Expedition to Easter Island and the East Pacific: Miscellaneous Papers*; Heyerdahl, T.; Ferdon, E., Jr., Eds.; School of American Research and Kon-Tiki Museum: Stockholm, 1965, Vol. 2; pp 469-495.
19. Michels J. W.; Stevenson, C. M.; Tsong, I. S. T.; Smith, G. A. In *Easter Island Archaeology: Research on Early Rapanui Culture*; Ayres, W. S.; Beardsley, F. R., Eds.; Monograph 1, Pacific Islands Studies Program, University of Oregon Press: Eugene, OR; in press.
20. Stevenson, C. M., Ph.D. thesis, Pennsylvania State University, 1984.
21. Stevenson, C. M.; Shaw, L.; Cristino, C. *Archaeology in Oceania* **1984**, *19*, 120-124.
22. Cristino, C.; Vargas, P.; Izaurieta, R. *Atlas Arqueologico de Isla de Pascua*; Corporacion Toesca: Santiago, 1981.
23. Leach, B. F.; Warren, S. E. In *Archaeological Studies of Pacific Stone Resources*; Leach, F.; Davidson, J., Eds.; British Archaeological Reports International Series 104, 1981; pp 151-166.
24. Goles, G. G. In *Physical Methods in Determinative Mineralogy*; Zussman, J., Ed.; Academic Press: London, 1978; pp 343-369.
25. Ayres, W. S. *Easter Island: Investigations in Prehistoric Cultural Dynamics*; University of South Carolina Press, Columbia, SC, 1975.

26. McKillop, H.; Jackson, L. J. In *Obsidian Dates IV: A Compendium of the Obsidian Hydration Determinations Made at the UCLA Obsidian Hydration Laboratory*; Meighan, C. W.; Scalise, J. L., Eds.; Monograph XXIX, Institute of Archaeology, University of California Press: Los Angeles, CA, 1988; pp130-141.

RECEIVED August 15, 1995

Chapter 6

Chrono-Cultural Technique Based on the Instrumental Neutron Activation Analysis of Copper-Based Artifacts from the "Contact" Period of Northeastern North America

J.-F. Moreau[1] and R. G. V. Hancock[2]

[1]Laboratoire d'archéologie et Départment des sciences humaines, Université du Québec à Chicoutimi, Chicoutimi, Québec G7H 2B1, Canada
[2]Slowpoke Reactor Facility and Department of Chemical Engineering and Applied Chemistry, University of Toronto, Toronto, Ontario M5S 1A4, Canada

The time period from 1500 AD to 1700 AD when the earliest "contacts" took place between Amerindians and Europeans in eastern Canada demands special absolute and relative dating techniques to establish the contemporaneity of artifacts and sites. Instrumental neutron activation analysis (INAA) of 500 copper-based artifacts from archaeological sites in Québec and Ontario has resulted in a database of 8 elements that allows us to assess the degree of homogeneity of these artifacts at the intra-site level as a means of establishing which parts of a site may be contemporaneous. The analytical data are also used in inter-site comparisons in an attempt to establish the chronology of the trading actions that link them.

The late 16th and the early 17th centuries witnessed European activities in the estuary of the Saint Lawrence River that included trading with Amerindians. Although current reflection on this subject has not yet weighed precisely the importance of this trading among other European activities (from whaling and fishing - mainly cod - to colonization), the continuity in commercial trading of European goods from about 1580 onwards attests that the cost involved, including the manufacturing of goods, their transport overseas and, on land, to the Amerindians, covered, at least, the overall return of the activity (whether it be in terms of economical return such as furs or in socio-cultural terms such as goods serving to shape alliances with Amerindians).

The earliest goods traded included ornamental items such as glass beads as well as goods perceived, at least at first glance, as utilitarian ones. Among these are those made of iron including adzes and knives, and those made of copper, specifically kettles that historical accounts indicate were often included as burial offerings. While this behavior has also been documented archaeologically for the MicMac (1) and the Huron and Neutral (2), most of the artifacts made of European copper have been fabricated from scraps presumably cut out of worn-out containers. These include objects serving a wide range of functions, from sustenance (projectile points) to decorative objects (tinkling cones, tubular beads).

Recent studies have shown that in non-burial sites, such artifacts, including unshaped scraps may vary in frequency from dozens to hundreds of specimens. See for example Neutral sites in Ontario (*2*), and in New York State, Onondaga (*3*) and Seneca (*4*) sites. These frequencies also characterize sites located in the southeastern fringe of the subarctic, especially in the Saguenay–Lac-St. Jean area (*5*).

Detailed INAA analyses of all the copper based artifacts from two sites of this region show that most artifacts exhibit very heterogeneous elemental composition. However, those that exhibit homogeneity support the archaeological hypothesis that, since they were found close together during archaeological excavation, they most probably are derived from the same container. This behavior is analogous to that regularly observed pattern where quite bulky, roughly flaked stones are thought to have been left on site for future stone tool production during subsequent occupations (*5*).

Attempts to use elemental compositions as a chronological tool for the early European-Amerindian contact era has been tested previously. Using emission spectroscopy, Fitzgerald and Ramsden (*6*) indicated that copper-based artifacts from Iroquoian sites in Ontario could be segregated between low (ca. 75±10% Cu) and high (ca. 96±2% Cu) copper content. This segregation was interpreted as a reflection of changing European suppliers from late 16th century Basques) to later French suppliers. Using INAA, Hancock et al. (*7*) proposed a seriation of northeastern North American archaeological sites based on the chemical analysis of copper artifacts.

On the basis of independent chronological markers such as morphological traits of artifacts [for glass beads traded by Europeans see (*8, 9*)] and contextual characteristics such as stratigraphy (*8*) and spatial distributions (*10, 11*), comparisons were drawn between three containers recovered from two non-burial sites from the Saguenay–Lac-St. Jean area [Chicoutimi (DcEs-1) and Ashuapmuchuan (DhFk-7)], and two containers associated with a burial from the Saint-Nicolas site (CeEu-12) near Québec (*12*). Inasmuch as these five containers could all be dated to the first half of the seventeenth century on the basis of their morphological traits and contextual characteristics, the extent of elemental heterogeneity they exhibit could be the result of kettles supplied from a larger number of smelting/shaping localities over a wider geographical range in Europe (*13*).

This paper has three objectives: (1) the presentation of the INAA results in terms of elemental composition of the multitesting of five artifacts from two Iroquoian sites (Tregunno and Milton Heights, also known as Gaetan) previously analyzed using emission spectroscopy (*6*); (2) the comparison of the results obtained by the two techniques; and (3) the comparison of the Tregunno-Gaetan data with some of those obtained from Québec.

Method

Manually abraded samples, weighing between 8 mg and 34 mg, were analyzed using INAA at the SLOWPOKE Reactor Facility of the University of Toronto. Individual samples were stored in 1.2 mL polyethylene vials and were analyzed as described in Hancock et al. (*14*). They were first irradiated serially for one minute at a neutron flux of 1.0×10^{11} neutrons cm^{-2} sec^{-1} and assayed for 200 seconds after a delay time of approximately one minute using a Ge detector gamma ray spectrometer, for copper (Cu), manganese (Mg), vanadium (V) and aluminum (Al). Elemental concentrations were calculated using the comparator method (*15*). After a delay time of approximately one hour, each sample was re-assayed for zinc (Zn), copper (Cu), manganese (Mn) and indium (In). Medium and long half-life radioisotope producing elements were analyzed by batch irradiating twenty to forty samples per irradiation container for 16 hours at 2.5×10^{11} neutrons cm^{-2} sec^{-1}. After 5 to 7 days, samples were serially assayed for 500 to 2000 seconds looking for gold (Au), copper (Cu), cadmium (Cd), arsenic (As), antimony (Sb), silver (Ag), zinc (Zn) and sodium (Na). A final counting was made after 10 to 14 days, at which time the samples were counted for 4000 to

10,000 seconds to determine the concentrations of tin (Sn), selenium (Se), mercury (Hg), gold (Au), arsenic (As), antimony (Sb), silver (Ag), nickel (Ni), scandium (Sc), iron (Fe), zinc (Zn) and cobalt (Co). This procedure produced replicate measurements of enough elements to guarantee no sample mix-ups. Analytical precisions ranged from ±1% to detection limits.

Pioneering papers on the application of the INAA method to copper-based artifacts found in North America enabled a clear distinction between the native copper artifacts made from source(s) in North America and the copper alloyed artifacts made in Europe (14-16). On this basis, subsequent studies were devoted to finer distinctions among the European material. These studies were based on the premise that a distinction could be drawn between elements occurring naturally with the basic element, copper, and those commonly added by alloying. Hence, there are elements in these artifacts that are of "natural" origin, namely copper and those associated with its ores: silver, gold and indium. Other elements, e.g., arsenic, antimony, tin and zinc, are known to have been added purposely to copper either to clean it (fluxing method), or to alloy it (change some aspect of its nature, such as its malleability), or both. As a general rule, elements present only in trace amounts (e.g., ppm order of magnitude) probably were present naturally in the copper or alloying ores, or both. Elements in the order of percentages were most probably added purposely. Palaeometallurgical studies have shown that only prior to the 19th century, these additions were made on the basis of the experience of the metalworker (17, 18). It seems, however, that these elements may also have been added in quantities, sometimes large, above that required by both fluxing and alloying. This observation has been pointed out by researchers (6), showing that there was a shift from high copper cauldrons in the time of earliest occurrences of European artifacts at the end of the 16th century to lower copper (brass) cauldrons during the first half of the 17th century. The over-loading in fluxing and alloying elements might illustrate the consumption of the artifacts made under a cheaper workmanship. Caution, however, should be given to these data as being uniquely of an economic nature: in fact, this shift does seem to correspond to that of a southern (Mediterranean or Basque) source of provenance of high copper containers in the last years of the 16th century as opposed to a later period when brass containers came presumably from Flanders.

In order to establish the differences between copper-based artifacts, we have tested several quantitative mathematical approaches. One based on percentage estimations allowed a limited number of specimens to be distinguished from one another (19). The use of weighted values, however, may be less appropriate than that of absolute ones. We thus sought a method based on a trigonometric approach. Although promising results have already been presented (20), some aspects of mathematical formulation remain to be tested. However, its graphical representation will be thoroughly used in this paper. Taking into account the palaeometallurgical aspects of copper working already referred to, these graphical representations are based on the grouping of "naturally" occurring elements (Cu, Ag, Au and In) and those (As, Sb, Sn and Zn) that, if they are not "purposely added," may interact with copper to change its physical properties. It should be mentioned that all eight of these elements may occur in ores. For example, arsenic is known to be part of some naturally occurring copper ores, while silver and gold are linked to ores of zinc (21, 22). In order to match the trigonometric approach currently under examination, the square root has been taken of the summation of the actual values of each of these two groups of chemical elements. A side, but not negligible, effect of this transformation is that it reduces, but does not eliminate, the discrepancies between the concentrations of the elements which differ by many orders of magnitude. Among our specimens, this is obviously the case for copper itself but also for one purposely added element, zinc.

Ontario and Québec Samples

Among the early contact sites whose copper-based scraps were analyzed by means of emission spectroscopy (*6*), Tregunno and Gaetan (Milton Heights) scraps of high and low level of Cu concentration were re-analyzed for this study. The presence of high copper specimens has been interpreted as having been derived from copper kettles of Basque origin, a group that was well established in the Saint Lawrence estuary in the late 16th century and early 17th century (*23*). The Tregunno and Gaetan sites have been dated on the nature of their European-manufactured glass beads ca. 1580 to 1600 (*2*: pp 170 ff). INAA data as well as the emission spectroscopy data are tabulated in Tables I and II, while the graphic distributions of eight (trace elements and Cu and Zn) and six elements (trace elements only) are illustrated in Figures 1 and 2.

The elemental composition of these five Ontario specimens can be compared with the published data of five "kettles" from Québec. Although we sampled the scraps assigned to these five kettles, their assignment to specific kettles was based on archaeological context. At the Chicoutimi site (DcEs-1), archaeological excavation produced two concentrations of scraps (451, 354). However, undertaken in the 1970's under the pressure of salvage archaeology, field data from the Chicoutimi site do not precisely report the nature of these concentrations. Exhaustive INAA of the Chicoutimi specimens exhibit a clear contrast between the chemical composition of the scraps from the concentrations and those found elsewhere on the site (*5*). The argument of the archaeological context is thus sustained by the homogeneity in elemental composition. But specimens at the "fringe" of the chemical distribution may not belong to the kettles despite their presence within the scrap concentration as noted in the course of archaeological fieldwork.

The assigment of nine scraps from the DhFk-7 site to a single kettle is based on provenance arguments similar to those used for the Chicoutimi site. In this instance, however, field techniques and ensuing data ensured that the nine scraps were retrieved from a spatially limited concentration (15 cm in diameter and 5 cm in depth). Like the Chicoutimi situation, the scraps from the DhFk-7 concentration also exhibit a high degree of elemental composition homogeneity in contrast to all other specimens found at the site (*5*). Still, however, some specimens may be improperly assigned to this "kettle".

Finally, two nearly complete cauldrons were found in a burial context at the St. Nicolas site (CeEu-12). INAA was performed on a series of tiny fragments from both kettles. While refitting to the kettles themselves of these tiny fragments was precluded by their minute dimensions, their respective assignment to one or the other of the kettles was performed on the basis of visual observation during the conservation process, a distinction that has been supported by elemental analysis. However, the overlap between the distributions of elemental composition of these two series of fragments may either constitute a real overlap in elemental composition of the two cauldrons, or it may be the result of incorrect visual assignment of a minority of these fragments.

In order to facilitate the comparison between the Ontario and the Québec cauldrons, the Québec data are represented in Figures 1 and 2 by ellipses that encompass 99% of the variation of the individual values as reckoned by SYSTAT (*24*). However, following a detailed study made for the Chicoutimi site, five anomalous samples were excluded from the original total of 39 samples that composed the DcEs-1 [451] concentration (*5*). A detailed study of the St. Nicolas samples also permits the exclusion of one specimen from the CeEu-12 [2] cauldron (*13*), illustrating a case of an unusual amount of Sn that could indicate a sample retrieved from a soldering join.

TABLE I. Elemental Composition of the Copper-based Artifacts from the Gaetan Site

Arch. Acc. #	Sample #	Ag ppm	Au ppm	As ppm	Cu %	In ppm	Sb ppm	Sn %	Zn %
MH1 (G1)	b1-15	730	23	480	99	11	2000	≤ 0.11	≤ 0.05
MH1 (G1)	b1-16	770	23	530	98	11	2100	≤ 0.13	≤ 0.07
MH1 (G1)	b1-17	760	26	520	99	11	2200	≤ 0.10	≤ 0.05
MH1 (G1)	b1-18	790	26	480	99	12	2000	≤ 0.11	≤ 0.05
MH1 (G1)	b1-19	760	24	510	97	12	2000	≤ 0.09	≤ 0.05
MH1 (G1)	b3-9	780	24	770	98	10.1	1900	≤ 0.05	≤ 0.01
MH1 (G1)	b3-10	770	26	620	98	11.2	1900	≤ 0.06	≤ 0.01
MH1 (G1)	b3-11	750	24	600	100	11.5	1700	≤ 0.06	≤ 0.0087
MH1 (G1)	b3-12	880	28	640	96	12.8	2100	≤ 0.05	≤ 0.0085
MH1 (G1)	b3-13	750	23	620	97	11.2	1800	≤ 0.04	≤ 0.0088
G1'	F & R	701		389	99.48		2100	0.026	0.0077
MH2 (G2)	b1-20	660	16	660	99	8.5	2200	≤ 0.18	≤ 0.06
MH2 (G2)	b1-21	690	16	630	97	8.3	2400	≤ 0.10	≤ 0.05
MH2 (G2)	b1-22	660	16	570	98	7.6	2300	≤ 0.21	≤ 0.09
MH2 (G2)	b1-23	680	17	580	99	8.6	2400	≤ 0.09	≤ 0.05
MH2 (G2)	b2-31	720	16	770	100	8,0	2200	≤ 0.11	≤ 0.018
MH2 (G2)	b2-32	690	16	730	99	8.6	2200	≤ 0.06	≤ 0.012
MH2 (G2)	b2-33	690	16	810	99	7.9	2200	≤ 0.06	≤ 0.012
MH2 (G2)	b3-14	720	16	740	98	7.28	2200	≤ 0.04	≤ 0.0088
MH2 (G2)	b3-15	690	16	730	97	7.61	2100	≤ 0.04	≤ 0.0082
MH2 (G2)	b3-16	750	17	940	100	8.28	2300	≤ 0.04	≤ 0.0089
G2'	F & R	709		480	99.36		2727	0.026	0.0028
MH3 (G3)	b1-11	440	3.1	540	77	2,0	150	2.1	21
MH3 (G3)	b1-12	430	3.0	620	79	1.7	160	2.0	19
MH3 (G3)	b1-13	440	2.7	540	75	2.3	140	2.5	21
MH3 (G3)	b1-14	410	2.7	590	77	≤ 1,0	150	2.1	21
NH3 (G3)	b2-34	350	2.8	530	76	≤ 0.9	150	1.5	23
NH3 (G3)	b2-35	370	2.8	580	76	≤ 0.7	160	1.7	22
MH3 (G3)	b3-1	310	2.7	580	76	≤ 0.8	130	1.6	21
MH3 (G3)	b3-2	370	3.1	590	73	≤ 1,0	150	1.8	24
MH3 (G3)	b3-3	320	2.6	580	75	≤ 0.8	140	1.7	21
MH3 (G3)	b3-4	380	3.2	590	75	1.7	150	2.1	21
G3'	F & R	375		820	81.29		1176	0.206	15.58

b1 to b3: batches of INAA data. F & R: emission spectroscopy data (6: 158)

TABLE II. Elemental Composition of the Copper-based Artifacts from the Tregunno Site

Arch. Acc. #	Sample #	Ag ppm	Au ppm	As ppm	Cu %	In ppm	Sb ppm	Sn %	Zn %
Tregunno 1	b1-24	1000	38	120	99	130	230	≤ 0.09	≤ 0.05
Tregunno 1	b1-25	990	38	130	99	110	230	≤ 0.08	≤ 0.04
Tregunno 1	b1-26	980	38	130	99	120	240	≤ 0.12	≤ 0.05
Tregunno 1	b1-27	930	36	130	99	130	230	≤ 0.09	≤ 0.04
Tregunno 1	b1-28	1010	38	140	99	130	230	≤ 0.19	≤ 0.04
Tregunno 1	b3-17	940	39	150	97	130	220	≤ 0.05	≤ 0.037
Tregunno 1	b3-18	980	38	150	98	120	220	≤ 0.06	≤ 0.0077
Tregunno 1	b3-19	1020	36	160	98	120	210	≤ 0.03	≤ 0.0046
Tregunno 1	b3-20	1090	43	150	98	120	250	≤ 0.04	≤ 0.037
Tregunno 1	b3-21	960	38	160	99	120	220	≤ 0.08	≤ 0.035
T1'	F & R	1156		180	99.63		210	0.09	0.039
Tregunno 2	b1-3	1060	11	1900	77	9.2	420	4.2	17
Tregunno 2	b1-4	1170	12	1940	81	8.7	460	4.1	17
Tregunno 2	b1-5	1010	11	1820	78	6.8	390	3.2	18
Tregunno 2	b1-6	1060	11	1880	79	7.4	420	3.7	18
Tregunno 2	b1-7	1120	11	1870	78	8,0	400	3.8	18
Tregunno 2	b1-8	1050	11	1860	79	8.3	390	3.3	18
Tregunno 2	b1-9	1020	12	1850	79	7.2	390	3.3	18
Tregunno 2	b1-10	1070	12	1920	83	7.5	400	3.5	18
Tregunno 2	b2-29	960	11	1910	77	8.3	390	3,0	20
Tregunno 2	b2-30	960	11	1830	77	8.2	390	3.1	19
T2'	F & R	1049		253	85.56		960	3.98	9.31

b1 to b3: batches of INAA data. F & R: emission spectroscopy data (6: 158)

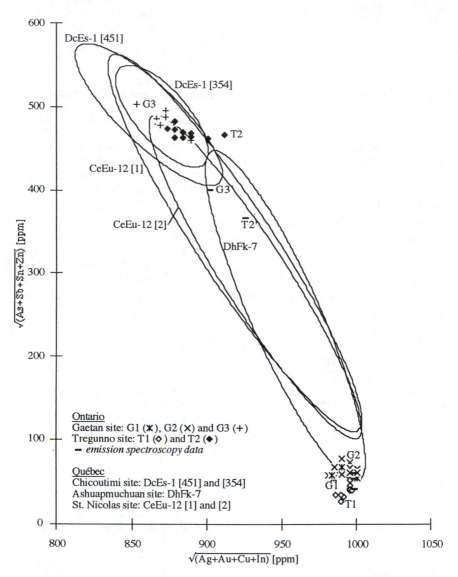

Figure 1. Elemental composition of Québec and Ontario copper-based artifacts for major and trace elements

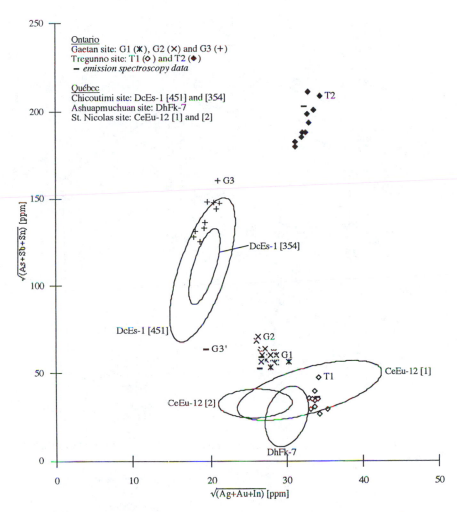

Figure 2. Elemental composition of Québec and Ontario copper-based artifacts for trace elements only

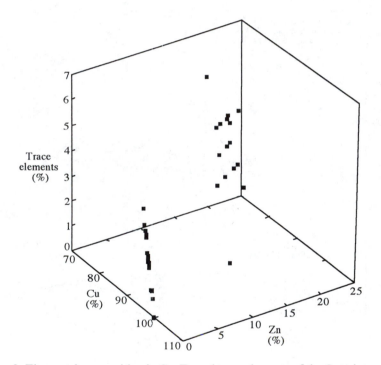

Figure 3. Elemental composition in Cu, Zn and trace elements of the Ontario copper-based artifacts

Results

The following results initially present the comparison between the Ontario and Québec cauldrons through a qualitative assessment based on the element distributions of eight and six elements, respectively. On the basis of the same representations, comparisons are then drawn between the elemental compositions obtained on the Ontario specimens using emission spectroscopy and INAA. This qualitative approach is then examined in the light of some statistical tests.

While the inverse relationship in content between Cu and Zn is obviously illustrated in the eight-elements graph (Figure 1), the three Ontario specimens with high copper content are grouped in a tight cluster toward the highest values of the abscissa, at the lower edge of the DhFk-7 distribution. In contrast, the individual values of the Gaetan 3 specimens and those of the Tregunno 2 specimens overlap in the relatively low-copper/high-zinc area of the graph, in the immediate vicinity of the two Chicoutimi cauldrons (DcEs-1 [451] and DcEs-1 [354]. The DhFk-7, CeEu-12 [1], and the CeEu-12 [2] cauldrons are located midway between the G1-G2-T1 cluster and the cluster formed by the Chicoutimi, G3 and T2 cauldrons.

When trace elements only are considered (Figure 2), the Ontario Gaetan 1 and 2 specimens remain clustered while the Tregunno 1 specimens which had been grouped with them under the eight element graph, are now separated from them, overlapping with the Québec St. Nicolas [1] cauldron. On the basis of the six trace elements, although Ontario G1, G2 and T1 specimens are high copper content cauldrons, they bear more resemblance to the Québec cauldrons from St. Nicolas and Ashuapmushuan, even though the latter are medium copper content cauldrons. All six of these cauldrons are, however, fairly sharply distinguished from the Ontario G3 and the Québec Chicoutimi cauldrons, all of them being low copper content containers. Finally the Tregunno 2 cauldron stands alone compared to either of the clusters.

Since the proportions of the different elements are expressed in relative values, Cu and Zn are bound to an inverse relationship expressed in the eight element graph of Figure 1 by the overall oblique slope followed by the data from all the cauldrons considered together. The question remains however as to the independence of the trace elements with Cu and Zn, the major elemental components of the cauldrons. The 3D graph of the 10 samples retrieved on each of the five Ontario kettles (Figure 3) shows clustering into low and high copper content specimens while these values are spread out on the plane of the inverse relationship between Cu and Zn. Figure 3, however, indicates no relationships of the trace elements to Cu and/or Zn. It may then be suggested that trace elements would be a useful and statistically independent tool to characterize the elemental composition.

The emission spectroscopy analysis undertaken for the five Ontario specimens did not test for Au or In (Tables I and II). Although we did not try to compensate for this in the representation of these data in Figures 1 and 2, the emission spectroscopy data did, nonetheless, fit within the INAA data. Hence for the eight elements graph, the three emission spectroscopy data for G1, G2 and T1 are clearly included in the cluster formed by the INAA data for these kettles. The G3´ and T2´ emission spectroscopy data, although outside the overall INAA variation for these two kettles, are still near this variation.

The overall resemblance of emission spectroscopy to INAA data is, however, better illustrated by the trace element graph (Figure 2): the G1, G2, T1 and T2 emission spectroscopy data are included inside the INAA variation. The G3 emission spectroscopy data is the only one for which the trace elements are outside the INAA variation. In order to take into account the absence of Au and In data for samples analyzed using emission spectroscopy, a three dimensional distribution of the three trace elements estimated through both techniques has been devised (Figure 4). Sn was

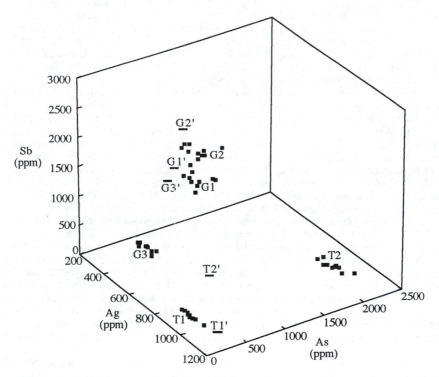

Figure 4. Elemental composition in Ag, Sb and As of the Ontario copper-based artifacts

not considered in this work given that it exhibits high values for low Cu cauldrons. Sn could thus have both status of an accidental (trace) and a voluntarily added (major) element. Figure 4 shows that for the three high copper specimens, the emission spectroscopy data is either close (G1, G2) or inside (T1) the INAA variation. However, for the two low Cu specimens (G3, T2), the emission spectroscopy data are clearly outside the variation of INAA.

In order to statistically assess these observations, the upper section of Table III reports the t-test results of the comparison between the single emission spectroscopy data and the ten INAA data retrieved from each of the five Ontario specimens. t-tests on the eight single elements are significant in 32 cases out of 40; t-tests on grouped elements still are significant in 12 out of 20 cases. In fact, except for T1 where the grouping of elements results in three unsignificant cases out of the four groupings, from a statistical point of view, the emission spectroscopy data do not correspond well with the INAA data. There are many reasons for these discrepancies. (1) The two techniques are fairly different from one another in that emission spectroscopy is a "surface" analysis technique while INAA is a "bulk" analysis technique. (2) Since both techniques result in relative values, the sum of both of which are identical, any discrepancy for one element in one technique compared to the other will exacerbate the discrepancies of the other elements. (3) Given that the concentrations of the different elements are not strictly independent, the t-test is perhaps not the best tool to statistically evaluate the similarity of the results obtained by both techniques. If the statistical estimations allow a prudent comparison of emission spectroscopy and INAA data, the qualitative appraisal from the graphs makes it clear, however, that the two techniques are not in complete contradiction.

Statistical tests of the differences in elemental mean concentrations were also systematically calculated between all five Ontario cauldrons (lower section of Table III). Except for the comparison of the G1 to G2 means for the grouped elements, almost all the differences in means are significant. This sustains the observation made from the graphs of Figures 1 and 2. Statistical tests of the differences in means were also calculated between the Ontario and Québec cauldrons (Table IV). As already suggested by Figures 1 and 2, the Chicoutimi [451] cauldron is completely identical to the G3 kettle: no difference in means is noted whether elements are considered on an individual or grouped basis. The only other statistical test that indicates similarity between Ontario and Québec specimens was the comparison between the CeEu-12 [1] cauldron and the T1 which exhibits three out of four not significant tests for the grouped elements while, however, their t-tests for single elements are unsignificant in only three out of eight cases. Of interest are also the five unsignificant t-tests for the comparison of the means of the single elements between DeEu-12 [2] and G1 while, however, grouping of the elements result in unsignificant cases only for the trace elements.

Conclusion

While INAA results seem to be reproducible, based on multiple testings of the five Ontario specimens, the comparison of results from the same samples with previous analysis using emission spectroscopy can be unreliable due to calibration problems. However, this discrepancy does not seem to be as great as the statistical tests would suggest. What seems to be in question is not so much the reliability of any of the two techniques but how to ascertain that comparison is made on the same kinds of data. This seems to be true at both the level of the acquisition of data (qualitative differences in the techniques) as well as at the level of graphical and statistical treatments.

On the basis of the INAA data, the Ontario specimens cluster into two opposite groups, one of low and the other of high Cu content, both situations being represented

TABLE III. t-tests of the Comparison Between the Ontario Copper-based Artifacts

	Ag	Au	As	Cu	In	Sb	Sn	Zn	$\sqrt{}$(Ag+Au +Cu+In)	$\sqrt{}$(As+Sb +Sn+Zn)	$\sqrt{}$(Ag+Au +In)	$\sqrt{}$(As+Sb +Sn)
G1'-G1	-5.710		-6.628	3.753		3.019	-5.560	-3.102	3.726	-6.586	-8.588	-5.640
G2'-G2	1.117		-8.266	2.087		10.630	-6.909	-3.763	2.087	-5.569	-0.775	-3.636
G3'-G3	-0.427		26.998	10.346		372.749	-17.125	-14.351	10.164	-21.617	-0.613	-21.887
T1'-T1	11.083		8.507	4.749		-4.885	0.428	0.986	4.746	1.196	0.665	0.848
T2'-T2	0.019		-131.394	11.063		77.665	3.498	-31.745	10.895	-53.233	-0.870	2.704
G1-G2	5.099	15.525	-3.001	-0.980	11.991	-5.133	-0.625	-0.015	-0.963	-2.393	5.818	-2.664
G1-G3	20.081	40.401	0.070	34.523	33.814	40.650	-18.303	-53.218	33.716	-104.322	19.845	-22.756
G2-G3	18.164	96.561	3.755	36.029	28.718	62.153	-17.942	-53.214	34.995	-99.102	17.299	-20.633
T1-T2	-2.348	44.501	-131.984	30.144	65.496	-21.432	-25.977	-65.235	28.967	-151.911	3.098	-40.700

T1-G1	10.778	16.961	-15.183	1.034	63.479	-38.763	0.191	0.276	1.115	-10.085	16.936	-10.925
T1-G2	16.748	37.015	-15.767	-0.117	65.796	-59.572	-0.417	0.226	-0.013	-11.010	25.686	-11.701
T1-G3	28.769	60.363	-42.608	39.792	69.635	17.310	-18.167	-53.218	38.025	-102.282	32.394	-26.525
T2-G1	11.274	-23.955	42.156	-26.799	-10.385	-34.560	26.086	65.230	-26.126	175.386	10.710	38.430
T2-G2	15.760	-23.811	30.565	-27.946	-0.390	-53.595	25.798	65.219	-27.124	155.148	16.390	35.573
T2-G3	26.336	46.988	84.849	3.573	22.514	33.232	9.832	-6.619	3.666	-3.213	26.337	10.926

Note: framed numbers indicate cases of probabilities NOT significant at the $p = 99\%$ (two tailed) level

TABLE IV. t-tests of the Comparison between the Ontario and Québec Copper-based Artifacts

Qué.	Ont.	Ag	Au	As	Cu	In	Sb	Sn	Zn	√(Ag+Au +Cu+In)	√(As+Sb +Sn+Zn)	√(Ag+Au +In)	√(As+Sb +Sn)
DcEs-1 [354]	G1	-26.771	-52.560	-1.117	-30.113	-39.509	-52.969	18.531	26.916	-28.526	51.648	-27.640	19.968
	G2	9.962	15.763	-18.610	-6.039	13.140	-88.065	-0.746	5.017	-5.883	6.925	11.031	-12.488
	G3	21.073	28.380	-37.116	10.463	25.466	-27.624	-24.619	-10.494	10.780	-8.867	23.970	-31.988
	T1	-1.470	-4.613	7.527	-6.067	-77.636	-45.692	-0.275	5.016	-5.937	8.047	-5.215	-0.964
	T2	-3.448	20.340	-142.686	8.270	12.956	-59.153	-35.283	-8.263	8.403	-8.379	-2.357	-48.440
DcEs-1 [451]	G1	-0.945	-49.022	0.614	-11.001	-7.871	-38.613	2.870	11.539	-9.201	20.273	-2.162	4.334
	G2	-0.575	-35.716	0.196	-11.232	-4.775	-47.354	2.827	11.532	-9.379	20.018	-1.573	4.038
	G3	0.873	1.806	0.621	-0.954	1.583	1.605	-0.368	0.192	-0.914	-0.134	0.985	-0.972
	T1	-1.935	-76.811	1.937	-11.229	-86.607	-0.437	2.845	11.531	-9.387	21.444	-4.076	5.842
	T2	-2.211	-21.579	-3.342	-2.242	-4.659	-4.793	-3.212	1.903	-2.064	0.534	-3.650	-4.389
DhFk-7	G1	6.596	7.238	-16.940	-5.649	6.762	-57.988	-0.097	5.018	-5.518	7.123	6.916	-11.208
	G2	9.962	15.763	-18.610	-6.039	13.140	-88.065	-0.746	5.017	-5.883	6.925	11.031	-12.488
	G3	21.073	28.380	-37.116	10.463	25.466	-27.624	-24.619	-10.494	10.780	-8.867	23.970	-31.988
	T1	-1.470	-4.613	7.527	-6.067	-77.636	-45.692	-0.275	5.016	-5.937	8.047	-5.215	-0.964
	T2	-3.448	20.340	-142.686	8.270	12.956	-59.153	-35.283	-8.263	8.403	-8.379	-2.357	-48.440

CeEu-12 G1 [1]	3.668	1.388	0.883	-1.960	1.849	-51.817	2.122	5.937	-0.975	7.260	2.339	-3.183
G2	4.330	4.272	-1.696	-2.038	4.340	-74.275	1.430	5.937	-1.012	7.119	2.914	-4.241
G3	6.888	8.744	0.989	1.446	9.405	-0.695	-24.400	-5.157	0.766	-4.210	5.387	-20.418
T1	1.912	-3.083	10.005	-2.032	-59.691	-5.954	1.897	5.936	-1.012	7.924	0.477	2.535
T2	1.416	5.923	-25.786	1.005	4.401	-16.573	-35.975	-3.507	0.523	-3.842	0.887	-32.068
DeEu-12 G1 [2]	-0.129	-2.437	1.814	-6.875	-1.541	-57.019	0.696	5.849	-6.669	9.453	-0.362	-1.548
G2	0.585	0.735	-2.990	-7.214	1.552	-86.339	0.611	5.848	-6.979	9.244	0.666	-2.082
G3	3.375	5.710	2.819	7.194	7.884	-3.845	-9.870	-7.553	7.343	-7.394	5.084	-10.923
T1	-2.036	-7.393	29.381	-7.237	-68.482	-17.566	0.674	5.847	-7.023	10.425	-3.704	1.240
T2	-2.561	2.573	-68.412	5.329	1.640	-38.019	-18.095	-5.586	5.350	-6.873	-2.951	-17.060

Note: framed numbers indicate cases of probabilities NOT significant at the $p = 99\%$ (two tailed) level in the Tregunno and Gaetan samples, confirming one of the overall conclusions reached by the emission spectroscopy study. Interestingly, while none of the Québec sites exhibit high Cu content cauldrons, the Ontario G3 cauldron exhibits a high degree of similarity with at least one cauldron from the Chicoutimi site. Such also is the case but in a more limited way for the Ontario T2 specimen compared to the Chicoutimi cauldrons.

Inasmuch as the five Ontario and the five Québec cauldrons all date to a relatively short period of time extending from 1580 to 1650, it could argued that factors other than a major transformation in technology through time explains the differences observed among these ten cauldrons. In the context of the comparison between the Chicoutimi, Ashupamuchuan and St. Nicolas cauldrons we hypothesized that the complete similarity of the two cauldrons from Chicoutimi indicated that the same sheet or at least the metal from the same smelting process was involved. Would that be the case for the G3 specimen?

We also proposed that the cluster made of the St. Nicolas and Ashuapmuchuan site indicated that the cauldrons may be related in some ways: were they the product of artisans that because they shared knowledge of craftsmanship would tend to fabricate cauldrons somewhat similar in terms of elemental composition? Again could this be applied to the G1 and G2 specimens? The quite strong elemental similarity between these specimens sustains the hypothesis of their provenance from a restricted era, presumably the Basque country. In this respect does the relative proximity in elemental composition of T1 compared to G1 and G2 indicate a cultural-geographical proximity of their stocking places in Europe?

Could we speculate further about the relationships between the Ontario group of cauldrons G1-G2-T1 and that of Québec CeEu-12 [1] and [2]-DhFk-7? First, we must remember that these two clusters share a fair degree of similarity in terms of trace elements while they tend to be different on the basis of their Cu-Zn content, the G1-G2-T1 being of high Cu content while the St. Nicolas and Ashuapmuchuan cauldrons are of medium Cu content. On this basis could their similarity of trace elements indicate some sharing in craftsmanship because of geographical proximity, namely cauldrons made in the Basque country and somewhere in Southeastern France? Otherwise, the level of Cu would have been maintained according to cultural patterning: the reddish color being that favored in the Mediterranean area while the yellowish color was especially appreciated in more temperate latitudes. In this respect, it is worthwhile stressing that our observation that the overall quantity of trace elements seems to be independent of the Cu-Zn content. Speculation about the sources of the cauldrons could also be expanded a little further. Hence, as opposed to this general "southern" cluster, could the Chicoutimi cauldrons and the G3 (and maybe T2) be characteristic of a more northern craftsmanship of cauldrons? Before any answer can be given to these questions, much more knowledge must be gained about the processes involved in smelting and the shaping of metal sheets into cauldrons as well as their implications for the spatio-temporal distribution of the European suppliers of kettles to the North American economy.

Acknowledgments

This work was supported in part by a joint grant made available to the Laboratoire d'archéologie of UQAC by R. Ouellet (Université Laval), principal investigator of a Programme équipes et séminaires grant from the Québec FCAR funds that includes J.-F. Moreau as co-researcher. Analytical work was also supported by the infrastructure grant from the Natural Sciences and Engineering Council of Canada to the SLOWPOKE Reactor Facility at the University of Toronto. Special thanks are extended to W. R. Fitzgerald, Department of Archaeology, Sir Wilfrid Laurier University,

Waterloo, Ontario, who provided the Tregunno and Gaetan samples previously analyzed by emission spectroscopy. W. R. Fitzgerald deserves special thanks for his thorough editorial assistance of the original text.

Literature Cited

1. Martin, C. *Ethnohistory* **1975**, *22*, 111-133.
2. Fitzgerald, W. *Chronology to Cultural Process : Lower Great Lakes Archaeology, 1500-1650*; Ph.D. dissertation, Department of Anthropology, McGill University: Montréal, 1990.
3. Bradley, J. W. *Evolution of the Onondaga Iroquois. Accommodating Change, 1500-1655*; Syracuse University Press: Syracuse, NY, 1987.
4. Wray, C. F.; Sempowski M. L.; Saunders, L. P. *Tram and Cameron. Two Early Contact Era Seneca Sites*; Research Records; Rochester Museum & Science Center: Rochester, NY, 1991; Vol. 21.
5. Moreau, J.-F.; Hancock, R. G. V. In *Proceedings of the Sixth Nordic Conference on the Application of Scientific Methods in Archaeology*; Mejdahl, V., Ed.; Esbjerg, Danemark, in press.
6. Fitzgerald, W.; Ramsden, P. *Can. Jour. Arch.* **1988**, *12*, 153-162.
7. Hancock, R. G. V.; Pavlish, L. A.; Fox, W. A.; Latta, M. A. *Archaeometry*, in press.
8. Moreau, J.-F. *Rech. Amérind. Qué.* **1994**, *24*, 31-48.
9. Moreau, J.-F. *Saguenayensia* **1993**, *35*, 21-29.
10. Moreau, J.-F. In *Transferts culturels en Amérique et ailleurs (XVIe-XXe siècles)*; Turgeon L.; Delâge D.; Ouellet R., Eds.; Les Presses de l'Université Laval: Québec, in press.
11. Moreau, J.-F. In *Mots, représentations. Enjeux dans les contacts interethniques et interculturels*; Fall, K.; Simeoni, D.; Vignaux D., Eds.; Actexpress, Les Presses de l'Université d'Ottawa: Ottawa, 1994; pp 343-374.
12. Chrétien, Y.; Bergeron, A.; Larocque R. In *Archéologies québécoises*, Balac, A.-M.; Chapdelaine, C.; Clermont, N; Duguay, F., Eds.; Recherches Amérindiennes au Québec: Montréal, 1995; pp 203-226.
13. Moreau, J.-F.; Hancock, R.G.V., In *Archéologies québécoises*, Balac, A.-M.; Chapdelaine, C.; Clermont, N; Duguay, F., Eds.; Recherches Amérindiennes au Québec: Montréal, 1995; pp 227-236.
14. Hancock, R. G. V.; Pavlish, L. A.; Farquhar, R. M.; Salloum, R.; Fox, W. A.; Wilson, G.C. *Archaeometry* **1991**, *33* , 69-86.
15. Hancock, R. G. V. *Analytical Chem.* **1976**, *48*, 1443-1445.
16. Hancock, R. G. V; Farquhar, R. M.; Pavlish, L. A.; Salloum, R.; Fox, W. A.; Wilson, G. C. In *Archaeometry '90*; Pernicka E.; Wagner, G. A., Eds; Birkhaüser Verlag: Boston, 1991, pp 173-182.
17. Smith, C. R.; Forbes, R. J. In *A History of Technology. Volume III, From the Renaissance to the Industrial Revolution. c.1500-c.1750*; Singer, C; Holmyard, E. J.; Hall, A. R.; Williams, T. I., Eds.; Clarendon Press : Oxford, 1957, pp 27-71.
18. Daumas, M. *Histoire générale des techniques. Tome II. Les premières étapes du machinisme*; Presses Universitaires de France: Paris, 1965.
19. Moreau, J.-F.; Hancock, R. G. V. In *The Archaeology of contact: processes and consequences*; Department of Archaeology, University of Calgary: Calgary, in press.

20. Moreau, J.-F.; Hancock, R. G. V.; Molhant, N. *Some advances in INAA of Cu based artifacts from the Saguenay-Lac-Saint-Jean*; paper presented to the 26th Annual Symposium of the Canadian Archaeological Association: Montréal, 5-9 May 1993.
21. Forbes, R. J. *Studies in Ancient Technology, Vol. IX*; Brill: Leiden, 1972.
22 Mohen, J. P *Métallurgie préhistorique. Introduction à la paléométallurgie*; Masson: Paris, 1990.
23. Fitzgerald, W. R.; Turgeon, L.; Whitehead, R. H.; Bradley, J. W. *Hist. Arch.* **1993**, *27*, 44-57.
24. Wilkinson, L. *SYGRAPH The System for Graphics*; Systat: Evanston, IL, 1989.

RECEIVED October 9, 1995

Chapter 7

Electron Microprobe and Neutron Activation Analysis of Gold Artifacts from a 1000 A.D. Peruvian Gravesite

Adon A. Gordus[1], Carl E. Henderson[2], and Izumi Shimada[3]

[1]Department of Chemistry and [2]Department of Geology, University of Michigan, Ann Arbor, MI 48109
[3]Department of Anthropology, Southern Illinois University, Carbondale, IL 62901

The gold-silver-copper ternary alloy contents of gold objects from an unlooted Middle Sicán (*ca.* 1000 AD) burial site on the north coast of Peru were determined by instrumental neutron activation analysis (INAA) using 10 µg metal rubbings from the interior of the objects. Electron microprobe analysis (EMPA) was also performed on a few of the objects allowed to be exported from Peru in order to measure the interior and surface alloy contents and to determine if the INAA samples are representative of the interior alloy. The near-perfect agreement between the probe data for the interior metal and INAA sample data confirms that the INAA sampling method results in metal characteristic of the interior alloy. The thin outer edges, which are only about five µm thick, are depleted markedly in copper (18 wt% interior copper, 7% edge copper, for example) and suggest that, for the high-quality gold objects (typical contents: 45 wt% Au, 37% Ag, 18% Cu), surface depletion of copper (and enrichment of gold and silver) may not have been deliberately induced but could have been the result of the repeated hammering-annealing required to produce the thin gold sheets used for the construction of the objects. Lower quality *tumbaga* gold sheets (typical interior contents: 25 wt% Au, 45% Ag, 30% Cu) also show marked surface depletion of copper.

The Burial Site

The northern coast of Peru was one of the primary centers of prehispanic metallurgy in the New World and, from the first millennium BC until the fourteenth century, The successive cultures inhabiting this region developed copper- and gold-based metallurgical expertise that was unmatched in innovative quality, sophistication, and overall production (*1, 2*). The Sicán Archaeological Project, under the direction of I. Shimada, is an interdisciplinary effort with some of the research directed toward an understanding of the technology and organization of metallurgical production and the role of the metal objects of the Middle Sicán culture that flourished on the northern coast of Peru during the 10th to 12th centuries AD.

0097–6156/96/0625–0083$12.00/0

The role of arsenical copper was first examined and excavations at various production centers in the Batán Grande region of Peru provided information on the technologies and uses of the arsenical copper products (*3, 4*). However, it is in the deep (*ca.* 11 meter) underground tombs of the Middle Sicán elite that unprecedented accumulations of precious metal objects were found (*5*) and between the 1930s to 1970s many of these tombs had been looted. Although many of the gold objects from these looted tombs found their way into private and public collections throughout the world, information about the specific gravesites, inventories of the sites, and other important contextual data are not available. Only a controlled archaeological excavation of an unlooted gravesite would provide the needed ancillary information.

In 1991-1992 Shimada excavated an intact shaft-tomb of a Middle Sicán nobleman at the base of an adobe-brick pyramid of Huaca Loro in the capital city of Sicán in the Batán Grande region of Peru. This excavation represents the first scientific documentation of such an elite tomb (*6*). The site contained about 1.2 tons of diverse grave goods, over 75% consisting of arsenical bronze and precious metal objects and scrap. Systematic examination and conservation of the grave contents was done by Shimada and coworkers in the Museo de la Nación in Lima where the objects will be on permanent display. The placement of the objects in the tomb and the positions of the main and sacrificial burials are illustrated elsewhere (*7*). Photographs of some of the objects are given in reference 6 and color photographs are given in reference 8.

The Gold Objects

About a hundred major gold objects (most constructed of separate pieces of gold) were contained in the tomb and included a gold mask with distinctive upturned eyes (which is characteristic of the Sicán style), a gold head ornament with an animal head, ceremonial *tumi* knives, crowns, various ornaments with attached bangles, dart throwers, ornate ear spools, long flat feathers, rattles, a pair of meter-long gold gloves, and piles of gold scraps and foils, including almost 2000 (*ca.* 1.4 cm square) pieces that had originally been sewn to a cloth mantle (now disintegrated) that had been laid beneath the principal burial.

Although all of the high-quality gold objects are very thin (usually only 0.1-0.5 mm thick) they are structurally sound due to the induced hardness which resulted from the repeated hammering that was required to produce the thin metal sheets used for fabrication of the final objects. The majority of the high-quality objects had interior gold contents in the range of about 40-75 wt% gold (*7*), the remainder being almost entirely copper and silver. Most of the gold scraps and foils are extremely thin (0.05 mm) and, hence, more flexible and fragile. As noted below, some of the scrap gold foil-fragments we analyzed have much lower gold contents in the range of 13-35 wt% gold and tend to be much more subject to corrosion.

Surface Depletion. The composition and method of fabrication of the gold objects is of interest. Based on the study of isolated pieces, it has been suggested that methods of deliberately removing copper and silver from the surface of ternary Au-Ag-Cu alloy objects were developed and used by some of the pre-Hispanic metalsmiths (*9-11*). (This process has sometimes been called depletion *gilding*, but modern goldsmiths limit the use of the word *gilding* to surface-additive, not subtractive processes.) Deliberate, induced surface depletion of silver and copper would allow the creation of an object with a surface of very high quality gold only a few μm thick, even though the bulk alloy starting material was of low gold content. Some of the analytical data presented in this study address directly the question of deliberate (as compared with unintentional) surface depletion of the Sicán gold objects.

Methods of Analysis

Two different methods of metal analysis were used. The first was based on instrumental neutron activation analysis (INAA) and involved taking very tiny (*ca.* 10 μg) metal rubbings of the interior alloy of the gold objects. The edge to be sampled was usually first stroked a few times using fine-grain emery paper in order to remove any surface corrosion products. This removed surface metal and oxides down to a depth of at least 20 μm. Then, a small (*ca.* 6 mm long, 2.5 mm diameter) piece of roughened high-purity silica tubing was rubbed against this cleaned area (*ca.* 2-4 mm^2), transferring about 10 μg of metal to the silica. A second sample was then taken from the same position and each sample packaged in a separate small polyethylene snap-cap vial for later irradiation and analysis. This method allowed the sampling of almost 400 separate gold pieces in the Museo de la Nación in Lima, transporting the rubbings to Ann Arbor, and irradiating the samples for 2.0 hours at a neutron flux of 3×10^{13} neutrons x cm^{-2} x sec^{-1} in The University of Michigan nuclear reactor. The second method involved electron microprobe analysis (EMPA) in Ann Arbor of fragments from a few of the gold objects that were allowed to be exported from Peru.

The INAA method has the advantage that sampling can be done in a museum and, if necessary, even while an object is mounted in a display case. The amount of metal required for each sample is so small (0.01 mg) that it is almost impossible to locate visually the place on the object where the metal rubbing was taken. Sampling can be done quickly; for example, 760 metal rubbings from 380 gold pieces were obtained during a three-day visit to Lima. A few hundred samples can be irradiated together with metal standards in the nuclear reactor. The activation analyses can be performed instrumentally using automatic sample changers followed by computer analysis of the resultant radioactivity spectra so that up to 100 samples can be analyzed in a day. The INAA method has the disadvantage that only the overall gross metal content of each object is measured. The INAA sampling and analysis method is described in detail in reference 7.

Since there can always be a question that the INAA samples are truly representative of the interior composition, it is important to measure directly the interior composition by EMPA. The near-perfect agreement in the data presented in this paper on the comparison between the INAA and the EMPA (interior) measured contents substantiates the conclusion that the INAA samples are representative of the interior composition.

The EMPA method has the advantage that very small areas of an artifact can be selectively analyzed to determine variations in the alloy content of a single object. This is especially important when determining the differences, if any, between the surface-edge and interior metal alloy. It has the disadvantage that the analysis procedure is more complicated, that it requires mounting a fragment of the metal object that weighs at least a few mg and more commonly 10-20 mg, and that it requires the loan of objects or the release of metal fragments. Only the few objects described in this paper, which were allowed to be exported from Peru, were analyzed by EMPA.

Small (*ca.* 2-7 mg) samples were mounted on edge, polished, coated with a very thin layer of electrically conducting carbon, and examined using a Cameca MBX microbeam electron microprobe analyzer. The electron beam was operated with a 15 kV accelerating voltage and a 20 mA beam current. Pure element standards were used to calibrate Au, Ag, and Cu. Compositional data were obtained for various areas of the thin surface edge and of the interior metal of each object. Areas ranging from a 1.0 μm spot to a 15 X 15 μm area were selected. The data for the 1 μm spot allowed, if desired, determination of the composition of selected, individual metal phases,

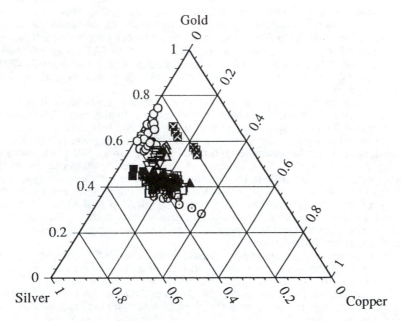

Figure 1. Weight fraction of gold, silver, and copper for major categories of Sicán gold objects listed in Table I: △ 31 2-cm Bangles, ▲ 12 Tumi Bangles, ▼ 28 Headdress Bangles, ⊙ 14 30-cm Disks, ○ 26 Ear Spool samples, □ 90 Headdress Feathers, ⊠ 11 Group-1 Feathers, ◘ 12 Group-2 Feathers, ■ 12 Group-3 Feathers, ◘ 11 Group-4 Feathers, ⊠ 6 Group-5 Feathers.

whereas the data for the larger areas provided compositional data representative of the metal content averaged over the metal phases. It is the latter phase-averaged data that can be compared with the INAA data. The only elements present in high concentrations were gold, silver, and copper with the exception of some of the corroded thin foil fragments which also showed high chloride concentrations.

The High-Quality Gold Objects

Shown in Figure 1 are the ternary alloy contents for a few of the major categories of gold objects as measured by INAA (7). Each category of objects tends to have its own characteristic alloy content as summarized in Table I. Ear spools, for example, which have greater thickness (*ca.* 0.5 mm), all have a copper content (1-6%) that is lower than all of the other objects. A few of the 30 cm disks and the square-foil scraps have low gold and high copper content. Except for Feather Group 6, all of the gold feathers have very consistent wt% Au:Ag ratios of about 1.1:1 unlike the groups of gold bangles which have Au:Ag ratios often approaching 2.0 (7). Other groups of objects also tend to exhibit their own characteristic ternary alloy content, each perhaps being chosen to result in desired mechanical properties. These various data illustrate the remarkable degree of understanding and control of the metallurgy of precious metals that was achieved by the Sicán goldsmiths.

One of two dart-throwers contained in the grave is suspected of being imported from coastal Ecuador based on its manufacturing details and repousséd decoration. The analysis data for the two portions of this dart thrower are clearly outside of the range of the other objects and serve to substantiate the supposition that this dart thrower was of a different origin (7).

Table I. Alloy Contents[a] of Major Categories of Sicán Gold Objects

Sample Type	N	Wt% Gold	Wt% Silver	Wt% Copper
2 cm Bangles	31	54.9 ± 0.2	32.2 ± 0.3	12.9 ± 0.1
Tumi Bangles	12	42.4 ± 0.4	33.8 ± 0.6	23.8 ± 0.6
Headdress Bangles	28	51.6 ± 0.5	26.7 ± 0.4	11.7 ± 0.3
30-cm Disks	14	36.3 ± 1.2	40.0 ± 1.1	23.7 ± 2.0
Ear Spools	26	63.2 ± 1.0	33.4 ± 0.8	3.4 ± 0.4
Headdress Feathers	90	41.4 ± 0.2	37.7 ± 0.3	20.9 ± 0.4
Feathers-Group 1	11	43.8 ± 0.5	40.7 ± 0.6	15.5 ± 0.8
Feathers-Group 2	12	43.3 ± 0.7	40.1 ± 0.8	16.6 ± 1.3
Feathers-Group 3	12	43.1 ± 0.9	42.4 ± 0.9	14.5 ± 1.4
Feathers-Group 4	11	42.8 ± 0.5	40.6 ± 0.7	16.6 ± 0.8
Feathers-Group 5	6	60.0 ± 2.0	21.6 ± 0.7	18.4 ± 2.6

[a]Data are given as the average ± s_m, the standard deviation of the mean; $s_m = s_x/\sqrt{N}$ where N is the number of separate metal pieces analyzed and s_x is the standard deviation.

Table II. Ternary Alloy Contents of a High-Quality Bangle and Support Wire[a]

Item	Position	Analysis	N	Wt% Gold	Wt% Silver	Wt% Copper	Ag/Au Ratio	Cu/Au Ratio
Bangle	Interior	EMPA	3	45.4 ± 0.3	37.0 ± 0.9	17.6 ± 0.6	0.82 ± 0.03	0.39 ± 0.01
Bangle	Interior	INAA	12	46.2 ± 0.5	36.3 ± 1.1	17.5 ± 0.9	0.79 ± 0.03	0.38 ± 0.02
Bangle	Surface	EMPA	4	51.1 ± 0.5	41.9 ± 3.2	7.0 ± 3.1	0.82 ± 0.07	0.14 ± 0.06
Wire	Interior	EMPA	3	44.8 ± 0.8	36.4 ± 1.3	18.8 ± 1.9	0.81 ± 0.02	0.42 ± 0.05
Wire	Interior	INAA	5	45.4 ± 0.7	36.1 ± 0.9	18.5 ± 0.7	0.80 ± 0.03	0.41 ± 0.02
Wire	Surface	EMPA	2	49.5 ± 0.9	46.3 ± 0.8	4.2 ± 0.1	0.93 ± 0.03	0.08 ± 0.01

[a]The ± is the standard deviation of the N replicate analyses.

INAA-EMPA Comparative Analyses of High-Quality Gold Objects. One small gold bangle and the gold wire used to attach the bangle were analyzed by both INAA and EMPA. The tear-shaped bangle measures 12 mm near the top tapering to 17 mm at the base, is 32 mm in length, and is 0.17 mm (170 μm) thick. It has a small (1.5 mm) hole at the top for inserting a gold wire to allow attaching the bangle to another object. The bent wire hook, if uncoiled, would measure about 17 mm in length, has a width of 0.9 mm, and a thickness of 0.20 mm (200 μm). EMPA examination of the edges of the bangle and wire showed surface layers about 5 μm thick having a composition different from that of the interior.

As was done with the objects in Lima, the INAA rubbings were taken so as to be representative of the interior metal alloy. Two series of six rubbings in succession were taken from two positions on the bangle and one series of five rubbings in succession was taken from the wire. As seen in Table II, the series of replicate INAA data were in extremely good agreement as shown by the very small standard deviations, s_x, associated with the average values for the (normalized to 100%) ternary metal contents.

Two or more replicate EMPA measurements were made at various positions in the surface edge-layer and in the interior of the bangle and wire samples. The individual replicate EMPA measurements were also normalized to %Au + %Ag + %Cu = 100%; the unnormalized sums ranged between 98 and 100% for the interior analyses and between 78 and 100% for the surface-edge analyses, the lower edge-values presumably including contributions from occluded copper(II) oxide. As seen in Table II, the normalized values for each EMPA set were also in very good agreement as shown by their small standard deviations. The near-perfect agreement between the data for the INAA rubbings and the EMPA data for the interior samples substantiates the assumption (7) that the INAA rubbing-samples were taken at a depth sufficient to ensure they are representative of the interior composition.

The interior alloy compositions of the bangle and the wire are, within the measurement uncertainties, the same and are similar to the alloy compositions of the 96 bangles and 8 wires sampled in Lima. These Lima samples have interior gold contents ranging from 43-56 wt%, silver from 31-37 wt%, and copper from 11-24 wt% (7). Ternary alloys of these compositions are very stable to corrosion (12), have melting points and metal-flow temperatures less than about 1000 °C (7, 13, 14), and while exhibiting a characteristic gold color (12), can be hardened to provide a degree of stability and flexibility not associated with thin pure gold which is relatively soft. Thus, the original alloy already has properties that would make it unnecessary to create a surface further enriched in gold and correspondingly depleted in copper and possibly also silver. Yet, the EMPA data of Table II indicate that the surface edges of the bangle and the wire are markedly depleted in copper. The Cu/Au ratio for the edge of the bangle relative to the Cu/Au ratio for the interior metal is 0.14/0.39 = 0.36. The corresponding comparison of Cu/Au ratios for the wire is 0.084/0.42 = 0.20. Almost two-thirds of the copper is depleted in the thin surface layer of the bangle and about 80% of copper in the wire is depleted in its thin surface layer. The comparative Ag/Au ratios for the surface-edge *vs.* the interior for the bangle and (within a few standard deviations) for the wire do *not* indicate surface depletion of silver.

These comparative edge *vs.* interior EMPA data allow us to comment on the manufacturing process. A gold-silver-copper ternary alloy ingot would be prepared using furnaces (with associated blowpipes to provide the necessary draft) available to the Sicán goldsmiths (3, 4). This ingot could have been as much as 10 mm thick and extensive hammering would be required to produce a final sheet as thin as 0.2 mm from which the desired gold object would be cut. After some degree of hammering, the slightly flattened ingot would become hard and require annealing in the furnace to soften it. As noted by Lechtman (11), the act of annealing would cause some of the surface copper to become oxidized creating a darkened appearance of the surface. The use of any of the numerous acids available to the Sicáns (for example, the acids resulting from the aging of fermented beverages) would permit dissolving and removing the oxidized copper layer. Repeated hammering, annealing, and removal of oxidized copper could then continue until the desired thinness was achieved. The thin, 1-5 μm thick surface layer would have less copper than the interior alloy mixture but the color and the mechanical and corrosion properties of the surface would not differ markedly (12) from that of the interior.

Since there would have been no need to alter deliberately the surface of the original alloy (to achieve a brighter gold color, for example), the observed copper depletion at the surface could have been simply the result of the method of creating the final thin sheet. Silver is much more stable and oxidizing acids would have been required if any surface silver were to be removed (10, 11).

In a separate study, Jo Ann Griffin and I. Shimada are conducting replicative experiments on the preparation of sheet gold having compositions similar to those of the Sicán gold objects. These studies will allow us to determine experimentally the amount of surface depletion of copper which results in the preparation of the sheets and should allow determining the degree to which the observed copper depletion at the surface of the Sicán objects is a coincidental byproduct of the hammering-annealing procedure. Analysis of the sheets will be by the same INAA and EMPA methods described in this paper.

INAA-EMPA Analyses of Lower-Quality Gold Foils. Twelve packets each containing a few gold foils and thin gold scraps were sent to Ann Arbor. Some were fragments that had broken from lower-quality gold objects; others were from scrap piles in the tomb; yet others were sheet fragments some of which, due to extensive corrosion, were adhering together in a layered array. In those cases where the foils had a visible distinct gold color, their measured thickness was only about 0.05 mm (50 μm). A typical foil-fragment weighed only about 7 mg, just enough for positioning in the EMPA mount so that the edge would be exposed to the electron beam. Some of the foils were highly corroded showing an extensive granular and porous surface layer presumably of copper (and possibly also silver) oxide and perhaps chloride.

During the 900 years that the objects were in the gravesite the water table would have often risen above the level of the grave goods (which were 11 m underground) and any chloride presumably represents corrosion that occurred during these wet periods. It is not surprising that a number of the foil fragments had visible surface corrosion. Many of these foil samples are alloys that apparently have relatively high copper and low gold contents and gold-silver-copper ternary alloys that are high in copper are especially subject to corrosion, unlike alloys of higher gold contents (characteristic of the high-quality gold objects) that are very stable to corrosion (11).

For a variety of reasons, it is much more difficult to obtain analysis data for the foils that matches the quality of the data for the high-quality gold objects. The reasons are: (a) The samples are so small that the same individual sample almost always cannot be subjected to both INAA and EMPA; a whole sample must be mounted for EMPA. (b) The samples are very fragile; the act of taking INAA rubbing samples can sometimes result in the foil fragment crumbling. (c) The extensive corrosion of some of the samples precludes obtaining analysis data that are representative of the original alloy. (d) The thinness of the foils (about 50 μm) often makes it difficult to obtain INAA rubbings that are devoid of a ca. 5 μm gold-enriched surface layer. (e) Although it is possible that the fragments from a single locale could have originated from the same (originally) larger piece of foil and have the same alloy composition, there is no prior assurance of this. In spite of these obstacles we have begun analysis of these foils and report here the initial data.

INAA of Gold Foils. It was extremely difficult to obtain foil rubbings for INAA that represent the interior metal alloy. The best that could be done was to take a series of rubbings in sequence and examine the analysis data to see if any obvious trend exists in the measured contents. If, for example, the first rubbing removed primarily surface alloy, subsequent rubbings removed increasing fractions of interior alloy, and the last rubbings were primarily of interior alloy, then increasing copper contents should be seen in the sequential data. No obvious trend in the sequential data existed for most of the foil samples examined by INAA. For some, the data were remarkably consistent; for others, the variation in copper content was random. These average data are shown in Table III. The standard deviations, s_x, provide an indication of the degree of scatter in the sequential replicate measurements.

EMPA of Gold Foils. Only a few of the foils, some showing extensive corrosion, have been examined in the electron microprobe. The unnormalized ternary alloy compositions ranged from 75-100%, the lower values presumably representing sampled areas containing some copper(II) oxide. Often only a single probe analysis was performed so that no averages or standard deviations are given. However, based on the total number of X-ray counts obtained (40,000 counts for each element in both the standards and the samples), Poisson statistics, and propagation-of-error methods, the standard deviation of the wt% values can be calculated to be about $\pm 1\%$ of the value and the standard deviation of the ratio values to be about $\pm 2\%$ of the value. Because of possible inhomogenities in these alloys, unlike the high-quality gold objects, a more realistic uncertainty might be $\pm 10\%$ of the % values and $\pm 20\%$ of the ratio values. These higher $\pm$ uncertainties actually are what are found for the few cases shown in Table III where replicate EMPA data were obtained.

In examining Table III it should be stressed that each set of analyses (except for the EMPA data) is for a separate sample. For example, five separate foil samples from a packet coded "PAS '94-M-79," noted in Table III as 79 a-e, were examined. As in other parts of the table, the EMPA analyses are for a single sample, 79-e, with the first analysis, EMPA-I, being for the interior alloy and the second, EMPA-E, being for the surface edge.

Where comparative edge:interior EMPA data are given they indicate a marked depletion of copper relative to gold as was the case for the high-quality gold objects. The silver:gold data for sample 79e (and possibly also 95c) suggest a possible surface silver-enrichment relative to gold, but future replicate analyses will confirm if this is truly the case. For those samples where replicate EMPA analyses were made (samples 100a and 109b), no such silver enrichment is apparent, at least when considering the associated uncertainties of the replicate measurements. Thus, the surface copper-depletion appears real whereas any indication of surface silver-enrichment appears very suspect.

In general, the INAA data for separate samples of each grouping result in similar data. Samples 79 a-d , samples 95 a, b, and samples 109 a-c all have less gold than silver and the average gold:silver ratios range from 0.4 to 0.8. This is opposite to what was found for the high-quality gold objects which have gold:silver ratios in the range of 1.0 to more than 2.0 (*7*). Samples 118 a-f have the reverse: much higher gold than silver and also very low copper. These series-118 samples have contents more characteristic of the very high-quality gold objects we have analyzed by INNA (*7*), the only difference being that much more extensive hammering would have been required to produce a final product as thin as these foils.

Table III. Ternary Alloy Contents of Lower-Quality Thin Gold Foils[a]

Foil No.	Identity	Analysis	N	Wt% Gold	Wt% Silver	Wt% Copper	Ag/Au Ratio	Cu/Au Ratio
79a	Foil	INAA	2	35.0 ± 3.5	43.2 ± 5.6	21.8 ± 9.1	1.23 ± 0.03	0.64 ± 0.33
79b	cluster	INAA	6	26.5 ± 9.6	35.0 ± 6.6	38.5 ± 15.4	1.39 ± 0.26	1.73 ± 0.97
79c	adjoining	INAA	6	26.5 ± 9.0	40.7 ± 9.7	32.8 ± 18.8	1.61 ± 0.27	1.71 ± 1.72
79d	amethyst	INAA	6	34.1 ± 16.2	35.3 ± 17.9	30.6 ± 17.9	1.22 ± 0.56	1.19 ± 1.05
79e	bead	EMPA-I	1	26.8	59.7	13.5	2.22	0.50
79e	cluster	EMPA-E	1	20.6	75.5	3.9	3.67	0.19
95a	Fragments	INAA	3	13.6 ± 5.1	32.7 ± 4.0	53.7 ± 9.0	2.54 ± 0.60	4.43 ± 1.98
95b	of Gold	INAA	3	28.4 ± 2.3	46.5 ± 5.3	25.1 ± 3.9	1.65 ± 0.31	0.88 ± 0.13
95c	Crown	EMPA-I	1	26.1	60.1	13.8	2.31	0.53
95c	Deposit 1	EMPA-E	1	25.1	69.7	5.2	2.78	0.21
100a	Scraps from	EMPA-I	2	16.7 ± 3.6	27.7 ± 1.4	55.6 ± 5.0	1.69 ± 0.28	3.43 ± 1.03
100a	Scrap Pile 4	EMPA-E	2	27.3 ± 2.6	39.5 ± 1.0	33.2 ± 1.6	1.46 ± 0.18	1.23 ± 0.18
109a	Scraps	INAA	6	23.7 ± 3.0	48.4 ± 7.4	27.9 ± 7.9	2.07 ± 0.42	1.21 ± 0.43
109b	from	INAA	6	25.8 ± 2.7	38.7 ± 3.8	35.5 ± 5.9	1.50 ± 0.13	1.41 ± 0.38
109c	Scrap	INAA	6	28.8 ± 4.8	51.5 ± 7.4	19.7 ± 12.1	1.79 ± 0.10	0.77 ± 0.68
109d	Pile 2	EMPA-I	3	30.4 ± 1.8	30.8 ± 9.0	38.8 ± 7.3	1.02 ± 0.37	1.27 ± 0.17
109d		EMPA-E	2	47.7 ± 4.3	46.9 ± 0.7	5.3 ± 3.6	0.99 ± 0.10	0.11 ± 0.09
118a	Fragments	INAA	6	69.0 ± 0.9	28.5 ± 1.3	2.5 ± 0.5	0.41 ± 0.02	0.04 ± 0.01
118b	of square	INAA	6	71.7 ± 13.1	23.5 ± 10.1	4.8 ± 9.2	0.35 ± 0.16	0.08 ± 0.18
118c	foils once	INAA	6	64.3 ± 2.6	21.0 ± 3.4	14.7 ± 4.8	0.33 ± 0.05	0.23 ± 0.08
118d	sewn to	INAA	2	56.8 ± 1.0	38.3 ± 0.1	4.9 ± 1.0	0.67 ± 0.01	0.09 ± 0.02
118e	cloth	INAA	2	61.1 ± 3.4	35.7 ± 3.3	3.2 ± 0.2	0.59 ± 0.09	0.05 ± 0.01
118f	mantle	INAA	2	59.2 ± 0.6	36.8 ± 0.1	4.0 ± 0.6	0.62 ± 0.01	0.07 ± 0.01

[a]The ± is the standard deviation of the N replicate analyses. EMPA-I are data for interior metal; EMPA-E are data for edge metal.

Future ongoing analyses will provide information about the extent of surface copper depletion of these foils. Similar EMPA data are being obtained by Shimada and Merkel on a different set of foils and fragments and their data indicate a pattern of surface depletion of copper similar to that found here (*15*).

Acknowledgments

We very much appreciate the cooperation and assistance provided by the University of Michigan Electron Microprobe Analysis Laboratory in making available the EMPA for analysis and the University of Michigan Memorial Phoenix Project in making available the Nuclear Reactor and associated radioactivity detection equipment for the INAA studies. Ed Birdsall of the reactor staff performed the radioactivity measurements. Karl Fisher and Scott Billecke assisted in the various sample preparations. Professor Wilbur Bigelow prepared the various EMPA mounted specimens. Their assistance is most appreciated.

References

1. Lechtman, H. In *Pre-Columbian Metallurgy of South America*; Benson, E. P., Ed.; Dumbarton Oaks: Washington, DC, 1979; pp 1-40.
2. Shimada, I. In *In Quest of Mineral Wealth: Aboriginal and Colonial Mining and Metallurgy In Spanish America*; West , R., and Craig, A. K., Eds.; Louisiana State Univ.: Baton Rouge, LA, 1994; pp 37-73.
3. Merkel, J.; Shimada, I.; Swann, C. P.; Doonan, R. In *Archaeometry of Pre-Columbian Sites and Artifacts;* Scott, D. A. and Meyers, P., Eds.; The Getty Conservation Inst.: Marina del Rey, CA, 1994; pp 199-227.
4. Shimada, I.; Merkel, J. F. *Scientific Am.* **July 1991**, *265*, 80-86.
5. Carcedo, P.; Shimada, I. In *Art of Pre-Columbian Gold: Jan Mitchell Collection*, Jones, J., Ed.; Weidenfield & Niccolson: London, 1985; pp 60-75.
6. Shimada, I.; Merkel, J. *Minerva*, **1993**,*4*, 18-25.
7. Gordus, A. A.; Shimada, I. In *Material Issues in Art and Archaeology IV*; Galvan, J. L.; Druzik, J.; Vandiver, P. B., Eds.; Materials. Res. Soc. Proc.: Pittsburgh, PA, 1995; Vol. 352, pp 127-142.
8. Shimada, I.; Griffin, J. A. *Scientific Am.*, **April 1994**, *270*, 82 -89.
9. Lechtman, H. In *Application of Science in Examination of Works of Art*; Young , W. J. , Ed.; Museum of Fine Arts: Boston, 1973; pp 38-52.
10. Lechtman, H. *Scientific Am.*, **June 1984**,*250*, 56-63.
11. Lechtman, H. In *The Beginning of the Use of Metals and Alloy* , Madden, R., Ed.; MIT Press: Cambridge, MA, 1988; pp 344-378.
12. Drost, F.; Hausselt, J. In *Interdisciplinary Science Reviews*, **17**, No. 3; Inst. of Metals: London, 1992; pp 271-280.
13. Chang, Y. A.; Neumann, J. P.; Mikula, A.; Goldberg, D. *Phase Diagrams and Thermodynamic Properties of Ternary Copper-Metal Systems*; Intl.
14. Scott, D. A.; Dorhne, E. *Archaeometry*, **1990**, *32*, 183-190 .
15. Merkel, J. F.; Seruya, A. I.; Griffiths, D.; Shimada, I. In *Material Issues in Art and Archaeology IV*; Galvan, J. L.; Druzik, J.; Vandiver, P. B., Eds.; Materials. Res. Soc. Proc.: Pittsburgh, PA, 1994; Vol. 352, pp 105-126.

RECEIVED August 15, 1995

Chapter 8

Chemical Compositions of Tiberian Asses, 15–23 A.D.

Giles F. Carter[1]

Eastern Michigan University, Ypsilanti, MI 48197

Twenty-eight Tiberian asses minted in Rome were analyzed for Fe, Co, Ni, Cu, Zn, As, Ag, Sn, Sb and Pb by wavelength dispersive X-ray fluorescence. The compositions of coins dated 15-16 A.D. were significantly different from those of coins dated 22-23 A.D. No asses were struck from 17 through 19 A.D. Undated coins were ascribed in the literature to 15-16 A.D. and to 22-23 A.D., and these dates are confirmed. Previously analyzed asses of Tiberius minted in 34-37 A.D. are less pure than the earlier Tiberian asses, which are copper usually containing less than 0.5% impurities. One Roman provincial coin of Tiberius contained much lead and tin, typical of provincial coins.

Beginning in about 16 B.C. the first Roman emperor, Augustus, struck a denomination of coins in almost pure copper. The "as" was a copper coin weighing about 10 to 13 grams. Asses had been struck many years before in the Roman Republic, but the earlier asses were larger coins, often cast rather than struck and made of copper alloyed with high concentrations of lead and sometimes appreciable amounts of antimony and arsenic (*1*). Augustus struck asses during three two-year periods: ca. 16-15 B.C., ca. 7-6 B.C., and 10-11 A.D. In each two-year period the chemical compositions were considerably different from those of the other periods, and usually the composition changed significantly during a given two-year period.

Tiberius, the successor to Augustus, struck asses in 15-16 and 20-23 A.D., and also in 34-37 A.D. (Only one very rare issue of asses was struck in 20-21 A.D., but none of these coins were available for analysis.) Dates have been obtained for several issues of coins from the titles appearing on the coins, such as the tribune power which usually was renewed annually using consecutive numbers.

Several issues of Tiberian asses are undated. Tentative dates have been assigned to these issues by Sutherland and Carson (*2*). Through a comparison of the chemical compositions of the undated series with the dated series this paper confirms that the dates assigned by Sutherland and Carson are correct.

[1]Current address: 243 Grove Drive, Clemson, SC 29631

0097–6156/96/0625–0094$12.00/0

Because the chemical compositions are significantly different for each two to four year period during which asses were struck, either by Augustus or by Tiberius, there is no indication that coins in the early Roman empire were remelted to mint asses in a later year. In this paper the chemical compositions of the early issues of Tiberius are compared with previous analyses of late issues of Tiberian asses.

Procedure

The coins were cleaned by a procedure described previously (*3*): first the surface oxides were reduced at the cathode in a hot solution of sodium carbonate. Then they were abraded with a pressurized air stream containing chemically pure aluminum oxide powder to remove any remaining loose oxides from the surface and a thin layer of metal about 10 fm thick. This procedure has been shown to produce surfaces that have essentially the same chemical composition as the coin interiors (*3*). Then the elements, Fe, Co, Ni, Cu, Zn, As, Ag, Sn, Sb, and Pb, were determined by wavelength dispersive X-ray fluorescence analysis according to the method described by Carter and Booth (*4*). The coins were identified according to the catalog numbers given by Sutherland and Carson (*2*). Photographs of typical coins are presented in Figure 1.

Chemical Compositions

The first group of six coins are dated to 15-16 A.D. because they bear the inscription, "TRIBUN POTEST XVII", or Tiberius's Tribune Power XVII. On the obverse is the bare head of Tiberius, usually facing left. On the reverse is a draped female figure, often attributed to be Livia, the widow of Augustus and mother of Tiberius. Hence these coins will be referred to as the "Livia" group for simplicity.

The six "Livia" coins are made of very pure copper for this period in history (see Table I). Most unusual are the extremely low Sb and Ag concentrations. Co, Ni, As, and Pb concentrations are also low, but these low concentrations are not as unusual in Roman coins as the extremely low concentrations of Sb and Ag. The "Livia" coins are quite different from the last issues of Augustus struck in 10-11 A.D., which contain relatively high concentrations of nickel and cobalt (*5*).

Lower limits of detectability for the various elements are approximately as follows: 0.0020% for Fe, Co, and Ni, 0.010% for Zn, 0.020% for As, 0.008% for Ag, 0.0010% for Sn and Sb, and 0.015% for Pb. The precision for each element present at low concentrations is about the same as the lower limit of detectability.

A second group of coins has the radiate (crown of divinity) head of Augustus and bears the title, "DIVUS AUGUSTUS PATER," Divine Augustus Father. On the reverse is a draped female figure, possibly Livia. This group is referred to here as the DAP "Livia" group. There is no title giving a clue to the date of these coins, but Sutherland and Carson, based on style and other information, have assigned these coins to ca. 15-16 A.D. (*2*). As shown in Table I, the chemical compositions of this group are essentially the same as the compositions of the "Livia" group of coins. The DAP "Livia" coins were most probably struck in 15-16 A.D., but, of course, these coins could have been struck either just before or just after the "Livia" group.

Two groups of coins are dated to 22-23 A.D.: the Tiberian "TRIBUN POTEST XXIIII" group and the Drusus group (Drusus was the only son of Tiberius, and Drusus was murdered soon after the coins bearing his image were struck). The coins of 22-23 are remarkable for their relatively high nickel and cobalt contents (see Table II). All twelve coins of Drusus, including six previously analyzed coins (*6*), contain from 0.089 to 0.37% Ni and 0.005 to 0.025% Co. The average nickel content of 'the more recent determinations is less than that of the old determinations. Improved techniques and equipment presumably make the recent determinations more accurate.

Figure 1. Representative Tiberian Asses (a) Coin No. 2066 obverse; (b) Coin No. 2066 reverse; (c) Coin No. 2118 obverse; (d) Coin No. 2118 reverse; (e) Coin No. 2075 obverse; (f) Coin No. 2075 reverse; (g) Coin No. 2083 obverse; (h) Coin No. 2083 reverse; (j) Coin No. 2200 obverse; (k) Coin No. 2200 reverse.

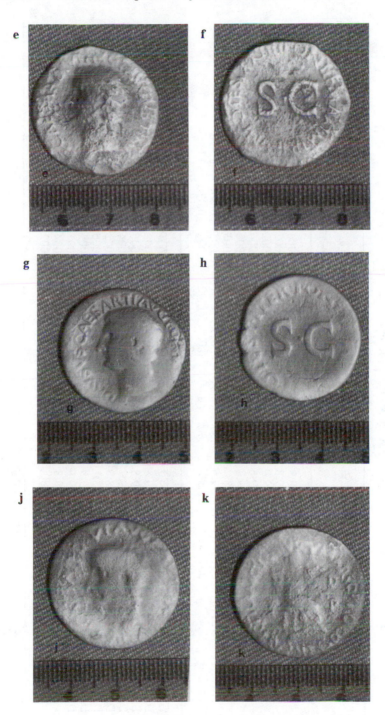

Figure 1. *Continued*

Table I. Chemical Compositions in Weight Percent of Tiberian Asses, ca. 15 – 16 A.D.

Coin Group	Fe	Co	Ni	Cu	Zn	As	Ag	Sn	Sb	Pb	Date, A.D.	RIC No.	Coin No.
"Livia"	0.050	0.000	0.000	99.9	0.000	0.010	0.033	0.001	0.004	0.000	15-16	33	2066
"Livia"	0.153	0.001	0.000	99.8	0.000	0.010	0.020	0.006	0.005	0.060	15-16	34 or 36	2067
"Livia"	0.098	0.000	0.008	99.9	0.000	0.010	0.018	0.003	0.002	0.010	15-16	34 or 36	2068
"Livia"	0.197	0.000	0.010	99.6	0.000	0.020	0.012	0.018	0.006	0.190	15-16	34 or 36	2069
"Livia"	0.179	0.001	0.000	99.7	0.000	0.010	0.034	0.023	0.000	0.080	15-16	34	2070
"Livia"	1.25	0.001	0.004	98.7	0.000	0.000	0.017	0.009	0.003	0.020	15-16	34 or 36	2071
Average	0.321	0.001	0.004	99.6	0.000	0.010	0.022	0.010	0.003	0.060			
Std. Dev.	0.457	0.001	0.004	0.500	0.000	0.006	0.009	0.009	0.002	0.070			
DAP "Livia"	0.144	0.001	0.000	99.7	0.000	0.000	0.020	0.009	0.001	0.110	ca. 15-16	71 or 72	2111
DAP "Livia"	0.091	0.000	0.015	99.8	0.010	0.020	0.017	0.026	0.003	0.010	ca. 15-16	71 or 72	2112
DAP "Livia"	0.116	0.001	0.000	99.8	0.000	0.010	0.030	0.009	0.006	0.010	ca. 15-16	71 or 72	2113
DAP "Livia"	0.114	0.000	0.015	99.7	0.000	0.020	0.017	0.006	0.004	0.080	ca. 15-16	72	2114
DAP "Livia"	1.88	0.001	0.009	98.0	0.000	0.020	0.022	0.018	0.002	0.050	ca. 15-16	72	2115
DAP "Livia"	1.89	0.001	0.002	98.1	0.000	0.000	0.022	0.007	0.006	0.000	ca. 15-16	72	2116
DAP "Livia"	0.109	0.000	0.000	99.9	0.000	0.000	0.029	0.005	0.005	0.010	ca. 15-16	72	2118
DAP "Livia"	0.223	0.001	0.011	99.6	0.000	0.010	0.007	0.020	0.001	0.130	ca. 15-16	72	2121
Average	0.570	0.001	0.006	99.3	0.000	0.010	0.020	0.012	0.004	0.050			
Std. Dev.	0.812	0.001	0.007	0.800	0.000	0.010	0.007	0.008	0.002	0.050			

DAP = Divus Augustus Pater; RIC = Roman Imperial Coinage (2).

The group of Tiberian TRP XXIIII coins, minted in 22-23 A.D., also have relatively high nickel and cobalt concentrations with one exception. As shown in Table II, one out of the nine coins, coin number 2077, has low nickel and low cobalt similar to the compositions found for the "Livia" coins. This coin, however, does not have the same catalog number as the other coins in the TRP XXIIII group. Two coins were analyzed many years ago, and these also have comparatively high nickel and cobalt similar to the recently analyzed coins (see coins numbered 250 and 251 in Table II).

A group of undated coins, which were analyzed many years ago (6), are distinguished by an altar on the reverse with the inscription, "SC PROVIDENT". This group will be referred to as the SC Provident group. Sutherland and Carson attribute this group to ca. 22-23 A.D., but possibly extending up to 30 A.D. Six out of eight of these coins contain relatively high nickel, and five of the eight coins contain relatively high cobalt (see Table II). The compositions are similar to those in the Tiberian XXIIII group, which were minted in 22-23 (see Table II). Interestingly, the two coins having low nickel contents also have relatively high silver and antimony concentrations, typical of later Tiberian asses. This suggests that some of these coins were, indeed, minted in the latter part of the reign of Tiberius.

Several Tiberian asses, assigned to 34-37 A.D. by Sutherland and Carson (2), were analyzed many years ago. Their compositions are presented in Table III. Usually these coins contain moderately low nickel concentrations. Cobalt was not determined in these coins. The silver and antimony concentrations are significantly higher than those found in the earlier Tiberian asses,

One Tiberian as, which was analyzed recently, was found to have unusually high tin, lead, and arsenic concentrations (coin number 2200, Table III). This coin was identified as a Roman provincial coin from Utica. Roman provincial coins, which of course were not minted in Rome, have chemical compositions deliberately different (apparently) from the coins minted in Rome (7).

Discriminant Analysis of the Chemical Compositions of Tiberian Asses

Carter and Frurip have developed a method using discriminant analysis which often enables the chronology of groups of ancient coins to be determined (8). When several variables change randomly over a period of time, the net change is usually larger for longer periods of time between two groups of coins. Often this enables the chronology of groups of coins to be determined. Several test cases, where the correct chronology is known, have demonstrated the effectiveness of the method (5, 9).

The method involves the use of Discriminant Analysis, which mathematically separates groups of variables, such as the concentrations of several elements, as far from each other as possible. The chemical composition of each coin is represented by a point in space, which is three dimensional for four groups of coins. Also each group of coins has a centroid, which is the mean of all the points for each coin in the group. The Mahalanobis Distance is the distance between a pair of group centroids.

If there are four groups of coins, A, B, C, and D, for example, then the correct chronology is one of the following: ABCD, ABDC, ACBD, ACDB, ADBC, ADCB, BACD, and so on. There are twenty-four possible chronologies for four groups of coins. The Carter-Frurip method involves summing the Mahalanobis Distances (MD) for each possible chronology. For instance, the chronology ABCD involves summing the MD for A-B, the MD for B-C, and the MD for C-D. This sum of MD's is called the Mahalanobis Drift Distance (MDD) because it represents the distance through which the centroids of coins passes over a period of time, somewhat like a drift through space as a function of time.

The Carter-Frurip hypothesis is that the chronology having the shortest MDD is the most probable chronology (8). Note that the MDD for the chronology ABCD is

Table II. Chemical Compositions in Weight Percent of Tiberian Asses, 22 – 23 A.D.

Coin Group	Fe	Co	Ni	Cu	Zn	As	Ag	Sn	Sb	Pb	Date, A.D.	RIC No.	Coin No.
Tiberius TRP XXIIII	0.052	0.008	0.151	99.7	0.000	0.010	0.019	0.008	0.000	0.020	22-23	44	2073
"	0.052	0.009	0.201	99.7	0.000	0.000	0.021	0.008	0.000	0.010	22-23	44	2074
"	0.068	0.011	0.205	99.6	0.000	0.010	0.019	0.009	0.000	0.050	22-23	44	2075
"	0.101	0.014	0.231	99.6	0.000	0.010	0.023	0.010	0.000	0.050	22-23	44	2076
"	0.219	0.011	0.224	99.4	0.000	0.000	0.017	0.008	0.000	0.080	22-23	44	2078
"	0.137	0.010	0.226	99.6	0.010	0.010	0.013	0.004	0.000	0.030	22-23	44	2079
Average	0.105	0.010	0.206	99.6	0.000	0.007	0.019	0.008	0.000	0.040	22-23	44	
Std. Dev.	0.065	0.002	0.030	0.100	0.000	0.005	0.003	0.002	0.000	0.030	22-23	44	
Tiberius TRP XXIIII	0.071	0.000	0.008	99.9	0.000	0.010	0.009	0.001	0.001	0.000	22-23	a	2077
Tiberius TRP XXIIII[b]	0.030	0.010	0.280	99.3	0.000	0.000	0.030	0.015	0.000	0.000	22-23	44	250
"	0.030	0.005	0.250	99.4	0.000	0.000	0.010	0.010	0.000	0.000	22-23	44	251
Drusus	0.504	0.015	0.194	99.0	0.000	0.010	0.020	0.010	0.003	0.260	22-23	45	2082
"	0.034	0.017	0.288	99.6	0.000	0.000	0.016	0.005	0.000	0.010	22-23	45	2083
"	0.060	0.006	0.089	99.6	0.000	0.020	0.026	0.011	0.001	0.220	22-23	45	2084
"	0.058	0.006	0.148	99.7	0.000	0.020	0.016	0.004	0.000	0.010	22-23	45	2085
"	2.272	0.006	0.165	97.5	0.000	0.000	0.023	0.004	0.000	0.020	22-23	45	2086
"	0.109	0.011	0.213	99.7	0.000	0.000	0.023	0.004	0.000	0.020	22-23	45	2087[c]
Average	0.506	0.010	0.183	99.2	0.000	0.010	0.020	0.006	0.001	0.090	22-23	45	2118
Std. Dev.	0.883	0.005	0.067	0.900	0.000	0.010	0.004	0.003	0.001	0.120	22-23	72	2121

Table II. *Continued*

Coin Group	Fe	Co	Ni	Cu	Zn	As	Ag	Sn	Sb	Pb	Date, A.D.	RIC No.	Coin No.
Drususᵇ	0.090	0.015	0.280	99.3	0.000	0.000	0.010	0.010	0.000	0.000	22-23	45	60
"	0.015	0.005	0.280	99.4	0.000	0.000	0.030	0.010	0.000	0.000	22-23	45	249
"	0.240	0.025	0.320	99.0	0.000	0.030	0.040	0.040	0.000	0.020	22-23	45	299
"	0.083	0.010	0.290	99.6	0.000	0.000	0.020	0.005	0.009	0.000	22-23	45	756
"	0.123	0.005	0.340	99.5	0.000	0.000	0.030	0.005	0.008	0.000	22-23	45	757
"	0.150	0.010	0.370	99.4	0.000	0.000	0.030	0.004	0.007	0.000	22-23	45	758
Average	0.117	0.012	0.310	99.4	0.000	0.005	0.027	0.012	0.004	0.000		45	
Std. Dev.	0.076	0.008	0.040	0.200	0.000	0.010	0.010	0.014	0.004	0.010		45	
SC Providentᵇ	0.085	0.015	0.270	99.3	0.000	0.000	0.020	0.010	0.000	0.000	ca. 22/23-30	81	96
"	0.030	0.025	0.310	99.3	0.000	0.000	0.040	0.010	0.000	0.020	"	81	252
"	0.100	0.010	0.300	99.3	0.000	0.000	0.030	0.015	0.010	0.040	"	81	605
"	0.057	0.005	0.340	99.6	0.000	0.000	0.030	0.002	0.004	0.000	"	81	760
"	0.064	0.000	0.230	99.7	0.000	0.000	0.030	0.002	0.005	0.000	"	81	761
"	0.070	0.010	0.290	99.5	0.000	0.000	0.030	0.005	0.009	0.030	"	81	762
Average	0.068	0.011	0.290	99.4	0.000	0.000	0.030	0.007	0.005	0.020		81	
Std. Dev.	0.024	0.009	0.040	0.200	0.000	0.000	0.010	0.005	0.004	0.020		81	
SC Providentᵇ	0.012	0.000	0.030	99.8	0.050	0.000	0.050	0.007	0.111	0.000	ca. 22/23-30	81	759
"	0.029	0.000	0.030	99.6	0.110	0.000	0.080	0.007	0.049	0.090	"	81	763

RIC = Roman Imperial Coinage (2); a = Similar to RIC 44, unknown coin, very worn; b = Analyses reported previously in Reference 6; c = Reverse contaminated with tin and lead.

Table III. Chemical Compositions in Weight Percent of Tiberian Asses, ca. 34 – 37 A.D.

Coin Group	Fe	Co	Ni	Cu	Zn	As	Ag	Sn	Sb	Pb	Date, A.D.	RIC No.	Coin No.
Eagle on globe[a]	0.253	nd	0.006	99.5	0.000	nd	0.095	0.007	0.041	0.100	ca. 34-37	82	79
"	0.132	nd	0.003	99.6	0.000	nd	0.089	0.010	0.055	0.090	"	82	260
"	0.096	nd	0.000	99.6	0.010	nd	0.056	0.018	0.064	0.100	"	82	518
"	0.121	nd	0.009	99.7	0.000	nd	0.046	0.009	0.086	0.040	"	82	767
Average	0.150	nd	0.004	99.6	0.000	nd	0.072	0.011	0.062	0.080		82	
Std. Dev.	0.070	nd	0.004	0.100	0.00	nd	0.024	0.005	0.019	0.030			
Winged Thunderbolt[a]	0.164	nd	0.008	99.5	0.000	nd	0.063	0.008	0.079	0.130	ca. 34-37	83	63
"	0.124	nd	0.003	99.6	0.000	nd	0.057	0.009	0.087	0.100	"	83	606
"	0.050	nd	0.023	99.8	0.000	nd	0.063	0.015	0.079	0.000	"	83	741
"	0.580	nd	0.014	99.1	0.000	nd	0.044	0.040	0.114	0.130	"	83	768
"	0.100	nd	0.012	99.5	0.000	nd	0.071	0.012	0.181	0.080	"	83	768
"	0.034	nd	0.006	99.8	0.000	nd	0.080	0.007	0.046	0.080	"	83 var.	95
Average	0.178	nd	0.011	99.6	0.000	nd	0.063	0.015	0.098	0.090		83	
Std. Dev.	0.204	nd	0.007	0.300	0.000	nd	0.012	0.013	0.046	0.050		83	
Winged Caduceus[a]	0.250	nd	0.000	99.6	0.000	nd	0.083	0.010	0.030	0.030	34-35	53	97
"	0.110	nd	0.010	99.6	0.000	nd	0.084	0.011	0.078	0.070	"	53?	766
Tiberius - Provincial	0.046	0.005	0.020	93.5	0.000	0.110	0.054	0.460	0.080	5.70	?	Sears 302[b]	2200

RIC = Roman Imperial Coinage (2); a = Analyses reported previously in Reference 10; b = Similar to Sears 302, Roman provincial coin from Utica (11).

Table IV. Chronology of Early Tiberian Asses

	Groups 1-2	Groups 1-3	Groups 1-4	Groups 2-3	Groups 2-4	Groups 3-4
Mahalanobis Distance	8.0	7.8	1.8a	1.4b	8.8	8.7

	1-2-3-4	1-2-4-3	1-3-2-4	1-3-4-2	1-4-2-3	1-4-3-2
Mahalanobis Drift Distance	18.1	25.5	18.0	25.3	12.1	11.9

	2-4-1-3	2-1-4-3	2-1-3-4	2-3-1-4	3-1-2-4	3-2-1-4
Mahalanobis Drift Distance	18.4	18.5	24.4	11.0	24.6	11.2

Coin groups 1 - 4 are respectively "Livia" asses (15-16 A.D.), Tiberius TRP XXIIII asses (22-23 A.D.), Drusus asses (22-23 A.D.) and DAP "Livia" asses (ca. 15-16 A.D.); [a]Groups 1 and 4 are very similar in composition (note the small Mahalanobis distance); [b]Groups 2 and 3 are extremely similar in their compositions. The preferred chronology is 2-3-1-4, which is the same as 4-1-3-2, i.e., DAP "Livia," Tiberius TRP XXIIII, Drusus. The chronology 4-1-2-3 is essentially indistinguishable from 4-1-3-2.

Table V. Chronology of Tiberian Asses, 15 – 37 A.D.

	Groups 1-2	Groups 1-3	Groups 1-4	Groups 2-3	Groups 2-4	Groups 3-4
Mahalanobis Distance	1.2	6.5	7.7	6.6	7.5	9.0

	1-2-3-4	1-2-4-3	1-3-2-4	1-3-4-2	1-4-2-3	1-4-3-2
Mahalanobis Drift Distance	16.8	17.7	20.6	23.0	21.8	23.3

	2-4-1-3	2-1-4-3	2-1-3-4	2-3-1-4	3-1-2-4	3-2-1-4
Mahalanobis Drift Distance	21.7	17.9	16.7	20.8	15.2	15.5

Coin groups 1 - 4 are respectively DAP "Livia" asses (ca. 15-16 A.D.), "Livia" asses (15-16 A.D.), Drusus asses (22-23 A.D.) and Winged Thunderbolt asses (ca. 34-37 A.D.); Groups 1 and 2 are very similar in composition (note the small Mahalanobis distance). The preferred chronology is 3-1-2-4, i.e., Drusus, DAP "Livia," "Livia," Winged Thunderbolt; Drusus is misplaced. The chronology 3-2-1-4 is essentially indistinguishable from 3-1-2-4.

identical to the MDD for the chronology DCBA. Therefore, the possible chronologies of four groups of coins have only twelve different MDD's. Frequently one group is known to be the oldest or the youngest, and then the unique chronology may be determined.

Discriminant analysis also provides relative measures of the dissimilarities of various groups. For instance when a Discriminant Analysis was run for the groups, "Livia," DAP "Livia," Tiberius TRP XXIIII, and Drusus, the two "Livia" group centroids were separated by the relatively small Mahalanobis distance of 1.8 (see Table IV). The two "Livia" groups, however, were separated very far from the groups struck in 22-23 A.D., as shown by the Mahalanobis distances ranging from 7.8 to 8.8. The Tiberius TRP XXIIII and Drusus groups also were extremely close together, as shown by the Mahalanobis distance of only 1.4 separating their group centroids. This indicates that they were struck at or nearly at the same time. These results confirm the visual comparisons of the means and standard deviations for the two pairs of groups: the undated DAP "Livia" group is confidently dated as essentially the same as the "Livia" group, and the Tiberius TRP XXIIII group is very near to the Drusus group in time.

A second discriminant analysis was run for the following groups of coins: DAP "Livia," "Livia," Drusus, and the Winged Thunderbolt, ascribed to ca. 34-37 A.D. by Sutherland and Carson (2). The preferred chronology according to the Carter-Frurip method is in error: Drusus, DAP "Livia," "Livia," and finally the Winged Thunderbolt (see Table V). Actually the method cannot clearly distinguish whether DAP "Livia" was struck before the dated "Livia" group or vice versa. The second most preferred chronology is correct: DAP "Livia" or "Livia" first, then Drusus, and finally the Winged Thunderbolt group. This is a rare example when the Carter-Frurip method gave the incorrect chronology, although the second best chronology is known to be correct.

Conclusion

The chemical compositions of Tiberian asses struck in 15-16 A.D. are significantly different from the compositions of asses struck in 22-23 A.D. The DAP "Livia" group, which is undated but which had been ascribed to ca. 15-16 A.D., is extremely close in composition to the "Livia" group, which is firmly dated to 15-16 A.D. Likewise the undated "SC Provident" group of asses has a composition close to that of the Drusus asses and the Tiberian TRP XXIIII asses, which are firmly dated to 2223 A.D. Because only a relatively few Tiberian asses from the latter part of his reign have been analyzed, more of these later asses should be analyzed. Also the rare Tiberian asses of 20-21 A.D. should be analyzed.

Acknowledgments

I am grateful to Clive Stannard for his assistance and also to Eastern Michigan University for the opportunity to use its equipment.

Literature Cited

1. Carter, G. F.; Razi, H. In *Archaeological Chemistry – IV;* Allen, R. O., Ed.; Advances in Chemistry Series 220; American Chemical Society: Washington, DC, 1989; pp 213-230.
2. Sutherland, C. H. V.; Carson, R. A. G. *The Roman Imperial Coinage;* Spink: London, 1984; Vol. 1, pp 9699.
3. Carter, G. F.; Kimiatek, M. H. In Proceedings of the *18th International Symposium on Archaeometry and Archaeological Prospection;* Scollar, I.,

Ed.; Archaeophysika 10; Rheinland-Verlag: Cologne, Germany, 1979; pp 82-96.

4. Carter, G. F.; Booth, M. M. In *Problems of Medieval Coinage in the Iberian Area;* Gomes Marques, M., Ed.; Polytechnic Institute: Santarem, Portugal, 1984; pp 49-69.
5. Carter, G. F. *J. Archaeological Sci.* **1993**, *20,* 101-115.
6. Carter, G. F.; Buttrey, T. V. *American Numismatic Soc. Museum Notes* **1977**, *22,* 49-65.
7. Carter, G. F. In *Science and Archaeology;* Brill, R. H., Ed.; MIT Press: Cambridge, MA, 1971; pp 114-130.
8. Carter, G. F.; Frurip, D. J. *Archaeometry* **1985**, *27,* 117-126.
9. Carter, G. F.; Powell, R. R.; Frurip, D. J.; In *Actas do III Congresso Nacional de Numismatica,* Clube Numis. de Portugal: Lisbon, 1985; pp 535-551.
10. Carter, G. F.; King, C. E. In *Metallurgy in Numismatics I;* Metcalf, D. M. and Oddy, W. A., Eds.; The Royal Numismatic Society: London, 1980; pp 157-167.
11. Sear, D. R. *Greek Imperial Coins;* B. A. Seaby Ltd.: London, 1982; p 27.

RECEIVED October 9, 1995

Chapter 9

Copper-Based Synthetic Medieval Blue Pigments

Mary Virginia Orna

Department of Chemistry, College of New Rochelle, New Rochelle, NY 10805

Blue pigments were in short supply in medieval times because only two naturally-occurring inorganic blue pigments, natural ultramarine and azurite, were known. Both pigments were difficult to obtain and consequently were very expensive. Medieval artists turned to the manufacture of copper-based blue pigments of surprising complexity in order to expand their palette. This paper outlines some of the methods used to obtain these complex pigments and elucidates their structures, when possible.

During the Middle Ages, the artist's interest in synthesizing blue pigments stemmed from the fact that blue, as a pigment, was in very short supply. An earlier study (*1*), which reviewed the availability of blue pigments from ancient times, showed that Egyptian blue, copper(II) sulfide, and azurite were the only inorganic blue pigments known in the ancient Roman empire. Indigo, an organic pigment and dye, was identified as the pigment used to decorate Roman parade shields, but its use in European painting and manuscript decoration was probably very limited until it began to be imported from India in the early 16th century (*2*). Egyptian blue was the first synthetic pigment: its earliest known occurrences can be dated to the Greek Bronze Age tombs in Knossos and Vergina (*3-5*) and to the third millenium B.C. in Egypt (*6*). Its subsequent use in tombs and in the manufacture of small objects has been amply documented (*2*).. Egyptian blue was also the first pigment to be subjected to modern chemical analysis, beginning with the trial preparations by Sir Humphry Davy (*7*), and continued by Chase (*8*) and Tite, et al. (*6*), who showed through laboratory reproduction that the ancient Egyptian blue mineral ($CaCuSi_4O_{10}$) was formed as a result of the solid state reaction between silica, lime and copper(II) oxide. On the other hand, copper(II) sulfide occurs naturally as the mineral covellite, although it too can be prepared synthetically (*9*). It was probably much used in antiquity, but it is not durable, decomposing slowly and spontaneously to black copper(II) oxide unless protected by varnish (*10*). The third ancient blue pigment, azurite, is a basic copper carbonate closely related to the green malachite and has the formula $2CuCO_3 \cdot Cu(OH)_2$. It occurs naturally as a monoclinic crystalline material throughout Europe and the former Soviet Union. Although its use may date from the fourth dynasty in Egypt, it is known that it

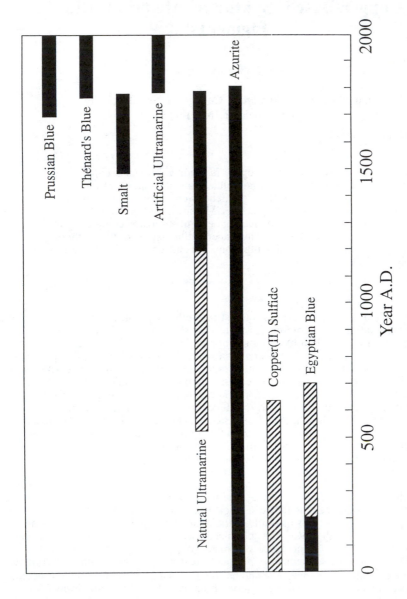

Figure 1. Chronological chart of some important artists' blue pigments. Solid areas signify periods of certain usage; striped areas signify periods of doubtful usage. Reproduced from reference 1 with permission.

was the most important and widely employed blue pigment throughout the Europe of the Middle Ages (*11*).

The use of a fourth blue pigment, natural ultramarine, derived from the semi-precious mineral lapis lazuli, can be traced to sixth and seventh century wall paintings in Afghanistan. It was introduced as a rare import to European artists in the thirteenth century, following the spectacular journeys of Marco Polo. During the subsequent three centuries, good quality ultramarine was as costly as gold, and patrons usually agreed to supply the pigment for a commissioned work or to pay for it separately at market price (*12, 13*).

Of these four inorganic blue pigments, only azurite and natural ultramarine seem to have survived into the Middle Ages. The secret of Egyptian blue manufacture was certainly lost by the ninth century, and the use of copper(II) sulfide fell into disuse in late Roman times.

Both azurite and ultramarine can be prepared synthetically. Azurite is made by precipitating a solution of copper(II) chloride with calcium hydroxide, and then heating the resulting precipitate with a mixture of potassium carbonate and calcium hydroxide. This procedure yields not only the basic copper(II) carbonate, but also a mixture of other products, notably the basic copper(II) chlorides (*14, 15*). Synthetic ultramarine was first observed as a byproduct in the soda furnaces at St. Gobain in 1814, and L. Gmelin was the first to synthesize it in 1822. Ultramarine blue, both natural and synthetic, is a clathrate complex consisting of a framework of $[(Si,Al)O_2]$ units enclosing a polysulfide, S_3^-, which is responsible for the blue color (*16*). The trisulfide radical anion was identified as the chromophore only as recently as the mid-1970's (*17*). Of these synthetic materials, only synthetic azurite, under the name "blue verditer" was available to the medieval artist-craftsman. The modern blue pigments, among which are smalt (a potassium cobaltous silicate of varying composition), Prussian blue (ferroferricyanide salts with a variety of cations), and Thénard's blue (cobaltous aluminate) were not available until the mid- to late sixteenth century and later.

From this review, one can see that European artists and craftsmen who were active between the ninth and sixteenth centuries had little choice in their use of blue pigments. Prohibitively costly, both natural ultramarine and azurite were hardly any choice at all. It is only natural, then, that much experimentation in the manufacture of synthetic inorganic blue pigments would have taken place throughout the Middle Ages. Figure 1 outlines the usage pattern of the major artists' blue pigments over the centuries.

Synthetic Blue Pigments: Recipes

Recipes for making artificial blue colors are very old. They are embedded in the literature of a technical tradition dating from the third century A.D. that managed to survive five centuries of "dark ages" to re-emerge in the late eighth or early ninth century in two Latin manuscripts, the so-called *Lucca Manuscript*, part of the *Liber Pontificalis* in the *Codex Lucensis 690* at Lucca, Italy, and the *Mappae Clavicula*, or a *Little Key to the World of Medieval Techniques*, that survives as a ninth century fragment and a more extended tenth century manuscript (*18*).

The recipes for synthetic blues contained in these and subsequent manuscripts that built upon the tradition of Lucca and the *Mappae* can be divided into three categories: (1) Pigments made from plant and animal materials; (2) Pigments based on synthesizing blue colors from silver, or the so-called "Silver Blue" recipes; and (3) Copper-based blue pigments.

Pigments made from plant materials are difficult to characterize. A recipe from the *Lucca Manuscript* tells us that an excellent azure-blue can be made from the petals of violet flowers ground with soap, and then heated with alum and urine. However, the identity of the plant from which the violet flowers came could be any one of several found in northern and southern Europe. Other plant recipes revolve around the production of indigo, which is easier to characterize, but much more so as a dye than as a pigment (*19*). This paper will concentrate on the blue mineral pigments manufactured by medieval artists since they seem to have much more definite starting materials and characteristics.

The "Silver Blue" Recipes

"Silver blue" recipes from the *Mappae Clavicula*, the *Strasburg Manuscript* (14th or 15th century) and the *Bolognese Manuscript* (15th century) call for subjecting sheets of the purest silver to must discarded from a wine press, to vinegar, and to vinegar in the presence of hot horse dung. In addition, the *Bolognese Manuscript* recommends alloying one part of copper with three parts of silver before carrying out the recipe. This latter recommendation would seem to suggest that by the 15th century, artists recognized the necessity of the presence of copper in order to obtain a reaction product, which one might assume was simply copper(II) acetate monohydrate. These recipes were carried out and reported on in a previous paper (*20*). The reaction products were analyzed by both the Debye-Scherrer X-ray powder diffraction method and by single-crystal X-ray crystallography and were found to be respectively copper(II) acetate monohydrate and tetra-μ-acetato-bisdiaqupcopper(II), a dinuclear crystal compound of copper(II) acetate (*21-23*).

Copper-Based Blue Pigments Made from Copper, Lime and Vinegar

The original *Mappae Clavicula* recipe (*1*) called for mixing copper with lime and vinegar. Many unpublished and unedited manuscripts from periods post-dating the *Mappae* contain this recipe since they are copies of the *Mappae* in whole or in part. One manuscript, Ms. Sloane 2584 (British Museum, London; 14th century) contains this recipe as the only remnant of the original *Mappae* recipes. Other compendia which contain it are the manuscript of Jehan LeBegue (*24*) and the *Bolognese Manuscript*. Another 14th century manuscript, Trinity College Ms. 1451, p. 7, contains the first mention of the addition of sal ammoniac (ammonium chloride) to the ingredients of the original *Mappae* recipe (*1*).

Synthesis of Blue Pigment Using the Original *Mappae Clavicula* Recipe. The original *Mappae Clavicula* recipe (*1*) tells us: "...take a flask of the purest copper and put lime into it halfway up, and then fill it with very strong vinegar. Cover it and seal it. Then put the flask in the earth or some other warm place and leave it there for one month; later uncover the flask. This azure is not as good as the other (made from silver), yet it is serviceable for painting on wood and plaster wall."
　　　　The synthesis carried out in our laboratory modified this recipe by using not a copper vessel, but copper plates, in a modern glass vessel. Since copper was the main reactant, this variation did not seem to be substantive. When the vessel was opened after a one-month incubation period, more than blue pigment was obtained. Since the mixture of the starting materials was not in stoichiometric amounts, large quantities of the starting materials remained unreacted, and many portions of the reaction vessel contained mixed crystals of green, blue and colorless hue. However, there was also a large amount of what looked like a pure blue crystalline compound. The blue color was unquestionably without traces of green, thus indicating that this compound was not either verdigris (copper(II) acetate) nor a form of azurite.
　　　　Elemental analysis of the product yielded 14.19% copper, 21.45% carbon, 5.40% hydrogen, 50.01% oxygen, and 8.95% calcium (obviously from the lime), by weight. The empirical formula from this elemental composition yielded a formula of $Cu(C_2H_3O_2)_2 \cdot Ca(C_2H_3O_2)_2 \cdot 6H_2O$ which was subsequently confirmed by X-ray crystallographic analysis. The d-values in Angstroms are given in Table I. Although the

crystals of this compound, calcium copper acetate hexahydrate, are a very pleasing deep blue, the crystals are soluble in water and the color fades to a pale blue when the material is ground to a fine powder. These factors call into question the value of this

Table I. d-Values (Å) for Calcium Copper Acetate Hexahydrate

8.29-7.58 (1)[a]	3.34-3.25	2.14	1.73
7.04	3.08-3.03	2.13	1.70
6.31	2.87-2.82	2.07	1.66
6.04	2.71-2.67	1.97	1.62
5.84-5.55 (2)	2.50	1.94-1.91 (4)	1.58
4.76	2.31	1.87	1.55
3.73-3.48 (3)	2.19	1.78-1.76	1.47

[a]Numbers in parentheses signify relative intensity.

compound as a pigment. The original *Mappae* recipe, we must recall, remarked that this pigment was not very good, but simply serviceable.

A literature search revealed that this compound can be prepared from solution by slow evaporation of an equimolar solution of calcium acetate and copper(II) acetate. After filtering off the pale green crystals that deposit initially, upon further evaporation, large, deep blue, tetragonal prisms crystallize out (25). The calcium copper acetate hexahydrate prepared in this manner in our laboratory exhibited the identical X-ray crystallographic characteristics as the compound prepared from the *Mappae* recipe by an entirely different route, confirming the fact that the two preparations are chemically identical. Langs and Hare (26) were the first to determine its crystal structure. They showed that the acetate anion acts as a bidentate bridging ligand between the calcium and the copper ions in such a way as to produce polymeric chains of alternate metal ions. They also showed that the water molecules coordinate only to the calcium ions and bind the polymeric chains together by hydrogen bonding. Billing et al. (25) determined the preferred assignment of the electronic spectrum of the compound and also determined that the chromophore was the four-coordinate Cu-O$_4$ moiety. The coordination polyhedra about the Ca and Cu atoms are depicted in Figure 2.

Synthesis of Blue Pigment Using the Recipe from Trinity College Ms. 1451. The Trinity College Ms. 1451, a 14th century manuscript, contains the first mention of the addition of sal ammoniac (ammonium chloride) to the ingredients of the original Mappae recipe: "...take of fyne vinegar half a pond, half a pond of lyme ontimeynt, iiij pond of sal armonyak...and poudre alle these thre materials, take of them by hitself aloone first and thanne aftrewardes alle to gidres, and thanne take and medle al the poudre with gode strong vinegar distilled or ell with ather gode winyer vynegre that be strong in the manere of past and thanne put it al in a pot of bras or of copper and close hit from the eyre and thanne lette hit ondre hote hors donge ly on .XV. dayes, and thanne take out and make hit up in to find the poudre as azure byzs..."

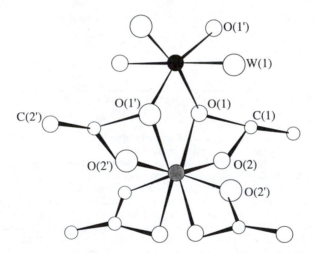

Figure 2. Coordination polyhedra about the Ca (●) and Cu (◉) atoms in calcium copper acetate hexahydrate. Reproduced from reference 26 with permission.

This recipe was reproduced in our laboratory in two ways. The recipe calls for sealing the container (closing it from the air) and then placing it under hot horse dung for fifteen days. If the pot were truly sealed so that it was a closed system, the presence of the hot horse dung would not affect the results, but might simply act as a hot water bath since a great deal of heat is generated as the dung ferments. On the other hand, if the closure of the container were not complete, that is, if the closure were porous (unbeknownst to the medieval craftsman), then gaseous products from the fermenting hot horse dung, notably carbon dioxide and ammonia, could indeed influence the outcome of the synthesis. Therefore, this recipe was carried out by placing the starting materials in both closed and open systems. The closed system, startlingly enough, did not yield a blue pigment after the fifteen prescribed days. The open system gave striking results: a mass of blue birefringent rosettes of varying diameters. Although these rosettes visually resemble rosettes of *cuprum carbonicum* (a basic copper carbonate from the Forbes Collection of Pigments, Conservation Center, Institute of Fine Arts, New York University), they have no counterpart in the X-ray powder diffraction file and remain uncharacterized. One of the problems in characterizing them is the fact that the crystals are intimately mixed with unreacted starting material and other colorless reaction products. A color photomicrograph of these crystals can be seen in Reference 27.

Copper-Based Blue Pigments Made from Verdigris, Lime, Sal Ammoniac and Oil of Tartar (Potassium Carbonate)

Manuscripts calling for the ingredients named above are characteristic of the late 15th century and the 16th century. The following three manuscripts contain similar recipes: Ms. Additional 12461 (British Museum, London; 16th century), Ms. Ashmole 750 (Bodleian Library, Oxford; 15th century); Ms. 1243 (Biblioteca Riccardiana, Florence; 15th century). The following translation from the Latin was made from folio 48r of the latter manuscript: "To make the most durable azure. R(ecipe): Mix well one part of sal ammoniac and three parts of verdigris with oil of tartar until it is soft and paste-like, or even softer. Then place it in a glass vessel under hot dung for a day; afterwards, you will find that the green has turned into the best blue. Another way of making the best blue. R(ecipe): Mix together three parts of sal ammoniac and six parts of verdigris with oil of tartar until soft and paste-like, or even softer. Then place the paste into a glass ampule, and when it is well-stoppered and sealed, place it in a hot oven and let it stand for some days; afterwards, take it out and you will find it turned into the best azure."

According to Crosland, the "oil of tartar" referred to in the recipe given above is a saturated solution of potassium carbonate (*28*). Trial and error with respect to starting materials probably led to the selection of this reagent since it provided the carbonate necessary to produce a blue pigment similar to the "*cuprum carbonicum*" referred to in the previous section of this paper. Synthesis in our laboratory, using both alternatives given in the recipe, yielded a bright blue pigment which, on microscopic inspection, consisted of blue needles or cylinders mixed with colorless crystals of varying morphologies. Again, these crystals have no counterpart in the X-ray powder diffraction file and remain uncharacterized. Reference 27 contains a color photomicrograph of these needle-like crystals.

Conclusion

Copper-based blue materials manufactured from recipes found in medieval artists' manuals present a very complicated chemical profile. Although pure compounds such as the tetra-μ-acetato-bisdiaquocopper(II) and calcium copper acetate hexahydrate were produced, other compounds and mixtures of compounds at the moment defy characterization. H. Kühn (*29*) has identified several additional pure compounds formed when acetic acid vapor, water vapor and air act on copper and copper alloys, and some forms of azurite have been identified by Gettens and Stout (2), but none of these resemble the copper-based blue pigments obtained from the recipes utilized in this study. It is well-known that copper and copper-silver alloys yield a variety of complex compounds depending upon the materials that react with them, and future research may reveal some new exotic compounds. Some of these compounds are definitely green (30), but some involving reactions of copper with such reagents as sour milk and cream of tartar can form some blue copper compounds which include lactates and tartrates (31). These, in addition to the acetates and putative carbonates discussed in this study, indicate the degree of sophistication attained by medieval artists and craftsmen long before the advent of the modern chemical theory that would provide a theoretical basis for these syntheses.

Acknowledgments

Thanks are due to the staff of the Conservation Center, Institute of Fine Arts, New York University, for its generous assistance, use of its facilities, and for access to the Forbes Collection of Old Pigments.

Literature Cited

1. Orna, M. V.; Low, M. J. D.; Baer, N. S. Studies in Conservation, 1980, 25, 53-63.
2. Gettens, R. J.; Stout, G. L. Painting Materials: A Short Encyclopedia; Dover Publications: Garden City, NY, 1966; pp 112-113.
3. Profi, S.; Weier, L; Filippakis, S. E. Studies in Conservation, 1976, 21, 34-39.
4. Filippakis, S. E.; Perdikatsis, B.; Paradellis, T. Studies in Conservation, 1976, 21, 143-153.
5. Filippakis, S. E.; Perdikatsis, B.; Assimenos, K. Studies in Conservation, 1979, 24, 54-58.
6. Tite, M. S.; Bimson, M.; Cowell, M. R. In Archaeological Chemistry - III; Lambert, J. B., Ed.; American Chemical Society: Washington, DC, 1984; pp 215-242.
7. Davy, H. Philosophical Transactions, 1815, 105, 97-124.
8. Chase, W. T. In Science and Archaeology; Brill, R. H., Ed.; MIT Press: Cambridge, MA, 1971; pp 80-90.
9. The Colour Index, 3rd Ed. Revised; Society of Dyers and Colourists: Bradford, U.K., 1975; p 4668.
10. Spitaels, P. Helinium, 1965, 5, 49-53.
11. Gettens, R. J.; FitzHugh, E. W. Studies in Conservation, 1966, 11, 54-61.
12. Plesters, J. Studies in Conservation, 1966, 11, 62-91.
13. Kühn, H. In Application of Science in Examination of Works of Art; Young, W. J., Ed.; Museum of Fine Arts: Boston, MA, 1973; pp 199-205.
14. Garrels, R. M.; Stine, L. P. Econ. Geol. 1947, 43, 21-30.
15. Garrels, R. M.; Dreyer, R. M. Bull. Geol. Soc. Am. 1952, 63, 325-380.
16. Orna, M. V. Color Research and Application 1978, 3, 189-196.
17. Cotton, F. A.; Harmon, J. B.; Hedger, R. M. J. Am. Chem. Soc. 1976, 98, 1417-1424.
18. Smith, C. S.; Hawthorne, J. G. Trans. Am. Phil. Soc. 1974, 64(4), 3-128.
19. Fuchs, R. "Blaufarbmittel in Illuminierten Handschriften und Drucken - Ihre Zerstorende Wirkungen und Restauratorische Konsequenzen," Paper presented at the 7th International Congress of Restorers of Graphic Art, 26th-30th August, 1991, Uppsala, Sweden.
20. Orna, M. V.; Low, M. J. D.; Julian, M. M. Studies in Conservation 1985, 30, 155-160.

21. Brown, G. M.; Chidambaram, R. *Acta Crystallographica* **1973**, *B29*, 2393-2403.
22. DeMeester, P.; Fletcher, S. R.; Skapski, A. C. *J. Chem. Soc. Dalton Trans.* **1973**, 2575-2578.
23. VanNiekerk, J. N.; Schoening, F. R. L. *Acta Crystallographica* **1953**, *6*, 227-232.
24. Merrifield, M. P. *Original Treatises on the Arts of Painting, Vol. I*; Dover Publications: Garden City, NY, 1967; pp 166-257.
25. Billing, D. E.; Hathaway, B. J.; Nicholls, P. *J. Chem. Soc. (A)*, **1970**, 1877-1881.
26. Langs, D. A.; Hare, C. R. *Chem. Comm.* **1967**, 890-891.
27. Orna, M. V. *Today's Chemist* **1991**, *4(6)*, 20-24.
28. Crosland, M. P. *Historical Studies in the Language of Chemistry*; Harvard University Press: Cambridge, MA, 1962.
29. Kühn, H. *Studies in Conservation* **1970**, *15*, 12-36.
30. Naumova, M. M.; Pisareva, S. A.; Nechiporenko, G. O. *Studies in Conservation* **1990**, *35*, 81-88.
31. Pisareva, S. A.; Grenberg, Y. I. "Russian 17-18th Centuries Production Formulas for Blue and Green Copper Pigments," paper presented at the 9th Triennial Meeting of ICOM Committee for Conservation, 26-31 August, 1990; Dresden, Germany. Grimstad, K., Ed.

RECEIVED August 15, 1995

Chapter 10

Archaeological Applications of Inductively Coupled Plasma–Mass Spectrometry

R. H. Tykot and S. M. M. Young

Archaeometry Laboratories, Department of Anthropology,
Harvard University, Cambridge, MA 02138

Inductively coupled plasma-mass spectrometry (ICP-MS) is a relatively new analytical technique increasingly used in the Earth Sciences in the last decade, and in "consumer" fields such as archaeology in the last few years. For archaeologists, ICP-MS has several important advantages over neutron activation and X-ray fluorescence analysis: (1) only a tiny powdered sample is required, so the technique is minimally destructive to valuable artifacts; (2) the large number of elements that can be accurately and precisely analyzed is particularly important for characterization and provenance studies; (3) isotope ratio measurements to three significant figures are possible without extensive sample preparation; and (4) the combination of small sample size and low per-sample cost allows assemblages of artifacts rather than individual objects to be studied. These advantages will be illustrated by the trace element characterization and source tracing of obsidian, the compositional analysis of copper-based artifacts, and the lead isotope ratio analysis of turquoise.

The chemical analysis of archaeological remains is now a widespread and fundamental part of any archaeological research project. The major element composition of metal tools, coins and other objects reflects the alloying technology employed, while the trace element composition of obsidian, copper, gold, marble, and clay has been used to match artifacts with the source of their raw materials. The stable isotope ratios and perhaps elemental content of skeletal remains are reflective of diet, and isotope ratios in metals, marble, ivory and obsidian have also been used to determine their provenance.

A now familiar range of instrumental techniques has been applied to archaeological materials in the last three decades, including neutron activation analysis (NAA), X-ray fluorescence (XRF), atomic absorption spectroscopy (AAS), proton-induced X-ray emission (PIXE), ICP-atomic emission spectroscopy (ICP-AES), and

0097–6156/96/0625–0116$12.00/0

electron probe microanalysis (with wavelength dispersive X-ray spectrometry) for elemental information, and stable isotope ratio analysis and thermal ionization mass spectrometry for isotopic data. Each of these techniques has its own sample requirements, preparation techniques, and data precision and accuracy which are relevant to particular archaeological materials and questions, and will not be reviewed here since there is an extensive literature on their use (see especially the journals *Archaeometry* and the *Journal of Archaeological Science*).

Inductively Coupled Plasma-Mass Spectrometry

Inductively coupled plasma-mass spectrometry (ICP-MS) is a relatively new technique in which samples are usually introduced as liquids, atomized and ionized at high temperatures, and the individual atomic species measured by a detector. In this sense, ICP-MS is similar to ICP-AES, except that individual mass units are measured rather than the energy or wavelength of the electromagnetic spectra produced. This technique therefore can be used to determine both elemental concentrations and isotope ratios in sample materials.

Although mass spectrometers were first used with an inductively coupled plasma (ICP) source in the 1980's (*1-7*), the increasing number of ICP-MS studies presented at analytical conferences confirms this technique as the "most important analytical system [developed] in the last decade of this century" (*8*). Several volumes on the applications of plasma source mass spectrometry have been published recently (*8-12*), and include, among others, elemental-compositional studies of metals (*13, 14*), rocks (*15*), plant materials (*16*), groundwaters (*17*), and automotive catalyst exhaust (*18*), and isotope ratio studies of metals (*19*), metabolic and environmental samples (*20*). A number of studies have even appeared already in the archaeological literature (*19, 21-30*).

Fisons, Finnegan-Mat and Perkin Elmer are the three major manufacturers of these instruments, of which there are now several hundred at research institutions worldwide. The analyses reported below were obtained using the Fisons PQ 2 Plus in the Department of Earth and Planetary Sciences at Harvard University. The ICP is used not as a source for producing uncharged atomic emission lines, as in the much more common ICP-AES technique (unfortunately, sometimes referred to simply as ICP or ICP-S), but as a source of charged ions for the mass spectrometer. The combination of a high-velocity argon flow, surrounded by a copper load coil, and initiated by an electrical spark, produces a plasma fireball of about 8000 °C at the mouth of the ICP torch. When the sample is injected into the torch through a capillary tube, it is both atomized and ionized into single-charged ions. Molecular and doubly-charged species are very few, reducing the otherwise significant problem of matrix effects. The ions are then passed into a quadrupole mass spectrometer for separation and measurement. Quadrupole mass filters utilize two pairs of oppositely charged rods, each superimposing a radio frequency AC signal on a DC voltage. This combination of RF and DC fields imparts a complex motion to the ion beam. For any set of operating conditions, there is a unique mass which can make its way through the exit slits and to the detector. Other ions will be pumped away by the vacuum system. The entire mass range may be scanned by the detector in about one second. Usually, up to 50 scans are accumulated, producing in a few minutes a simple mass spectrum, theoretically representing all elements (and their isotopes) in the periodic table except H, He, C, N, O, F, Ne, Cl and Ar. Solution detection limits are much lower than for other analytical methods, typically on the order of 1 ppb for light elements, and 50 parts per trillion for heavy elements. Some ICP mass spectrometers are now equipped with graphite furnaces, lowering detection limits into the low parts per trillion range. Precision is on the order of ± 2-4% for most elements (*7, 31*), and extensive analyses of U.S.G.S. basalt and andesite reference materials have accuracies better than 3% for 18 of 28 elements tested, with only 1 element differing by more than 7% from published values (*6*). More detailed information about ICP-MS may be found in a number of recent surveys of the technique (*4, 9-11, 31*).

Finally, there is also a laser ablation device, which can remove from a solid specimen a tiny sample from a spot much less than 1 mm in diameter, sending it directly to the ICP source (*32*). Although this method saves all the preparation time involved in dissolving a silicate sample, results are not nearly as precise, and detection limits are "only" in the ppm range, although the precision may be improved for homogeneous materials like glass. The laser has already found an application in the spot analysis of metals, siliceous materials, marble crusts, and bone (*22, 30*), its detection limits being superior to that of the microprobe (*2*), and in the rapid microanalysis of uranium compounds (*33*). While the small amount of sample ablated makes this technique minimally destructive - an important consideration for archaeological materials - samples must nevertheless fit inside the ablation chamber, which is less than 4 cm in diameter.

Comparison with Other Techniques

In addition to its ability to provide isotope data, ICP-MS has a fundamental advantage over other instrumental methods. NAA, XRF, AAS, and ICP-AES all rely on the interaction of electromagnetic radiation with the nuclei and electron shells of individual atoms. More specifically, excitation of a sample will produce peaks of characteristic wavelength and energy, with the peak heights proportional to the concentration of each element in the sample. Problems of spectral overlap and matrix interferences plague all of these techniques, and may obscure elements of interest. X-ray techniques generally provide only a surface analysis, and the absorption by air of lower energy wavelengths usually precludes the analysis of elements less than atomic number 12. In ICP-AES, for example, many individual elements have more discrete optical lines in their spectra than the total of 211 possible mass lines for all elements in ICP-MS spectra. ICP-MS gives excellent performance for rare earth elements, platinum group elements, and Ag, Au, Th, and U.

The few potential interferences in ICP-MS may be predicted and in most cases resolved by measuring alternative isotopes of the same element. For example, the argon plasma itself coincides with the major isotope of calcium (^{40}Ca), but one can measure ^{44}Ca instead. $^{40}Ar^{16}O$ also interferes with ^{56}Fe (92% natural abundance), requiring the analysis of minor iron isotopes at significantly reduced precision. The only elements without alternative isotopes are fluorine and phosphorus.

ICP-MS (and ICP-AES) response curve calibrations are linear over several orders of magnitude, so that the same sample solution may be used for measuring a range of sample concentrations. In contrast, AAS has very poor linearity, and interference problems; along with lamp changes, it can be quite tedious if analyzing for more than a few elements (*34*). When equipped with a graphite furnace, however, AAS does have comparable detection limits to ICP-MS for many elements.

ICP-AES (*35, 36*) has excellent precision (± 1%), and solution detection limits (0.2-25 ppb) several orders of magnitude higher than ICP-MS (certainly sufficient for most archaeological applications), but is limited to the routine determination of 35-40 elements (including rare earth elements, B, Be, S, P, Ti, V), many of which are subject to spectral interferences. ICP-AES has superior precision to ICP-MS for Fe analysis because of the mass spectral overlap described above.

NAA (*37*) can be non-destructive, otherwise requiring only a powdered sample, and access to a nuclear reactor. For the 30-35 elements that can be routinely measured, it is particularly sensitive for the rare earth elements, and Sc, Co, Cr, Cs, Hf, Ta, Th and U. Matrix effects are not a problem, although there are considerable spectral overlap interferences. Precision is approximately ± 2-5% for rare earth elements, but may be several times that for solid samples, depending on the artifact's shape. The concentrations of 20 elements in a groundwater sample determined by ICP-MS were extremely close to those measured by NAA (*17*), and were produced much more quickly and less expensively.

XRF (*38, 39*) is also a solid sample technique, is capable of better analytical precision (± 0.5%) than ICP-MS, but the matrix interference corrections required for XRF result in similar accuracy to ICP-MS. XRF is the preferred technique for major-element determinations in homogeneous and/or highly refractory materials. Related X-ray techniques like electron probe microanalysis and PIXE are used for spot analyses of thin sections and mounted samples, but at considerably higher elemental detection limits. Laser ablation ICP-MS has superior detection limits (but inferior precision) and can be applied directly to small artifacts without sample removal.

Archaeological Applications

ICP-MS has now been used a number of times in archaeological contexts, for determining the major and trace element composition of metals (*25, 27-28*), marble (*21*), and obsidian (*24*). Laser ablation has been applied to coins, glass, ceramics, bone (*22*), and marble (*30*), while provenance information has also been obtained from lead isotope ratio analysis of metals (*19, 23*) and turquoise (*40*).

ICP-MS appears to offer several important advantages over other analytical techniques: (1) Only a few milligrams of sample are required. Small or precious artifacts (e.g. coins, sculptures) may therefore be analyzed with minimal sample destruction, although larger samples are warranted if the material is heterogeneous, for example   ceramics; (2) A very large number of elements can be accurately and precisely analyzed. This is particularly useful for characterization and provenance studies of obsidian, ceramics, jade, and marble; (3) Isotope ratio measurements to three significant figures are possible without extensive sample preparation, allowing preliminary assessment of the provenance of metals, turquoise, and ivory; and (4) The combination of small sample size and low per sample cost allows assemblages of artifacts rather than individual objects to be studied. This is critical for the study of patterns of resource exploitation or technological production in the archaeological record.

We present here a few examples of our own research using ICP-MS, in order to emphasize the advantages and disadvantages of this technique in answering specific archaeological questions about particular types of artifact materials.

Sample Preparation

Specific procedures were used to create sample solutions from different materials (*41*). Copper and iron typically dissolve in nitric acid at room temperature; tin, lead, gold and silver require hydrochloric acid as well as nitric acid. Rock samples require concentrated acid attack: refractory silicates (metal slag, glass, obsidian) and non-siliclastic minerals were dissolved using alkali fusion, or using elevated temperature and pressure digestions. Analytical precision is considerably reduced for rock samples because of the high concentration of flux and silicon in the sample solution, but this problem can be overcome through careful matching of standards to samples, the averaging of replicate samples (ideally at different concentrations), or the use of standard addition.

All sample solutions were diluted so that elements of interest were present in concentrations of about 500 ppb or less, and 100 µL of a 10 ppm ^{115}In or ^{186}Rh solution was added to each sample as an internal spike, to correct for drift in the mass count signal.

Metals. Samples were removed from archaeological artifacts either by careful drilling to minimize the visual impact of sampling, or by cutting with an emery disk. Surface corrosion was either discarded or removed by soaking in nitric acid for a few minutes. Copper samples weighing 5-10 mg were then dissolved at room temperature in ca. 10 mL concentrated HNO_3 and brought up to 100 mL volume; 20 mL of aqua regia (1:3 v/v conc. HNO_3/conc. HCl) were also added to dissolve iron, gold, tin and tin-bronze artifacts. Samples containing gold or tin must remain in concentrated acid for 24 hours before dilution. For analysis, a 1 mL aliquot of the sample solution plus 2 mL concentrated HNO_3 was again diluted to 100 mL.

Calibration curves were produced using commercially available multi-element standards. The average value of several acid blanks was subtracted from that of all sample solutions, and the resulting solution concentration values were multiplied by the dilution factor to reproduce the elemental concentrations in the solid copper sample.

Obsidian and Slag. Samples were removed from archaeological artifacts either by flaking or using a high-speed diamond saw, ultrasonically cleaned in distilled and deionized water, and dried at 60 °C. Each sample was then pulverized at liquid nitrogen temperatures for 2.5 minutes using a SPEX 6700 freezer mill. We have then employed two different sample dissolution procedures.

For sealed-vessel digestion, 100 mg of powdered sample were weighed into the inner Teflon beaker of an acid digestion bomb (Parr Instrument Co., Moline, Illinois). 0.2 mL of aqua regia were used to wet each sample; 2.5 mL of hydrofluoric acid were added and the bomb sealed. Bombs were heated at 120 °C for 2 hours, increasing the internal pressure to several tens of atmospheres, causing minerals which are resistant to attack at normal pressures to decompose. After cooling, the digestion bombs were carefully opened, their contents quantitatively transferred to a 250 mL polypropylene volumetric flask already half-filled with twice-distilled and deionized water, and the sample solution brought up to volume. Samples of less than 100 mg were diluted proportionally so that solution concentrations were similar for all samples. The Teflon containers were cleaned by soaking in a concentrated nitric acid bath for several hours before reuse.

For alkali fusion, 100 mg of powdered sample were mixed with 200 mg of lithium metaborate flux in a high-purity graphite crucible (SPEX Industries). The crucible was heated at 1050 °C for 5 minutes, gently shaken twice, and the hot glass melt poured into a polypropylene sample bottle containing 50 mL of 10% nitric acid. Bottles were vigorously shaken for 5 minutes by hand, and agitated overnight using a shaking table. For analysis, 1 mL aliquots of the sample solution were diluted to 10 mL with twice-distilled and deionized water.

Calibration curves for obsidian were produced using accepted values for the NIST standard obsidian SRM-278, prepared in the same way as the obsidian samples. The average value of several acid blanks was subtracted from that of all sample solutions, and the resulting solution concentration values were multiplied by the dilution factor to reproduce the elemental concentrations in the solid obsidian sample.

Turquoise. All additional rock matrix was removed from the mineral ore samples and then samples were removed by flaking. Samples weighing 50-100 mg were dissolved at 60 °C in 5 mL HNO_3, 2 mL HCl, 0.33 mL HF, and 10 mL distilled and deionized water in Teflon perfluoroalkoxy (PFA) bottles, and later brought up to a 50 mL volume.

Turquoise samples were analyzed for ^{204}Pb, ^{206}Pb, ^{207}Pb, and ^{208}Pb. Care was taken to ensure that all four lead isotopes were in the correct detection range for the instrument. Acid blanks were subtracted from all samples, and isotope ratios of $^{206/204}Pb$, $^{206/207}Pb$, and $^{208/207}Pb$ were calculated and calibrated against NIST lead standards 981 and 982. It must be noted that while thermal ionization mass spectrometry (TIMS), the most common method of isotope ratio analysis, fractionates samples because of differences in volatility, ionization, and space charge effects, ICP-MS fractionates samples only because of mass. For this reason, it is critical to normalize data against frequently run isotopic standards. Accuracy can also be improved by performing separation chemistry on the samples to isolate the element of interest. While three-significant figure precision is insufficient to distinguish among many of the copper and lead ore sources in the Mediterranean, for example, this precision may be sufficient to distinguish among source regions, and perhaps even some sources within those regions.

Analytical Results

Metals. 14 copper and arsenical bronze artifacts from Gonur in Turkmenistan (*29*), and one arsenical bronze ingot from near Rooiberg in the Northern Transvaal of southern Africa (*42*) were analyzed by ICP-MS for major and trace elements. The Gonur material, all from the Bronze Age Margiana civilization, includes axes, spears, bracelets, pins, a small vessel, and a mirror. The Rooiberg ingot is a typical example of the form in which metal was traded in the African Iron Age. Results from NAA for a sample from the same ingot (Grant, M.; Miller, D.; Young, S. M. M., unpublished report) are also available. Although reasonable agreement for both major and trace elements was obtained (Table I), comparison of these data are not a fair assessment of the relative accuracy of ICP-MS and NAA since metals are potentially heterogeneous. A more extensive comparison of ICP-MS with NAA has produced excellent results for groundwaters (17).

Table I. Comparison of ICP-MS and NAA Data for an Arsenical Bronze Ingot

Element	ICP-MS	NAA
Sn (%)	1.24	1.78
Fe (%)	0.94	2.11
Ni (%)	0.046	0.065
Co (%)	0.009	0.012
Sb (%)	0.027	0.022
Cu (%)	78.5	76.5
Ag (%)	0.0030	0.0035
As (%)	19.1	18.0

Electron probe microanalysis data are also available for the Turkmenistan artifacts (Killick, D. J.; Young, S. M. M.; Hiebert, F., unpublished report), producing general agreement with ICP-MS (Table II). Once again, metal objects do not provide a fair test of these methods since only a small surface area is analyzed by the probe. The detection limits of the probe are also inferior, and subject to spectral interferences.

Nineteen bronze statuettes of Iron Age date from Spain and Italy, now in the collections of the Peabody Museum (PM), Harvard University, and the Department of Classics, Tufts University (TU) were analyzed for their major element composition using the procedure described above. The statuettes were presumably cast by the lost-wax technique in clay molds which were then broken open and discarded (43-45). Knowledge of the alloys used provides information on the types of ores exploited, as well as the technological nature of the production process itself (e.g. was lead added to improve casting properties? Were standardized recipes used, suggesting centralized production? Was fresh metal or scrap bronze used?). Lead isotope ratio analysis by ICP-MS of 11 Sardinian bronze figurines has already suggested the use of local ore deposits, in contrast to the apparent Cypriot origin of the copper oxhide ingots found in Sardinia (19). Table III gives the results of the analyses, normalized to 100%. In addition to the elements shown, Cr, Mn, Co, Se, Cd, Sb and Ba were also determined but were never present in more than trace quantities. The 17 Spanish figurines average about 5% Sn and nearly 8% Pb, although there is considerable individual variability in composition (Sn ranges from 1% to more than 10%, Pb from 0% to 23%). These results are entirely consistent with previous analyses of Iberian bronze figurines (*45*). The two Italian figurines have 7-8% Sn and 2-4% Pb, also consistent with previous analyses (*44, 46, 47*). Fe was only significant in one figurine, and none of the figurines had more zinc (maximum 1.4%) than is likely to have been present in the copper ore(s) used.

For these particular studies, the use of ICP-MS permitted a quantitative, minimally destructive analysis for both major and trace elements, while minimizing sample preparation and analysis time and cost.

Obsidian. One hundred and eighty-six geological samples from western Mediterranean sources and 33 archaeological artifacts from sites in Italy and France were prepared and analyzed for 38 major and trace elements using the procedures described above. The chemical source-tracing of obsidian in the western Mediterranean has been pursued for more than thirty years (*48-53*); it is only recently, however, that the multiple flows of the Monte Arci volcanic complex in Sardinia were thoroughly surveyed, chemically characterized, and followed by the analysis of significant numbers of artifacts (*24, 54, 55*). The analyses confirm the existence of 7 chemically distinct Monte Arci sources, 5 of which are represented among archaeological artifacts. Stepwise multivariate discriminant analysis of the elemental data was performed to determine the statistical probability at which each artifact matched a particular source. A bivariate plot of the first two discriminant functions derived from the concentration data for Ba, Mn, Sc, Rb, Sr, Y, La, Ce, Pr, Nd, Eu, Gd, Tb, Dy and Ho (Figure 1) illustrates the clear source differentiation provided by multi-element analysis. Among archaeological artifacts, geological sources SA, SB2 and SC1 are represented most often, but their relative usage varies considerably from site to site, in some cases in chronological or geographic patterns. These patterns may reflect particular cultural affinities, maritime routes and capabilities, and the exchange of items that are invisible in the archaeological record; chronological changes in distribution may be related to the adoption of a neolithic way of life, and later, to increasing social complexity.

For this particular study, the use of ICP-MS permitted a quantitative analysis of a large selection of elements not technically or monetarily feasible with any other single technique.

Turquoise. The lead isotope ratios of turquoise from more than 20 mines in the southwestern United States and Mexico were determined to three significant figures. The preliminary results (Figure 2) show that the Cerrillos mine can be distinguished from the other turquoise sources tested using ICP-MS (*26, 40*). Based on the limited number of analyses shown here, it appears that inter-source differences are greater than both intra-source heterogeneity and the analytical resolution of ICP-MS. Many more turquoise sources are currently known in the Southwestern US than in Mexico, and it is likely that additional, undocumented sources exist in Mexico which may have also supplied turquoise for archaeological artifacts. Nevertheless, it has been suggested that the residents of Chaco Canyon, where extensive workshops have been found, controlled much of the turquoise market both in the southwest US and in Mexico (*56*). The carrying capacity of the Chaco area may have been insufficient to

Table II. Comparison of ICP-MS and Electron Probe Microanalysis Data for Copper and Arsenical Bronze Artifacts from Gonur, Turkmenistan

Sample	As (%)		Sb (%)		Fe (%)		Pb (%)		Sn (%)		Ag (%)		Zn (%)	
	ICP	Probe	ICP	Probe	ICP	Probe	ICP	Probe	ICP	Probe	ICP	Probe	ICP	Probe
3	2.26	2.19	0.003	nd	0.54	0.96	0.22	0.12	0.29	0.29	0.10	0.11	0.11	nd
4	1.89	1.57	0.001	nd	0.60	0.61	0.43	0.24	0.09	nd	0.03	nd	0.07	nd
6	2.66	2.56	0.013	nd	0.97	1.47	0.14	0.10	0.19	0.12	0.03	nd	0.08	nd
9	1.61	2.92	0.014	nd	0.38	0.36	0.12	nd	0.03	nd	0.01	nd	0.16	nd
10	1.14	1.10	0.010	nd	0.51	0.76	1.04	0.47	0.08	0.06	0.03	nd	0.36	nd
12	1.31	1.37	0.004	nd	3.12	0.42	0.17	0.14	0.09	nd	0.04	nd	1.08	nd
13	4.89	5.52	0.66	0.32	0.13	nd	0.04	nd	0.05	nd	0.19	0.22	0.22	nd
16	2.85	2.82	0.011	nd	0.51	0.32	0.28	0.22	0.62	0.67	0.05	nd	0.18	nd
18	3.28	4.70	0.008	nd	0.44	0.40	0.06	nd	0.03	nd	0.08	nd	0.38	nd
22	0.40	2.55	0.03	0.08	0.29	0.07	0.12	0.14	0.15	0.16	0.04	0.06	2.67	nd
23	1.56	1.67	0.016	nd	0.50	0.45	0.14	0.09	1.14	0.83	0.05	nd	0.16	nd
24	2.55	2.82	0.06	0.07	0.94	1.34	0.78	0.99	1.54	1.69	0.06	nd	0.20	nd
26	1.79	1.76	0.011	nd	0.67	0.65	0.16	0.14	0.09	0.05	0.03	nd	0.13	nd
27	1.46	1.76	0.25	0.32	0.19	nd	0.23	0.23	0.94	0.95	0.07	0.06	0.15	nd

nd = not detected

Table III. ICP-MS Data for Bronze Figurines from Spain and Italy

Museum Number	Cu (%)	Sn (%)	Pb (%)	Ni (%)	Ag (%)	As (%)	Zn (%)	Fe (%)	Total (%)
PM 33-67-40/113	80.2	6.95	10.6	0.13	0.06	0.14	1.31	0.60	100.0
PM 33-67-40/114	83.2	10.5	4.68	0.03	0.09	0.10	1.36	nd	100.0
PM 33-67-40/118	87.9	3.50	7.47	0.01	0.05	0.02	1.04	nd	100.0
PM 33-67-40/120	94.6	2.62	1.53	0.03	0.06	0.03	1.11	nd	100.0
PM 33-67-40/121	94.5	5.16	0.26	0.00	0.03	0.01	nd	nd	100.0
PM 33-67-40/122	83.2	4.62	11.6	0.06	0.04	0.19	0.33	nd	100.0
PM 33-67-40/123	82.8	5.67	10.6	0.03	0.08	0.07	0.84	nd	100.0
PM 33-67-40/124	70.9	5.54	23.1	0.02	0.06	0.05	0.32	nd	100.0
PM 33-67-40/125	95.0	1.41	3.48	0.03	0.05	0.01	nd	nd	100.0
PM 33-67-40/126	92.4	2.80	4.29	0.01	0.06	0.05	0.36	nd	100.0
PM 33-67-40/127	88.8	5.46	5.61	0.02	0.06	0.05	nd	nd	100.0
PM 33-67-40/128	89.0	9.95	0.97	0.01	0.06	0.03	nd	nd	100.0
PM 33-67-40/129	91.5	1.29	6.68	0.02	0.05	0.09	0.41	nd	100.0
PM 33-67-40/130	89.7	5.05	4.64	0.03	0.05	0.11	0.39	nd	100.0
PM 33-67-40/134	76.2	4.16	18.7	0.03	0.09	0.10	0.74	nd	100.0
PM 33-67-40/135	80.2	5.00	13.8	0.04	0.06	0.09	0.85	nd	100.0
PM 33-67-40/136	89.7	6.92	2.84	0.03	0.06	0.05	0.37	nd	100.0
TU 1985.9.3	89.4	8.38	1.89	0.03	0.08	0.26	nd	nd	100.0
TU 1986.11.2	88.3	7.21	4.14	0.06	0.07	0.17	0.06	nd	100.0

nd = not detected

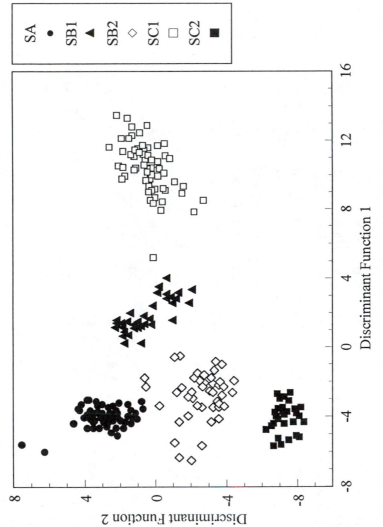

Figure 1. Bivariate Plot of First Two Discriminant Functions for Obsidian Samples Analyzed by ICP-MS

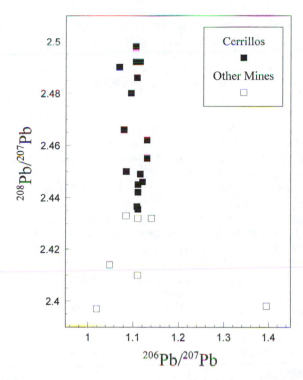

Figure 2.   Bivariate Plot of Lead Isotope Ratios for Turquoise Sources in the Southwestern United States and Mesoamerica Analyzed by ICP-MS

support the local population, and turquoise may have been exchanged for agricultural produce. The Cerrillos mine - the closest to Chaco Canyon and far larger than regional consumption would have demanded - could have supplied much of the turquoise worked at Chaco Canyon and traded to other sites. A lead isotope database for all the turquoise sources in this vast region is now being constructed so that these hypotheses may be tested.

Conclusions

ICP-MS is rapidly becoming an important technique for the analysis of archaeological materials, with particular advantages over other techniques in terms of the number of elements that can be analyzed, the amount of time required for sample preparation and analysis, the minimal sample size required, the ability to do isotope ratio analysis, and the relatively low cost of analysis. As with other solution analysis techniques, careful selection of dissolution procedures, meticulous sample preparation, and close matrix matching of standards are necessary to insure good results. Archaeologists should continue to use appropriate analytical techniques to answer specific archaeological questions; ICP-MS is now an established method that is available towards this end.

Literature Cited

1   Jarvis, I.; Jarvis, K. E. *Chemical Geology* **1985**, *53*, 335-344.
2   Hutton, R. C. *Journal of Analytical Atomic Spectrometry* **1986**, *1*, 259-263.
3   Potts, P. J. *A Handbook of Silicate Rock Analysis*. Chapman and Hall: New York, 1987.
4   Gray, A. L. In *Inorganic Mass Spectrometry*; Adams, F.; Gijbels, R.; Van Grieken, R. Eds.; John Wiley & Sons: New York, 1988; pp 257-300.
5   Jarvis, K. E. *Chemical Geology* **1988**, *68*, 31-39.
6   Jenner, G. A.; Longerich, H. P.; Jackson, S. E.; Fryer, B. J. *Chemical Geology* **1990**, *83*, 133-148.
7   Longerich, H. P.; Jenner, G. A.; Fryer, B. J.; Jackson, S. E. *Chemical Geology* **1990**, *83*, 105-118.
8   *Applications of Plasma Source Mass Spectrometry II*; Holland, G.; Eaton, A. N., Eds.; The Royal Society of Chemistry: Cambridge, 1993.
9   Date, A. R.; Gray, A. L. *Applications of Inductively Coupled Plasma Mass Spectrometry*. Blackie: Glasgow, 1989.
10   Jarvis, K. E.; Gray, A. L. *Plasma Source Mass Spectrometry*. Royal Society of Chemistry Special Publication 85: London, 1990.
11   Jarvis, K. E.; Gray, A. L.; Houk, R. S. *Handbook of Inductively Coupled Plasma-Mass Spectrometry*; Blackie: Glasgow, 1991.
12   *Applications of Plasma Source Mass Spectrometry*; Holland, G.; Eaton, A.N., Eds.; The Royal Society of Chemistry: Cambridge, 1991.
13   Ekstroem, H.; Gustavsson, I. In *Applications of Plasma Source Mass Spectrometry II*; Holland, G.; Eaton, A. N., Eds.; The Royal Society of Chemistry: Cambridge, 1993; pp 150-157.
14   Rigby, I. A.; Stephen, S. C. In *Applications of Plasma Source Mass Spectrometry II*; Holland, G.; Eaton, A. N., Eds.; The Royal Society of Chemistry: Cambridge, 1993; pp 158-164.
15   Perry, B. J.; Barefoot, R. R.; Van Loon, J. C.; Naldrett, A. J.; Speller, D. V. In *Applications of Plasma Source Mass Spectrometry II*; Holland, G.; Eaton, A. N., Eds.; The Royal Society of Chemistry: Cambridge, 1993; pp 91-101.
16   Ming, Y.; Ningwan, Y. In *Applications of Plasma Source Mass Spectrometry II*; Holland, G.; Eaton, A. N., Eds.; The Royal Society of Chemistry: Cambridge, 1993; pp 115-123.
17   Probst, T.; Zeh, P.; Kim, J. I. In *Applications of Plasma Source Mass Spectrometry II*; Holland, G.; Eaton, A. N., Eds.; The Royal Society of Chemistry: Cambridge, 1993; pp 29-37.
18   Knobloch, S.; König, H. In *Applications of Plasma Source Mass Spectrometry II*; Holland, G.; Eaton, A. N., Eds.; The Royal Society of Chemistry: Cambridge, 1993; pp 108-114.
19   Angelini, E.; Rosalbino, F.; Atzeni, C.; Virdis, P. F.; Bianco, P. In *Applications of Plasma Source Mass Spectrometry II*; Holland, G.; Eaton, A. N., Eds.; The Royal Society of Chemistry: Cambridge, 1993; pp 165-174.
20   Turner, P. J. In *Applications of Plasma Source Mass Spectrometry II*; Holland, G.; Eaton, A. N., Eds.; The Royal Society of Chemistry: Cambridge, 1993; pp 175-185.

21   Özdemir, N.; Moens, L.; Güçer, S. Determination of Rare Earth Elements in White Marbles by Inductively Coupled Plasma Source Mass Spectrometry. Poster paper presented at the 29th International Symposium on Archaeometry, 9-14 May 1994, Ankara, Turkey.

22   Gratuze, B.; Giovagnoli, A.; Barrandon, J. N.; Telouk, P.; Imbert, J. L. Characterisation of Archaeological Materials by Laser Ablation-Inductively Coupled Plasma-Mass Spectrometry. Poster paper presented at the 29th International Symposium on Archaeometry, 9-14 May 1994, Ankara, Turkey.

23   İnce, A. T.; Çukur, A.; Kunç, Ş. A Comparative Study of Lead Isotope Ratio Analysis. Poster paper presented at the 29th International Symposium on Archaeometry, 9-14 May 1994, Ankara, Turkey.

24   Tykot, R. H. *Prehistoric Trade in the Western Mediterranean: The Sources and Distribution of Sardinian Obsidian*. Ph.D. dissertation, Department of Anthropology, Harvard University, Cambridge, Massachusetts, 1995.

25   Young, S. M. M. In *Interregional Contacts in the Later Prehistory of Northeastern Africa*; Krzyzaniak, L.; Kroeper, K., Eds.;   Museum Archeologiczne W. Poznaniu: Poznan, 1992.

26   Young, S. M. M.; Phillips, D. A., Jr.; Matthien, F. J. In *Proceedings of the 29th International Archaeometry Symposium, 9-14 May 1994, Ankara, Turkey*; Melek Özer, A., Ed.; Middle East Technical University: Ankara, 1995.

27   Young, S. M. M.; Miller, D. In *10th Congress of the Pan African Association for Prehistory and Related Studies*; Pwiti, G., Ed.; University of Zimbabwe: Harare, 1995.

28   Miller, D.; Young, S. M. M.; Sandelowsky, B. In *The Prehistory of Mining and Metallurgy*; Craddock, P.T., Ed.; The British Museum: London, 1995.

29   Hiebert, F.; Killick, D.; In *New Studies in Bronze Age Margiana (Turkmenistan)*; Hiebert, F., Ed.; Information Bulletin 19; International Association for the Study of the Culture of Central Asia: Moscow, 1993, pp 186-204.

30   Ulens, K.; Moens, L.; Dams, R.; Vanwinckel, S.; Vandevelde, L. *Journal of Analytical Atomic Spectrometry* **1994**, *9*, 1243-1248.

31   Jarvis, I.; Jarvis, K. E. *Chemical Geology* **1992**, *95*, 1-33.

32   Arrowsmith, P. *Analytical Chemistry* **1987**, *59*, 1437-44.

33   Machuron-Mandard, X.; Birolleau, J. C. In *Applications of Plasma Source Mass Spectrometry II*; Holland, G.; Eaton, A. N., Eds.; The Royal Society of Chemistry: Cambridge, 1993; pp 186-192.

34   Hughes, M. J.; Cowell, M. R.; Craddock, P. T. *Archaeometry* **1976**, *18*(1), 19-37.

35   Walsh, J. N.; Howie, R. A. *Applied Geochemistry* **1986**, *1*, 161-171.

36   Hatcher, H.; Tite, M. S.; Walsh, J. N. *Archaeometry* **1995**, *37*(1), 83-94.

37   Harbottle, G. In *Radiochemistry: A Specialist Periodical Report*; Newton, G., Ed.; The Chemical Society, Burlington House: London, 1976; pp 33-72.

38   Jenkins, R. *X-Ray Fluorescence Spectrometry*. J. Wiley: New York, 1988.

39   Williams, K. L. *An Introduction to X-Ray Spectrometry: X-Ray Fluorescence and Electron Microprobe Analysis*. Allen & Unwin: Boston, 1987.

40   Young, S. M. M.; Pflaum, R. In *Pittcon, 1995*; Barrel, H. A.; Perry, M. B., Eds.; The Spectroscopy Society of Pittsburgh: Pittsburgh, 1995.

41   Totland, M.; Jarvis, I.; Jarvis, K. E. *Chemical Geology* **1992**, *95*, 35-62.
42   Grant, M. R. *Journal of Archaeological Science* **1994**, *21*, 455-460.
43   Gallin, L. J.; Tykot, R. H. *Journal of Field Archaeology* **1993**, *20*(3), 335-345.
44   Atzeni, C.; Massidda, L.; Sanna, U.; Virdis, P. In *Sardinia in the Mediterranean: A Footprint in the Sea*; Tykot, R. H.; Andrews, T. K., Eds.; Sheffield Academic Press: Sheffield, 1992; pp 347-354.
45   Prados Torreira, L. *Exvotos Ibericos de Bronce del Museo Arqueológico Nacional*; Ministerio de Cultura: Madrid, 1992.
46   Balmuth, M. S. *Studi Sardi* **1978**, *24*, 145-156.
47   Craddock, P. T. In *Italian Iron Age Artefacts in the British Museum*; Swadding, J., Ed.; The British Museum: London, 1986; pp 143-152.
48   Cann, J. R.; Renfrew, A. C. *Proceedings of the Prehistoric Society* **1964**, *30*, 111-133.
49   Hallam, B. R.; Warren, S. E.; Renfrew, A. C. *Proceedings of the Prehistoric Society* **1976**, *42*, 85-110.
50   Mackey, M.; Warren, S. E. In *Proceedings of the 22nd Symposium on Archaeometry, University of Bradford, Bradford, U.K. March 30th - April 3rd 1982*; Aspinall, A.; Warren, S.E., Eds.; University of Bradford: Bradford, 1983; pp 420-431.
51   Francaviglia, V. *Preistoria Alpina* **1984**, *20*, 311-332.
52   Herold, G. *Mineralogische, chemische und physikalische Untersuchungen an den Obsidianen Sardiniens und Palmarolas. Grundlagen zur Rekonstruktion Prähistorischer Handelswege im Mittelmeerraum.* Ph.D. dissertation, Universität (TH) Fridericiana Karlsruhe, 1986.
53   Francaviglia, V. *Journal of Archaeological Science* **1988**, *15*, 109-122.
54   Tykot, R. H. In *Sardinia in the Mediterranean: A Footprint in the Sea*; Tykot, R. H.; Andrews, T. K., Eds.; Sheffield Academic Press: Sheffield, 1992; pp 57-70.
55   Tykot, R. H. In *Method and Theory in Archaeological Obsidian Studies*; Shackley, M.S., Ed.; Plenum Press: New York, in press.
56   Mathien, F. J. In *The American Southwest and Mesoamerica: Systems of Prehistoric Exchange*; Ericson, J. E.; Baugh, T. G., Eds.; Plenum Press: New York, 1992; pp 27-63.

RECEIVED August 15, 1995

Chapter 11

Studies of Soils from an Aleutian Island Site

Henry J. Chaya

Department of Chemistry, Vassar College, Poughkeepsie, NY 12601

The study of properties of archaeological soils may give valuable information as to the location of middens and previously settled areas. A study is described of a site on Chernabura Island in the Aleutian Islands which was occupied by marine hunter-gatherers for at least 1500 years. Relatively simple analytical procedures for phosphorus and organic content were used to locate possible middens and places and manner of occupation. Phosphorus determination utilizing the molybdovanado-phosphate colorimetric method rather than the molybdenum blue method is described. Methods used for successfully dealing with interference for the determination of total phosphorus, organic phosphorus, and inorganic phosphorus are discussed. Most of the analytical steps for phosphorus and organic carbon determinations were performed by students using simple laboratory equipment. The methods used were simple and rapid.

High soil phosphorus content may indicate past human settlement activity (1). Phosphorus content in soils may indicate past human settlement even after 2500 years. Therefore, there is a need among archaeologists for a simple, rapid, economical, and reasonably accurate means of analyzing phosphorus in soils. Several authors have published modifications of procedures for the analysis of soil phosphorus (2 – 6), and portable field tests for phosphorus (2, 6, 7). Another consideration is the fractional separation of different forms of phosphorus that relate to aspects of previous human occupation (4). This paper will deal with fractional separations of soil phosphorus in terms of total phosphorus, inorganic phosphorus, and organic phosphorus as recommended by Bethel, et al. (8) as a useful breakdown of soil phosphorus.

The determination of total phosphorus by wet chemical extraction is considered one of the most accurate measures of archaeological soil phosphorus. Wet chemical extraction of total phosphorus from soils and sediments is time consuming and in some cases requires the use of dangerous reagents such as perchloric acid. There is however, a more practical, if slightly less accurate means of preparing the

0097–6156/96/0625–0131$12.00/0

soil sample for the determination of total phosphorus. This preparation involves heating the sample to 550 °C followed by acid extraction of the total phosphorus. Ordinarily, inorganic phosphorus determination involves acid extraction without the ignition step. For soil sample preparation, the ignition method is easier and safer than a chemical digestion. It permits the determination of total, organic and inorganic phosphorus and as a bonus, total organic matter (9 – 11). Organic phosphorus is calculated as the difference between total and inorganic phosphorus (12).

In the situation described here, extracted phosphorus is determined colorimetrically using the molybdovanado-phosphate (MVP) color procedure. In our experience, this color complex is more stable and measurements more reproducible than the molybdenum-blue colorimetric procedure commonly reported in the archaeological literature. The MVP colorimetric determination is recommended (13, 14) for its stability, freedom from temperature and time factors and ease of color development.

Phosphorus in the Soil

Calcium, aluminum, and iron are present in most soils. At high pH, inorganic phosphate is formed by bonding to calcium in an ionic manner and as the pH decreases, phosphate bonds to aluminum and iron. Organic phosphate on the other hand is bonded as an ester. Organic phosphate is found in all living things and it is known that some mineralization to the inorganic form has taken place over time. Mineralization takes place relatively slowly in temperate regions and somewhat more rapidly in tropical regions. Bone which contains phosphorus will be preserved for long periods of time in soils with basic pH, but decomposes fairly quickly in acidic soils (3, 15). Phosphorus found in organic refuse accumulates in the soil and even after several thousand years may still be detected.

The soil samples analyzed for phosphorus in this report were from Archaeological Site XSI-040 located on Chernabura Island in the Shumagin Islands which are a part of the Aleutian Islands of Alaska. It is a place that was occupied by marine hunter-gatherers at various times probably over a period of at least 1500 years as indicated by radiocarbon dating. The area, a strip of land on a cliff overlooking a northern part of the beach, is approximately 130 m long and extends inland for approximately 50 m. Soil samples were taken in a systematic manner utilizing a grid to locate middens or areas where refuse was deposited. Soil samples were also taken at surface points of interest, such as signs indicating probable former underground dwellings (16}.

Methodology

Our previous use of the analytical procedure briefly described above involved taking two portions of the same soil sample. The first portion was ignited to 550 °C, extracted with 2 M HCl and then subjected to "total phosphorus" determination using the molybdovanado-phosphate colorimetric procedure. The second portion was directly extracted with 2 M HCl for the determination of "inorganic phosphorus" using the same colorimetric procedure. Any interference due to brown soil coloration was satisfactorily accounted for by running a blank with each sample consisting of just the acid solution without the color reagent. The net colorimetric instrument reading between the two was taken as that due to the phosphorus content. The procedure of using a sample blank to account for brown coloration was checked by removing the brown coloration with activated carbon for several samples. No interference was ever encountered in the acid extract of the ignited portion. The difference between the total and inorganic phosphorus was regarded as that due to organic phosphorus.

Table I represents total phosphorus content of a group of sediments and soils that were analyzed by electron microprobe (*17*) and then by the analytical procedure described above.

Table I. Total Phosphorus of Sediment and Soils Analysis

Samples	% P Electron Microprobe	% P MVP Colorimetric
A	0.022	0.023
B	0.052	0.051
C	0.054	0.052
D	0.073	0.075
E	0.022	0.023
F	0.057	0.057
G	0.057	0.058
H	0.086	0.086

When the above analytical procedure was applied to the archaeological soils from Site XSI-040, no interference was experienced for the determination of total phosphorus. Serious interference was encountered in the determination of inorganic phosphorus in the form of dark brown colors in the (unignited) acid solutions that masked the developed color. Use of activated charcoal (Norite™) effectively removed the brown soil coloration, but a disadvantage was that it gave a high blank for phosphorus. For some samples, even after the brown color had been removed, a greenish color developed instead of a yellow one. According to the literature (*12*), the greenish color indicated the presence of Fe(II), an interference factor. The greenish color was removed by adding a small amount of hydrogen peroxide to each sample, which did not interfere with the MVP determination.

In order to overcome the interference problem described above, it was necessary to analyze for total phosphorus and inorganic phosphorus in different portions of the same archaeological sample. This was done in the following manner: a sample of soil was taken and its total phosphorus was determined as previously described. Another portion of the same sample was taken and extracted with 2 M HCl. The acid extract was separated from the solid portion by either centrifuging or Büchner funnel filtration with ashless filter paper. The solid contained organic phosphorus, while the acid extraction contained inorganic phosphorus. Each separated portion was analyzed following the procedure for total phosphorus content. In the solid portion, the total phosphorus represented the organic phosphorus of the original soil sample, while the total phosphorus from the acid extraction represented the inorganic phosphorus of the original sample. The two forms of phosphorus from this second portion should add up to the value for total phosphorus determined directly on the first portion of the sample.

Table II represents a number of soil samples analyzed from Archaeological Site XSI-040. A comparison is shown between total phosphorus expressed as the sum of individually determined organic and inorganic phosphorus, and total phosphorus determined directly. The sums for total phosphorus compare well with the directly determined total phosphorus values.

In practice, therefore, total phosphorus may be determined directly and in another portion only the organic phosphorus as described above. The inorganic phosphorus may be calculated by difference.

Table II. Total Phosphorus as a Sum and as a Direct Determination

Sample Number	% Inorganic P	% Organic P	% Total P (Sum)	% Total P Direct Determination
1	0.86	0.20	1.06	1.02
2	1.40	0.22	1.62	1.67
3	0.08	0.01	0.09	0.10
4	0.05	0.06	0.11	0.13
5	0.10	0.07	0.17	0.17
6	0.16	0.10	0.26	0.24
7	0.11	0.13	0.24	0.26
8	0.27	0.14	0.41	0.35
9	0.11	0.15	0.26	0.26
10	0.52	0.17	0.69	0.68

Experimental Procedure

The soil samples were broken into small particles by percussion without grinding and allowed to dry at room temperature for at least four days. The determinations were based on the weight of sample at equilibrium with room temperature and ambient humidity. The soil samples were then ground with a mortar and pestle and sieved using 1 mm mesh screen. For the determination of total phosphorus, 1 g of each of the prepared samples was weighed into a 20 mL glass scintillation vial with a Teflon™-lined cap which was included in the tare. The vial should be capable of withstanding a temperature of at least 550 oC. A reagent blank was included with each set of samples. The uncapped samples were placed in a drying oven at 105 oC for one hour, then capped, cooled and weighed to 0.1 mg. The uncapped samples were then placed in a muffle furnace that was slowly brought up to 550 oC over 1.5 h and maintained at that temperature for an additional 0.5 h. The samples were capped, cooled, and weighed again. The measured weight difference, the "loss on ignition" or LOI, has been shown to have a linear relationship to organic carbon content and will be discussed in the next section.

After addition of 10 mL 2 M HCl, each sample was weighed to 0.01 g and heated in a 90 oC water bath for 1 h with occasional swirling of each vial. The samples were weighed again, and additional distilled water was added to make up for any lost through evaporation. The samples were then removed from the water bath and any insoluble materials were allowed to settle over a period of several hours.

After the insoluble materials in the sample vials had settled, each sample was treated in the following manner: an aliquot not exceeding two mL of the supernatant liquid was quantitatively transferred to a 25 mL volumetric flask. Five mL of MVP color reagent (13, 18) were added and the solution was diluted to the mark with distilled water. The color was allowed to develop for at least ten minutes. The absorbance of the solution was measured using distilled water as a reference at a wavelength of 470 nm in a colorimeter. A cell with a pathlength of 2 cm is recommended.

The amount of phosphorus the absorbance represents was read from a calibration curve prepared from standard phosphorus solutions treated in the same manner as the samples.

Results and Discussion

Total phosphorus content in soil is a well-accepted form for indicating places of human occupation for an archaeological site. Locations of high phosphorus content help an archaeologist decide where to excavate. Figure 1 is a map of Site XS1-040 showing the distribution of total phosphorus content in terms of three value ranges. The distribution of phosphorus content indicates the location of probable middens and underground houses called barabaras which are of primary interest here. The various circles represent soil samples from test pits of surface points of interest in terms of definite and probable barabara locations. The distribution clearly indicates several areas of high soil phosphorus concentration. Since middens usually have high soil phosphorus content, it appears that the previous settlers threw much of their refuse over the cliff to the beach below. Other areas of high phosphorus concentration are in the middle and southeastern parts of the site.

About 150 soil samples were analyzed for total phosphorus and corresponding loss on ignition. The samples were run usually in groups of 20. Each group included a reagent blank, and three control soil samples to monitor reproducibility, which was within five percent. Some samples were collected off the site to determine the total phosphorus in "unlived" areas away from the site. At six different sampling points, two soil samples were taken at depths of 20 to 40 cm. and 40 to 60 cm. The range in concentrations of total phosphorus was 0.09 to 0.13 %.

A soil sample is dried at 105 °C and weighed and then ignited at 550 °C and reweighed. The weight difference or loss on ignition represents the organic carbon ignited. A linear relationship between organic carbon content and percent loss on ignition (LOI) has been demonstrated in the literature (*19*). In this work the LOI was determined as a routine step for each total soil phosphorus determination. Just as Figure 1 shows a distribution of total phosphorus content in three ranges of concentration, Figure 2 shows the LOI for the same samples in three ranges of values. There are similarities in the distribution of both soil properties.

Conclusions

We have described herein a method for analyzing phosphorus in archaeological soils in the laboratory. This method is rapid, simple, and agrees well with results from other methods in terms of total, organic, and inorganic phosphorus. The use of the stable molybdovanado-phosphate (MVP) colorimetric procedure for phosphorus determination was evaluated. Interference in the soil samples prevented the direct determination of inorganic phosphorus. In turn, the organic phosphorus could not be determined by difference. The problem was solved, using modifications that were relatively simple and straightforward.

A major objective in this work was to provide the archaeologist with a rapid and reliable method for obtaining archaeological soil phosphorus data. Using the ignition method, total phosphorus content and organic carbon content were determined in soil samples from an Aleutian Island archaeological site. The results have provided useful information on dwelling construction and on house and midden locations.

This type of information is particularly advantageous in soil-survey archaeology, and since the color development and reagents used are stable over long periods of time, this method has the potential, with modification, to become useful as a field test.

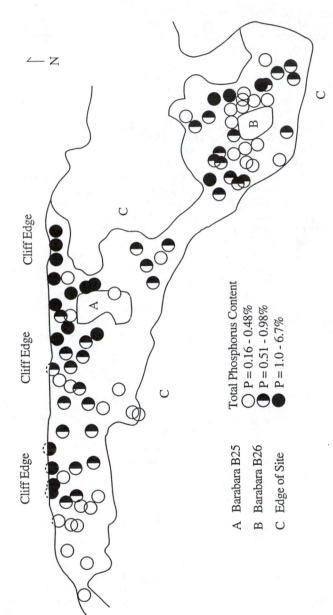

A Barabara B25   Total Phosphorus Content
B Barabara B26   ○ P = 0.16 - 0.48%
C Edge of Site   ◑ P = 0.51 - 0.98%
                       ● P = 1.0 - 6.7%

Figure 1. Distribution of total soil phosphorus content from shovel test pits, Site XSI-040.

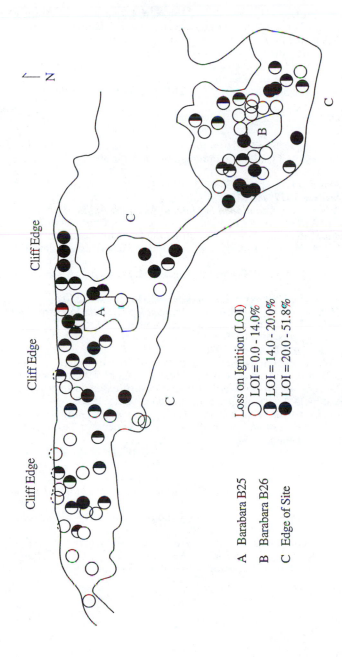

Figure 2. Distribution of soil loss on ignition from shovel test pits, Site XSI-040.

Acknowledgments

The author thanks Professor Lucille Lewis Johnson, Department of Anthropology, Vassar College, and her students, Mindy Perilla, Rachel Goddard, and Jonathan Howard who participated in the excavations and also helped in the analysis of the soil samples, and Adrienne Fowler who prepared the SITE XSI-040 maps. Thanks are also extended to the members of the Chemistry Department of Vassar College for their advice and for the use of laboratory facilities. This research was supported by National Science Foundation Grant OPP-9223473 and by Vassar College.

Literature Cited

1. Cook, S. F.; Heizer, R. F. *Studies on the Chemical Analysis of Archaeological Sites*; University of California Publications in Anthropology 2: Berkeley, CA, 1965.
2. Eidt, R. C. *American Antiquity* **1972**, *38*, 206-210.
3. Eidt, R. C. *Science* **1977**, *197*, 1327-1333.
4. Eidt, R. C. *Advances in Abandoned Settlement Analysis: Application in Prehistoric Anthrosols in Colombia, South America*; University of Wisconsin/Milwaukee, Center for Latin America: Milwaukee, 1984.
5. Cavanagh, W. G.; Hirst, S; Litton, C. D. *Journal of Field Archaeology*; **1988**, *15*, 67-83.
6. Craddock, P. T.; Gurney, D.; Pryor, F.; Hughes, M. J. *Journal of Archaeology*; **1985**, *142*, 361-376.
7. Overstreet, D. F. *The Wisconsin Archaeologist* **1974**, *55*, 262-270.
8. Bethel, P.; Maté, I. *Scientific Analysis in Archaeology and Its Interpretation*; Henderson, J., Ed.; Oxford University Committee for Archaeology, Institute of Archaeology: Oxford, 1989; pp 1-29.
9. Andersen, J. M. *Water Research* **1976**, *10*, 329-31.
10. Hamond, F. W. *Landscape Archaeology in Ireland*; Reeves-Smyth, T.; Hamond, F., Eds.; Bureau of Archaeological Research: Oxford, 1983; pp 47-80.
11. Gurney, D. A. *A Guide for the Field Archaeologist. Technical Paper No. 3*; Institute of Field Archaeologists: Birmingham, (UK), 1985.
12. *Water and Environmental Technology, "Total Recoverable Phosphorus and Organic Phosphorus In Sediments"*; American Society for Testing of Materials: Philadelphia, 1983; *11.02*, pp 716-720.
13. Gee, A.; Deetz, V. R. *Anal. Chem.* **1953**, *25*, 1320-1324.
14. Kitson, R. E.; Mellon, M. G. *Ind. Eng. Chem.* **1944**, *16*, 379-383.
15. Provan, D. M. J. *Norwegian Archaeology Review* **1970**, *4*, 37-50.
16. Johnson, L. L.; Perilla, M. J.; Chaya, H. J. *Journal of Archaeological Science*, in press.
17. McKinley, T. D.; Heinrick, K. F. J.; Wittry, D. B. *The Electronmicroprobe*; John Wiley & Sons: New York, NY, 1966.
18. *Standard Methods for the Examination of Water and Wastewater*, 13th Ed.; American Public Health Association, Washington DC, 1971, 527-530.
19. Dean, W. E. Jr. *Journal of Sedimentary Petrology* **1974**, *44*, 242-248.

RECEIVED October 9, 1995

Chapter 12

Chemical Analysis of Residues in Floors and the Reconstruction of Ritual Activities at the Templo Mayor, Mexico

L. A. Barba[1], A. Ortiz[1], K. F. Link[1], L. López Luján[2], and L. Lazos[1]

[1]Laboratorio de Prospección Arqueológica, Instituto de Investigaciones Antropológicas, Universidad Nacional Autónoma de México, Ciudad Universitaria, Del. Coyoacán 04510, México, D.F., México
[2]Museo del Templo Mayor, Instituto Nacional de Antropologia e Historia, Guatemala 60, Col. Centro, Del. Cuauhtémoc, 06060 México, D.F., México

Lime plaster floors in the "Hall of the Eagle Warriors," Templo Mayor of Mexico-Tenochtitlan, have been studied intensively using chemical techniques after a detailed archaeological excavation. The simultaneous interpretation of chemical, historical and archaeological data has provided new and detailed information concerning the ritual activities performed in this unique structure during Aztec times. Given that the firing process used to produce lime minimizes chemical contamination, then newly plastered floors can be considered to be a clean homogeneous surface on which extraneous chemical compounds are deposited as a result of human activities that left residues in specific areas. This style of research confirms that a significant quantity of potential information is contained in lime plaster floors and illustrates how simple chemical techniques can be applied to the study of ritual activities in the past.

Since its inception in early 1978, the Proyecto Templo Mayor of the INAH has recovered part of one of the most prominent ceremonial complexes of the Mesoamerican world: the Sacred Precinct of Mexico-Tenochtitlan. Some of the most outstanding discoveries include the remains of 15 structures (almost all with various sub-structures), 132 rich offerings with more than 8,000 elements, along with a large number of sizable sculptures, bas-reliefs, mural paintings and ceramic fragments.

Among the most impressive discoveries made during the first season of fieldwork was Building "E" better known as the "Hall of the Eagle Warriors" (*1*). The structure is approximately 52 meters in length by 24 meters in width and corresponds to Phase VI of the Templo Mayor (ca. 1486-1502 A.D.). There are two flights of steps leading into the building at its western end, one facing west and the other south. The stairs are flanked by double inclined moldings in the form of knots. A pair of polychromatic sculptures in the form of eagle heads emerge from the west-facing *alfardas* (Figure 1).

In 1981, after excavating the interior of this building layer, an earlier structure that corresponds to Phase V of the Templo Mayor (ca. 1481-1486 A.D.)

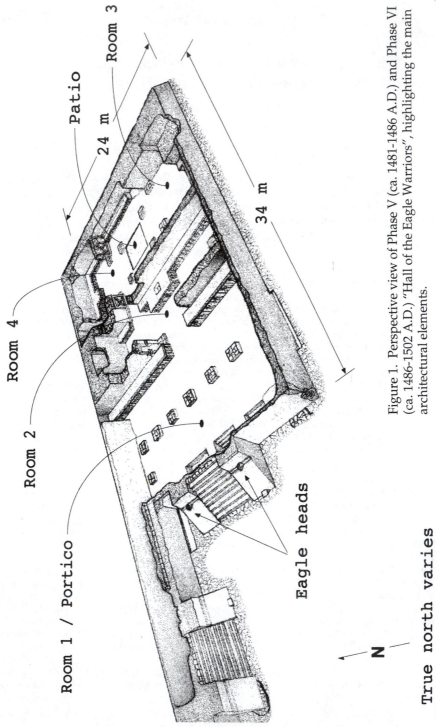

Figure 1. Perspective view of Phase V (ca. 1481-1486 A.D.) and Phase VI (ca. 1486-1502 A.D.) "Hall of the Eagle Warriors", highlighting the main architectural elements.

was uncovered and found to contain a series of interior rooms that were in an excellent state of conservation. From the exterior plaza level rise two flights of steps which lead to a large portico which was probably covered by a flat roof, of which a series of column bases distributed in the form of an *L* still remains. Two life-size ceramic statues which represent full-bodied Eagle Warriors flanked the entryway to the main room. A doorway protected by a pair of ceramic human skeletons leads to an open-air patio with two additional adjoining rooms at its northern and southern ends. Almost all of the interior walls of the hall conserve the remains of painted murals depicting temples, priests, shields and diminutive warriors. The lower parts of the walls are furnished with long benches which are composed of two vertical sections of bas-relief basalt. The lower and larger section of these polychrome carvings represent various processions of armed warriors which converge on a small altar where the *zacatapayolli* was kept. This was a plaited grass ball and held the *maguey* (*Agave sp.*) thorns that were bloodied during the ritual of self-mortification. The upper section, detailed as a frieze, was decorated with undulating feathered-serpents in bas-relief. Moreover, a total of eight ceramic braziers containing carbonized plant material were found in front of these benches.

It is noteworthy to mention that the physical layout, form, proportions and decorations of the benches of the "Hall of the Eagle Warriors" have many features in common with the Burnt Palace of Tula, as well as the Market, the Temple of the Warriors and the northern and north-eastern colonnades of Chichén Itzá. The strong analogies which can be made between the "Hall of the Eagle Warriors" and other older structures from Toltec and Mayan cultures, along with the ceramic representations of Eagle Warriors, and the continual allusion to acts of ritual self-sacrifice on the benches, have been the subject of many controversies concerning the function and significance of this building (*2 -5*).

To be precise, our current research seeks to obtain archaeological, iconographic, and ethno-historical information which will allow us to determine the functions for which this building was used, its relation to other structures within the ceremonial complex, its religious significance and certain aspects of the military order to which it was dedicated.

The Hall of the Eagle Warriors comprises five main areas which will be denominated rooms 1-4, and the patio. Room 1 is a porticoed space with a series of columns that supported a flat roof. This area leads to Room 2, the first chamber located at the center of the structure, through a doorway guarded by two sculptures of Eagle Warriors. Room 2 contains an altar whose characteristics suggest that it was the most important feature within the enclosure. In front of this altar, two braziers have been found. To the south of this room, there was a stone masonry pit containing large amounts of charcoal and ashes. Towards the north of the room, there is a further doorway, this one being guarded by two fleshless sculptures representing *Mictlantecuhtli* (Death God). This leads to an open space with a sunken patio or *impluvium*, framed by columns. On the eastern side of this patio, there is a bench with two rough braziers. To the north and south of this space, there are two other rooms denominated 4 and 3 respectively. There is an altar in each room, each having two braziers. All the braziers bear an image of *Tlaloc* (Rain God) except for the ones at the patio. The floors of the entire hall are made of lime plaster, a mixture of hydrated lime and ground tezontle (a volcanic scoriaceous rock).

Previous studies and analogies.

Through the study of floors samples we have believe that stuccoed floors are the best substrate for the study of chemical in floors. Stuccoed floors can be considered chemically "clean" just after their construction and gradually they accumulate chemical compounds, mainly in form of solutions spilled on them, which are fixed into the pourous matrix.

It has been demonstrated that "soil from archaeological contexts retains markers of anthropogenic activity" (6,7). Research of modern household floors has demonstrated that the distribution of chemical compounds in floors is not uniform nor random. The relationship between human activities and the distribution of chemical elements and compounds has been confirmed. Based on ethnographic analogies and ethno-archaeological experiments, areas of food storage, preparation and consumption, as well as areas of frequent transit have all been located (8). Parallel applications of the same methodology to the study of archaeological households have shown that the same patterns observed in ethnographic cases appear can be observed in archaeological floors (9-10). The most recent applications of study of activity areas has been in ritual spaces. Until now most of the chemical research into activity areas within archaeological sites had been oriented towards domestic dwellings. Activities performed and materials involved are different in domestic and ritual activities, nevertheless, results from ritual spaces have shown that the distribution chemical compounds is susceptible to archaeological interpretation, as can be seen in the present work. Previous studies associated with ritual activities have analyzed a Aztec altar or *momoztli* in the center of Mexico City (11), a Mayan laberinthic structure called Satunsat in Oxkintok, Yucatán (12), and a domestic shrine in Oztoyahualco, Teotihuacan (13).

Ritual ceremonies are considered here as individual or group acts of symbolic nature which are repeated and stereotyped according to a set of rules. The same ritual ceremony may present prayers, offerings, taboos, games, immolations, self-mortification, magic, or mythical representations (14).

The contaminating liquids (blood, sweat, foodstuffs, etc.) which were repeatedly absorbed by the floors during the rituals, allows us to chemically identify the areas where the acts took place and perhaps in the future characterize the perishable materials that were utilized.

We propose that via a detailed investigation integrating diverse fields of study such as archaeology, chemistry, history, historical philology, and iconography it will be possible to identify the kinds of ritual activities which took place within the "Hall of the Eagle Warriors."

In this particular case, there is a large number of ancient documents which testify to the kind of activities and materials involved. Rooms 2,3, and 4 all present elements in common: an altar, the iconographic representation of a *zacatapayolli*, and the *Tlaloc* braziers. All these elements can be recognized on page 79r of the Magliabechiano Codex (15), where several individuals are practicing self-mortification in front of a deity, inserting the blooded spines in a *zacatapayolli* and offering *copal (Bursera jorrullensis)* to the braziers.

This approach highlights the potential information contained in lime plaster floors and illustrates how simple chemical tests can be used to study human activities.

Materials and Methods

Materials. High purity chemical reactives were used. All the glassware was rinsed with concentrated nitric acid, deionized water and organic solvent before use. Screw-topped glass vials were used to store the samples to eliminate contamination risk.

Samples. Approximately 500 samples with an average weight of 20 grams were obtained from the stuccoed floors. Based on a 24 x 22 m grid originally laid out during excavation, extraction points were established at the intersection of every grid line. Samples were extracted using a drill bit 2 cm in diameter and 3-5 cm deep (Figure 2).

Analysis. Once in the laboratory, each sample was subjected to a series of simple tests such as Munsell color comparison, phosphate assay, carbonate assay and pH determination.   The results of these analyses showed notable differences of concentrations in certain areas which motivated us to apply a further series of tests to determine the concentration of total carbonates and phosphates as well as organic residues (such as albumin, carbohydrates and fatty acids).

Inorganic tests were performed on all of the samples.   Taking into consideration chemical results, archaeological and iconographic data we decided to only apply the organic tests to rooms 2,3,4, and patio since they were the most promising ritual areas. The results from the inorganic tests are the first approach in determining contamination patterns, as a result, we identify the more promising areas in which to apply the rest of the tests.

Three stages can be distinguished in the analysis of floor samples.   1) Application of semi-quantitive tests, allowing the differentiation of the defined zones.   2) Application of quantitative techniques to determine phosphate and carbonate levels giving greater accuracy than at the previous stage.   3) Based on these results, a selection of 228 samples was taken and semi-quantitative analysis were performed to determine organic compounds.

In Table I, the techniques applied at each of these stages are mentioned, semi-cuantitative test are described in detail by Barba, Rodríguez and Córdova (*16*).

Table I. Analytical Techniques

Stage	Substance or Property Analyzed	Method
1. Semi-quantitative inorganic	Color	Munsell Soil Color Chart
	Phosphates	Ring chromatography
	Carbonates	Effervescence with HCl
2. Quantitative	Phosphates	Method of phospho-vanado-molybdate
	Carbonates	Carbonate evaluation
	pH	
3. Semi-quantitative organic	Fatty acids	Hydrolysis with ammonia
	Albumin	Reaction with alkaline oxides and detection of ammonia
	Carbohydrates	Hydrolysis, dehydration and furfuraldehyde detection

During the first stage, color was determined with 500 mg of the sample which was compared with the Munsell Soil Color Chart (*17*). The pH of each sample was determined by adding 5 mL of distilled water and then measuring with a Beckman 3500 digital pH meter after 2 minutes. Phosphate determination was

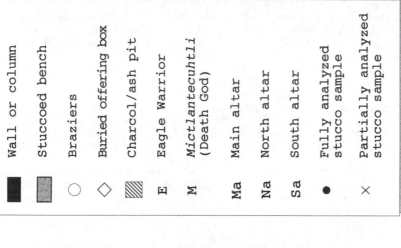

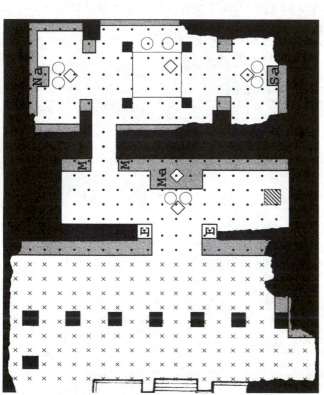

Figure 2. Location of the extraction points and main features within the "Hall of the Eagle Warriors."

performed by placing 50 mg of the sample in the center of a Whatman No. 42 filter paper. Two drops of 5% w/v ammonium molybdate (dissolved in 5 M HCl) were then added. After 30 seconds, two drops of 0.5% ascorbic acid were added and 90 seconds later saturated sodium citrate solution was added. The intensity of the blue color that developed and the diameter of the ring were compared (graded on a scale one to six) (*18*). Carbonate determination was performed by adding 3 mL of 10% HCl to 50 mg of sample and comparing the difference in effervescence between the samples (graded on a scale one to six) (*19*).

At the second stage, quantitative tests were performed to evaluate the presence of phosphates and carbonates. Phosphates were detected colormetrically through the formation of a phospho-vanado-molybdate complex. 10 mL of 6N HNO_3 were added to 500 mg of the sample. After 15 minutes, 10 mL of 0.5 % ammonium vanadate and 10 mL of 5% ammonium molybdate were added. The volume was then increased to 100 mL. After 15 minutes, colorimetry was performed at 460 nm. Carbonate analysis: 100 mg of the sample were placed in an Erlenmeyer flask and 25 mL of distilled water and 50 mL of 0.1 M HCl were added. The mixture was heated to remove CO_2 and when cool, the solution was back-titrated with standardized NaOH to indirectly determine the quantity of carbonate in the sample (*20*). The correlation between the qualitative and the quantitative values for phosphate and carbonate can be seen in Table II.

Table II.   Correlation values.

Semiquantitative scale	Quantitative scale for Phosphates (mgP/g of sample)	Quantitative scale for Carbonates (%)
0	0	0
1	<0.3	1-34
2	0.3-0.6	35-44
3	0.7-0.9	45-54
4	1.0-1.3	55-64
5	1.4-1.7	65-74
6	1.8-2.1	75-80

Organic compounds were determined using the following tests:
(a) Fatty acids were determined by placing 10 mg of the sample in a test tube and 3 mL of chloroform was added. This was then warmed to form a concentrate. One drop of the concentrate was then placed on a flat glass slide and two drops of 28.7% ammonia were added. After two minutes, two drops of 20% hydrogen peroxide were added. The resulting differences in quantity and stability of foam were compared. (graded on a scale one to three)
(b) Carbohydrate determination was performed by mixing 10 mg of the sample with 10 mg of oxalic acid and placing the mixture in a crucible. Five drops of sulfuric acid (1:3) were added and the mixture was then heated. The resulting furfuraldehyde was determined by observing the reaction with O-dianizidine impregnated on filter paper and exposing it to the resulting vapors. The change in the color of the paper was observed (graded on a scale one to six)
(c) Albumin determination was performed by placing 10 mg of the sample in a test tube and adding 0.1 g of CaO and two drops of water. The mixture was heated for

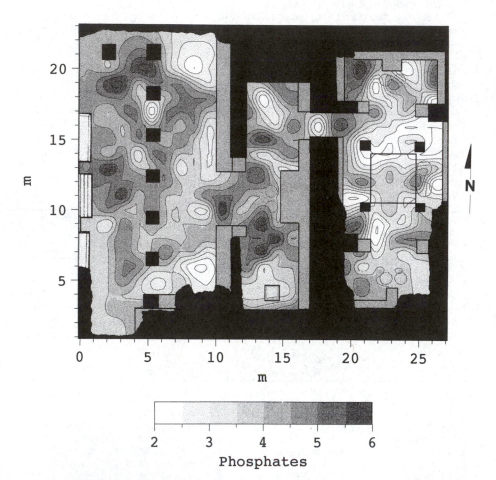

Figure 3. Distribution map of phosphates.

60 seconds and the resulting vapors were exposed to test paper to determine the presence of NH_3 via the change in the color of Whatman full range pH paper .

It should be noted that the lack of iconographic and archaeological elements (altars, braziers, etc.) related with self-mortification rituals in Room 1 was the main reason for excluding the samples of this area from the organic tests.

Visualization. The results from each of the analysis was represented a pseudo-colored raster image to facilitate initial interpretation. Missing data points were interpolated using the Kriging method. Data was then resampled using a bi-cubicly weighted filter and contextualized within the site by overlaying a geo-referenced plan of the excavation. The final representation in gray scales for publication was overlaid with black iso-lines to help distinguish the transitions between values.

Results

Phosphates. Residues of solutions rich in phosphates accumulated and became fixed in the abundant calcium carbonate present in the construction material of the floor, forming calcium phosphates.

In Room 1, two zones can be observed with different phosphate concentrations. The first is an area of high values and is found between the stairs and the columns. The area of low values is found between the columns and the entrance to Room 2. In this, there is a patch of high values close to the doorway which contrasts with the low values around it. In Room 2 high values were observed in two areas, a very extensive one to the south which begins near the main altar and finishes at the charcol and ash pit; the other area is near the altar and extends towards the north. In this room the low values are in the extreme north, however, at the entrance to the patio, another patch of high values can be noted. Rooms 3 and 4 as well as the patio show low values in general except for the zone to the west in Room 4 (Figure 3).

Carbonates. The distribution of carbonates in the floor is directly related to the construction material of the floor. The relative proportions of lime and sand are the principal determinants of the prescence of carbonates, but wear to the surface through constant use modifies the proportion.

In this map, one can again note that the higher values are to be found in the portico (Room 1) and tend to diminish towards the inner rooms though not to the same marked degree as previously. Once again, there is a contrast in Room 2, in front of the main altar, where values show a noticeable decline. Near the braziers, the values diminish slightly and this also occurs in the NW corner of Room 1 (Figure 4).

pH determination. pH values were determined by the levels of hydroxides which are residues of ash. However, in stuccoed floors, the base value for measurements is the equilibrium value of calcium carbonate in water, that is, approximately 8.2. All the differing values will therefore be the result of varying uses of the surface.

Room 1 contains the areas with the highest values of pH in the entire enclosure. Specifically, these are in the NW corner and around the central columns. In Room 2 high values are only to be found near the main altar and at the entrance to the patio. The rest of the room presents low values. In Room 4, the increase is only noticeable in front of the braziers. In the patio, high values are to be found around the *impluvium* and it should be noted that in this case the braziers are not associated

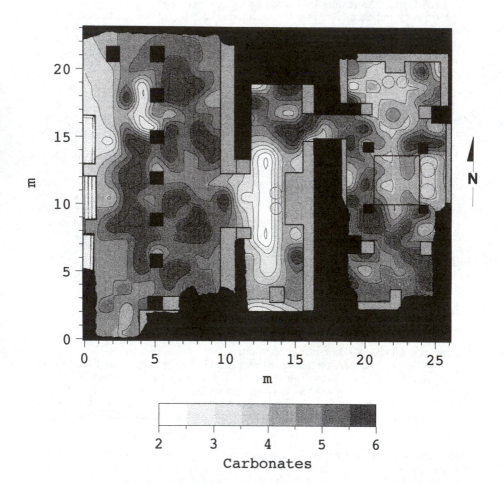

Figure 4. Distribution map of carbonates.

with a high pH level. In Room 3, the highest values are found in the SE corner (Figure 5).

Color comparison. As in the case of carbonates, color is determined by the mixture used to make the floor. Color is homogeneously distributed particularly in the two interior rooms, but in the porticoed area (Room 1) there is a broad zone of a redder color. In a number of samples the color is noticably darker. This may be due to contamination with a small quantity of sub-floor matrix (Figure 6).

Fatty acids. Fatty acids are the residue of substances formed by oils, fats or resins which were spilt on the floor and became impregnated in it. The fact that the distribution of the concentrations is very symmetrical is to be noted.

Room 2 presents high values directly related with the main altar and the braziers. Other zones of high values are the entrances and the area to the north of the charcol and ash pit. In Rooms 3 and 4 and the patio, the highest values are directly associated with the location of the braziers and altars (Figure 7).

Carbohydrates. Carbohydrates are the residues of substances with high starch and sugar levels.

In Room 2, specific areas of high admixture can be noted, particularly to the south of the room directly to the north of the charcol and ash pit. The levels also increase at the two entrances and in front of the braziers of the main altar. In Room 4, there are zones of high values behind the braziers and in three of the corners of the room. In the patio there are two zones of high values directly to the north and south of the *impluvium*, and in Room 3, there is only one occurence in the NW corner (Figure 8).

Albumin. Residues of albumin come from solutions containing proteins that were spilt onto the surface.

In Room 2, there are three main zones: to the south of the room, associated with the charcol and ash pit, and again to the north and center, associated with the entrances. In the patio the residues of albumin were only found in the northeast behind the braziers. In Room 3, two principal zones are observed: in the SE corner and in the entrance to the room (Figure 9).

Discussion

The maps of color and carbonates indicate that there was a change in the proportions used to make the lime plaster mix with which the floor of the Room 1 was constructed, because the color becomes reddened as a result of the *tezontle* present and the increased values of carbonates, indicating a greater proportion of lime in the mix. This might indicate changes in construction methods, going from the use of differently colored *tezontle* to remodelling or the building of extensions.

The results demonstrate that one of the principal areas of activity was in front of the main altar located in Room 2. This is characterized by a reduction in the concentration of carbonates due to wear to the floor, which would have been caused by heating and constant coming and going. This matches the semi-circle of high pH values formed around the altar, the two patches of high concentration of phosphates, as well as the presence of organic residues such as carbohydrates and fatty acids.

The next zones of interest, in terms of their chemical residues, are the areas at the entrances to Room 2 and the patio. These zones present high values of phosphate, pH, albumin, carbohydrates and fatty acids that are similar to the results obtained for the main altar. It is possible that the higher chemical content of this area is due to ritual activities associated with the sculptures guarding these entrances.

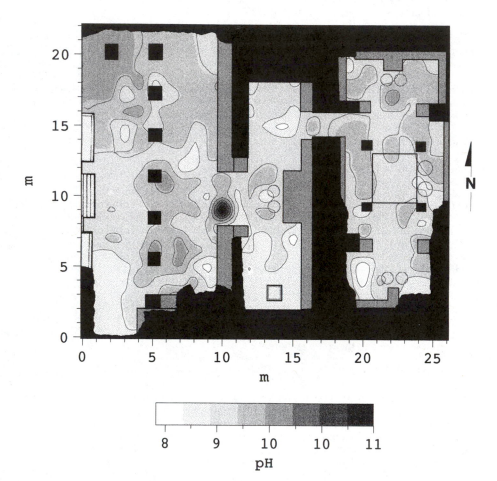

Figure 5. Distribution map of pH.

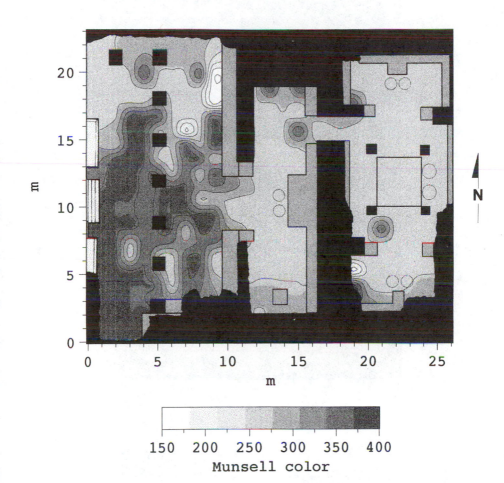

Figure 6. Distribution map of Munsell color.

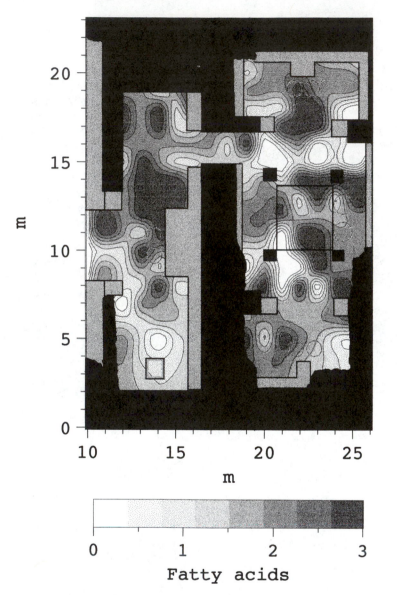

Figure 7. Distribution map of fatty acids.

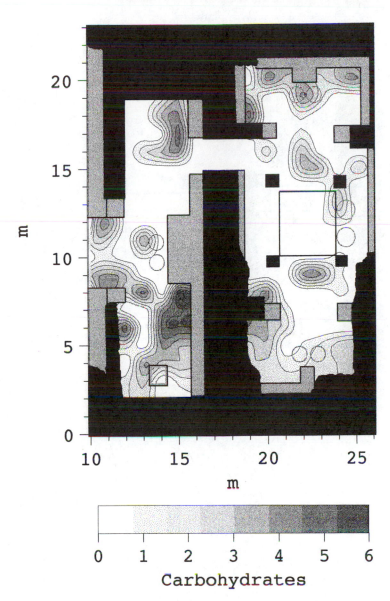

Figure 8. Distribution map of carbohydrates.

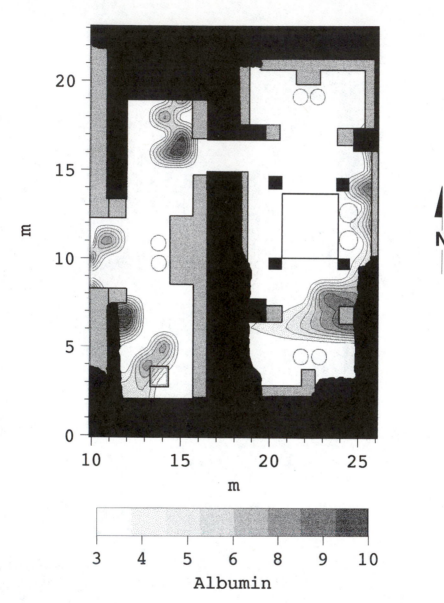

Figure 9. Distribution map of albumin.

The following zone of interest is an area in the angle formed between the two lines of columns in Room 1. In this area again, low carbonate values coincide with high pH and phosphate values which denote higher levels of activity.

The final areas of interest are those located immediately in front of the north and south altars, near the braziers. These results indicate activities clearly associated with rituals. Strangely, the concentrations of chemical compounds behind the braziers in the patio indicate that the activities took place between the benches and the braziers and not on the patio side, as had previously been thought.

In general terms, it is clearly noticeable that the patio with the *impluvium* is the area where the lowest concentrations of compounds were found, and in contrast, the areas in front of the altars are the richest.

The highest levels of chemical indicators were found in the ritual areas around the braziers, the ceramic sculptures and the representations of *zacatapayolli*.

In the Nahuatl literature there are numerous descriptions concerning ritual, sacrifice and ceremonies. León Portilla reports a great diversity of such practices and mentions that in addition to human victims, it was common to offer several species of small animal, fruits, *copal*, *pulque* (fermented juice of *Agave sp.*), as well as paper and spines sprinkled with blood, just to mention a few (*21*). The rituals were oriented towards deities at altars and braziers. Fluids from these bodies and materials produced the chemical compound enrichment of the floors .

According to Leach (*22*), there are three spatial components in any ritual scene, the first being the sacred area itself. Normally there is some sort of symbolic icon which indicates the materialization of the sacred. The second area is adjacent to the first, and is where most of the ritual takes place. Access to the first and second areas is reserved for the priests and religious servants. Finally, we find the faithful in a third area which is separated in turn from the sacred area by the sector of ritual activity.

Both the first and second areas can be identified in the "Hall of the Eagle Warriors": the first is represented by a series of iconographic elements, such as the *zacatapayolli* (placed on the altars and physically located in the interior or most hidden areas of the hall) and the ceramic sculptures of Eagle Warriors and *Mictlantecutli* (flanking two entrances). The second type of area, where most of the rituals took place, was located in the central and back parts of the hall and can be distinguished by the presence of a series of elements such as braziers and altars.

In conclusion, the Hall of the Eagle Warriors chemically exhibits two of the three key areas for a ritual to take place. The areas of ceremonial activity that stand out are: in front of altars, directly around the braziers, and the entryways where clay figures representing Eagle Warriors and *Mictlantecuhtli* were found. In those places ritual activities spill fluids on the floor whose chemical compounds were fixed and permited their recovery and analysis to reconstruct past human activities performed in this hall.

Literature Cited

1. López Luján, L. *The Offerings of the Templo Mayor of Tenochtitlan*; University Press of Colorado: Niwot, 1994.
2. de la Fuente, B. *Artes de México*, Nueva Época. **1990**, *7*, 36-53 .
3. Bonifaz Nuño, R. *Artes de México*, Nueva Época. **1990**, *7*, 26-35 .
4. Matos Moctezuma, E. *The Great Temple of the Aztecs*; Thames and Hudson: London, 1988, pp 19-20.
5. Klein, C. In *The Aztec Templo Mayor*, Boone, E. H., Ed.; Dumbarton Oaks: Washington, D.C., 1987; pp 293-370.

6. Knights, B. A.; Dickson, C. A.; Dickson, J. H.; Breeze, D. J. *Journal of Archaeological Science.* **1983**, *10*, 139-152.
7. Pepe, C., Dizabo P., Scribe, P., Dagaut, J., Fillaux, J.; Saliot. A. *Revue d'Archeometrie.* **1989,** *13*, 1-11.
8. Barba, L.; Ortiz, A. *Latin American Antiquity* **1992**, *3*, 63-82 .
9. Barba, L. In *Unidades habitacionales mesoamericanas y sus áreas de actividad* ; Manzanilla, L., Ed.; IIA-UNAM: México, 1986; pp 21-39.
10. Manzanilla, L.; Barba, L. *Ancient Mesoamerica* . **1990**, *1*, 41-49 .
11. Getino, F.; Ortiz, A. *Das Altertum* . **1990**, 2, 126.
12. Ortiz, A.; Barba, L. In *Oxkintok* ; Misión Arqueológica de España en México: Madrid, 1992, Vol. 4; pp 119-126.
13. Ortiz, A.; Barba, L. In *Anatomía de un conjunto residencial teotihuacano en Oztoyahualco;* Manzanilla, L., Ed.; IIA-UNAM: México, 1993, Vol. 2, pp 617-660.
14 Cazeneuve, J. *Sociología del rito*; Amorrortu: Buenos Aires, 1972.
15 *Codex Maglabechiano.*; Hill Boone E., Ed.; University of California Press: Berkeley, 1983; Vol. 2.
16. Barba, L.; Rodríguez, R.; Córdova, J. L. *Manual de técnicas microquímicas de campo para la arqueología*; IIA-UNAM: México,1991.
17. Barba, L. *Antropológicas*, **1989**, 3, 98-107.
18 Eidt, R. C. *Science.* **1977**, 197, 1327-1333.
19. Dent, D.; Young, A. *Soil Survey and Land Evaluation*; George Allen and Unwin: London, 1981.
20. Jackson, M. L. *Análisis químico de suelos;* Omega:Madrid, 1982.
21. León-Portilla, M. *Ritos, sacerdotes y atavíos de los dioses. Fuentes indígenas de la cultura Nahuatl. Textos de los informantes de Sahagún*:; UNAM: México, 1958; Vol. 1, pp 31.
22. Leach, E. *Cultura y comunicación. La lógica de la conexión de los símbolos, una introducción al uso del análisis estructuralista en la antropología social;* Siglo XXI:Madrid, 1978.

RECEIVED August 15, 1995

Chapter 13

Application of Multimolecular Biomarker Techniques to the Identification of Fecal Material in Archaeological Soils and Sediments

Richard P. Evershed and Philip H. Bethell

School of Chemistry, University of Bristol, Cantock's Close, Bristol BS8 1TS, United Kingdom

This paper focuses on the chemical analysis of organic residues in archaeological soils, particularly on the development and application of lipid "biomarker" techniques for the detection and characterization of disaggregated fecal matter. The work has been directed towards the study of soils from experimental and archaeological sites to assess the possibility of employing "biomarkers" characteristic of feces. Initial efforts in this area focused on building on the principle that 5β-stanols (which arise in feces from Δ^5-stenols by microbial reduction in the gut) may be of use in assessing fecal inputs into archaeological soils and sediments. The analytical methods employed are based on the use of gas chromatography/mass spectrometry with selected ion monitoring (GC/MS-SIM) to provide a very sensitive and selective means of analyzing for the characteristic steroidal marker compounds. Enhanced selectivity can be achieved by use of GC/MS/MS employing selected reaction monitoring (SRM). The use of 5β-cholestan-3β-ol (coprostanol and its congeners) to identify the sites of ancient cesspits has been demonstrated. In an investigation of a set of soil samples taken at intervals across an experimental field only the variation in the concentration of 5β-stanols reflected manure addition, together with the dynamic effects of soil erosion and possibly bioterbation. This paper summarizes our recent findings in this area, placing special emphasis on developments in the use of multi-component mixtures of biomarkers, i.e., 5β-stanols and bile acids, to identify the origin of fecal inputs into soils with a high degree of specificity.

Aged feces, either in the form of intact coprolites or dispersed in soil, have attracted considerable attention as a result of their potential to yield unique archaeological information relating to early human behavior, diet, parasites, disease, ecological adaptation, resource utilization, etc. (*1-3*). Intact fecal remains have been found in arid caves in desert areas, in intact cadavers, such as mummies and bog bodies, and in other extreme preservation environments such as frozen or waterlogged deposits. In temperate regions fecal remains are more commonly encountered in a disaggregated

0097–6156/96/0625–0157$12.00/0

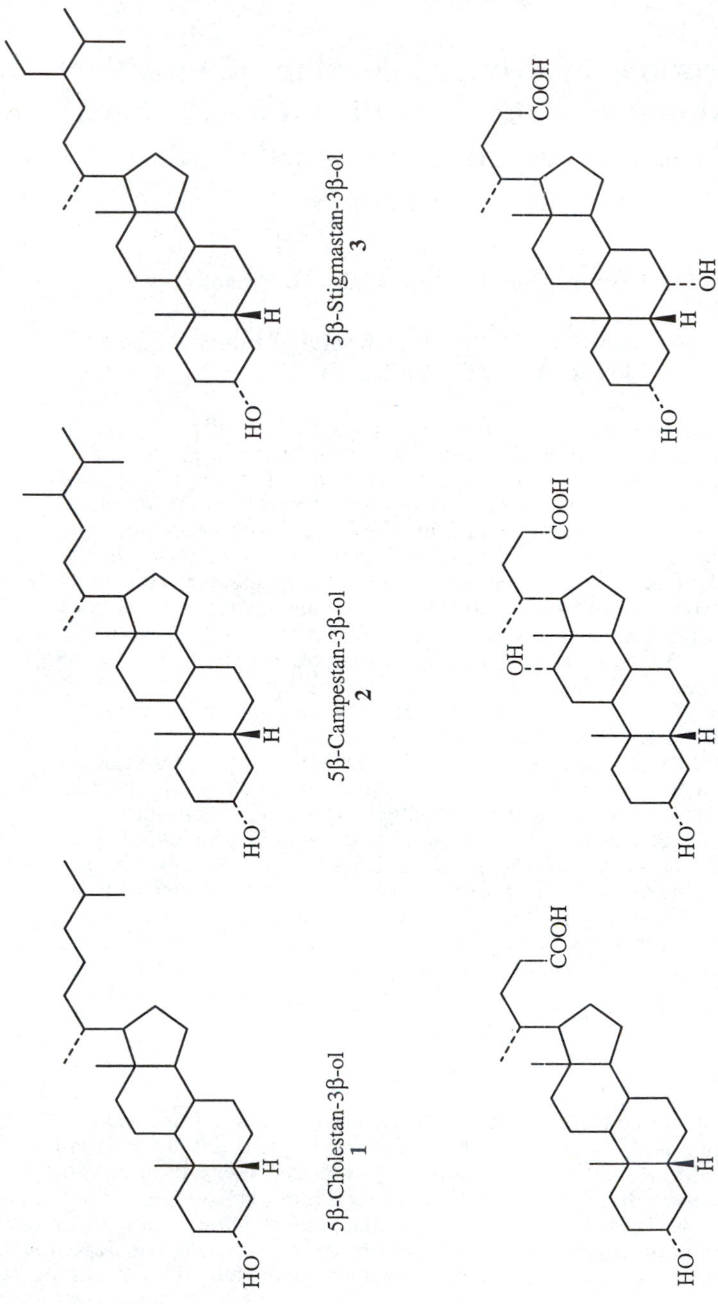

Figure 1. Steroid biomarkers referred to in the text by boldface numbers.

state, concentrated in such features as latrines, cesspits and midden deposits, or widely scattered, as in the case of ditch-fills or manuring.

Established methods of studying fecal remains include mainly the use of microscopy to identify morphologically intact fragments of food, e.g., plant fibers, pollen, bone fragments, and fish scales as well as parasite eggs and body parts, and inorganic materials, e.g., grits, derived from the tools or vessels used in the preparation of food (2, 4, 5). Chemical analysis of aged fecal remains, until recently, received only scant attention. Analyses for inorganic ions were among the first chemical analyses to be performed with a view to gaining dietary information (6). Problems exist in drawing conclusions from analyses of fecal remains recovered from waterlogged sites due to the dissolution of many inorganic components. In more recent years, the preservation of the organic constituents of ancient fecal remains has been demonstrated by use of GC/MS. Probably the most detailed study reported to date is that of Lin et al. (7) who investigated the survival of steroidal components in 2000-year-old human coprolites from dry deposits in Lovelock Cave, Nevada. All six samples that were analyzed using gas chromatography (GC) contained steroidal compounds. Although the concentrations of sterol, stanol and bile acids were somewhat lower than those of fresh human stool, the relative abundances of the individual components were very similar. In another study of cave coprolites from Danger Cave and Glen Canyon, Utah, amino acids, bile acids and lipids were readily detected (8, 9). The GC analyses performed in this latter study revealed no significant differences in the distributions of various lipids compared to desiccated modern samples. The potential for the chemical analysis of coprolites has also been noted by Wales and Evans (10).

While these findings are important in demonstrating the survival of fecal lipids in intact coprolites, substantial scope exists for the investigation of lipids from disaggregated feces in soils. Earlier examples include the study of the contents of the fill of a ditch associated with a Roman fort at Bearsden, Scotland. Conventional palaeobotanical and zoological techniques were used in conjunction with GC and GC/MS (11). Different strata of the ditch fill were distinguished by their sterol content. One layer was deemed to represent a sewage deposit based on the detection of coprostanol, 5β-cholestan-3β-ol (1; for structures see Figure 1) and its C_{28} (2) and C_{29} (3) homologs and related oxo-steroids, and bile acids (e.g., 4 and 5). The finding of coprostanol was taken to be an indicator of human fecal material, and hence, the use of the ditch as a latrine drain. Pepe and co-workers (12, 13) analyzed archaeological sediments for their lipid content. Their findings of coprostanol in high abundance in one horizon was taken to indicate cess deposition. Related studies of lipids in soils and sediments include the analysis of fatty acids from an Eskimo midden, which provided evidence consistent with a high input of debris from marine animals (14), and the analysis of fatty acids in soils from a buffalo jump site in Canada (15).

The above studies have established that specific chemical components, notably the steroidal compounds that occur ubiquitously in feces, possess the necessary chemical stability to survive for several millennia in a range of contrasting environments that are encountered at archaeological sites. Even where the physical evidence of feces is heavily degraded, e.g., through acid soil conditions, or is widely distributed as in the case of manuring, the detection of these specific markers should allow the location of fecal deposition. Our research has concentrated on the development of the application of biomarkers to the detection of disaggregated fecal matter in archaeological soils, with the following specific aims: (1) to assess the usefulness of different classes of chemical substance as indicators of inputs of fecal matter into archaeological soils and sediments; (2) to provide effective analytical methods for the detection of specific biomarkers in soils and sediments; (3) to identify the sites of ancient latrines, cesspits and midden dumps; (4) to distinguish human fecal matter from that of other animals; (5) to

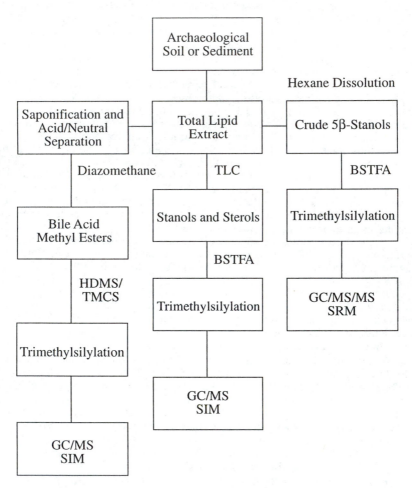

Figure 2. Scheme of extraction of fecal steroid biomarkers from archaeological soils. Abbreviations: BSTFA = *N,O*-bis(trimethylsilyl)trifluoroacetamide; HMDS = hexamethyldisilazane; TMCS = trimethylchlorosilane. (See text for all other abbreviations.)

distinguish fecal matter from decayed vegetable matter; and (6) to establish sites of animal penning enclosures or stables.

Our aim is therefore to be able to demonstrate the survival of molecular markers of fecal matter in soils and sediments at archaeological sites. Although the primary goal of this investigation was to develop the use of lipid biomarkers, particularly steroidal compounds, as indicators of fecal matter inputs into archaeological soils and sediments, throughout these studies we have compared and contrasted the information provided by lipid biomarkers to that derived from other techniques that are used in the study of soils and sediments at archaeological sites. Where comparisons have been drawn they will be referred to in the appropriate sections below. (See also Bethell, P., et al. submitted for publication to *J. Archaeol. Science*).

Analytical Method Development

The analytical protocols employed in this work focused primarily upon the use of 5β-stanols and bile acids (e.g. **1 – 5**) preserved in archaeological soils and sediments. The overall analytical scheme used to assesses archaeological soils and sediments for the presence of these compounds is summarized in Figure 2.

More detailed accounts containing the precise practical information can be found in References 16-18. Although long-chain alcohols, fatty acids, alkanes and wax esters were the major components of the soils studied these are believed to represent the natural background input from plants, fungi and microorganisms associated with the soil. The 5β-stanols and bile acids were chosen as biomarkers for use in this work due to their ubiquitous occurrence in mammalian feces and the fact that their structures can be used diagnostically to determine their original source with a high degree of certainty. Moreover, these compounds are not the normal products of the decay of plant or animal matter in the soil. Although 5β-stanols and bile acids occur only in very low concentrations in soils and sediments the use of combined GC/MS/SIM affords the high sensitivities and selectivities necessary to detect the target compounds in the majority of archaeological and experimental materials studied. The electron ionization (70 eV) mass spectra of the most commonly occurring animal and plant derived Δ^5-stenols and 5β-stanols are shown in Figure 3 and the ions used to detect characteristic marker compounds (stanols and stenols) are listed in Table I. Specific examples of the application of these techniques are discussed in the remainder of this paper.

An alternative approach that has been explored uses the new technique of tandem mass spectrometry, specifically GC/MS/MS using selected reaction monitoring (SRM). Figure 4 shows the product ion spectrum produced by the collision induced dissociation of the m/z 370 ion ($[M-TMSOH]^+$) of coprostanol in the collision cell of a triple stage quadrupole mass spectrometer. The decomposition, m/z 370 → 215, which arises via cleavage across the D-ring of the steroid nucleus, accompanied by a proton transfer, raises possibilities for the highly selective detection of 5β-stanols by GC/MS/MS employing SRM as shown schematically in Figure 5. Indeed in practice this new technique was found to have considerable potential for the trace analysis of 5β-stanols, allowing highly selective analyses with a significantly simplified sample preparation strategy to be adopted compared to that required when conventional GC/MS was employed. The hexane-soluble fraction of the total lipid extracts (chloroform/methanol, 2:1 v/v) were trimethylsilylated and analyzed directly by GC/MS/MS, using a triple stage quadrupole instrument, without further purification. Although the hexane soluble fraction comprises a complex mixture of cyclic and acyclic lipid species 5β-stanols (and 5α-stanols) were readily detected with high selectivity. The use of the method was demonstrated in the analysis of 5β-stanols in various archaeological and experimental agricultural soils. The method greatly simplifies

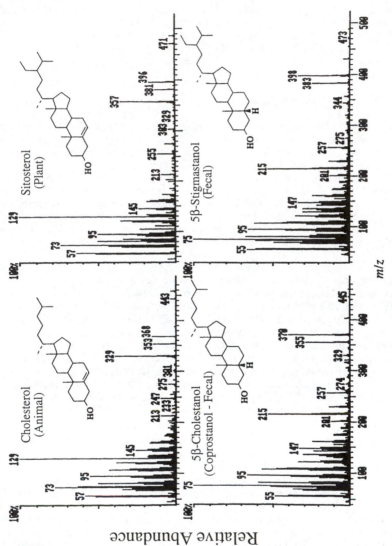

Figure 3. 70 eV electron ionization mass spectra of the trimethylsilane derivatives of Δ^5-stenols and the corresponding 5β-stanols.

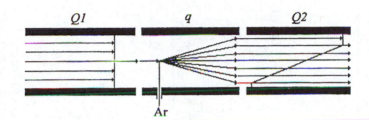

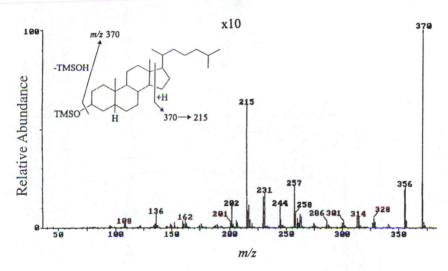

Figure 4. Unimolecular decomposition product ion spectrum obtained from the [M-TMSOH]+ ion of coprostanol.

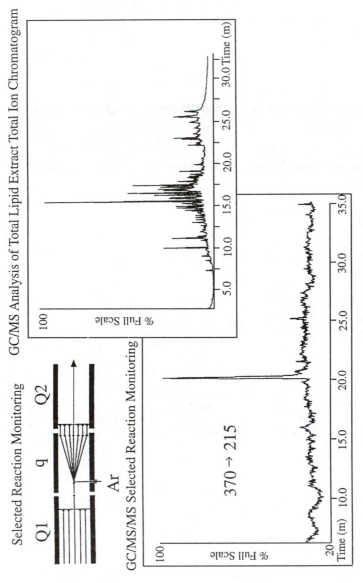

Figure 5. Chromatograms showing the high selectivity (lower left) that can be achieved in the analysis of 5β-stanols in complex lipid extracts (upper right) by means of tandem mass spectrometry. The inset shows schematically the mode of operation of the GC/MS/MS SRM technique.

analyses by increasing sample throughput by eliminating the need for time-consuming adsorption chromatographic steps, e.g., thin layer chromatography, prior to GC/MS.

Table I. Diagnostic Fragment Ions Used to Detect Target Stanols and Sterols by GC/MS with Selected Ion Monitoring

m/z	Parent Compound	Origins of Fragment Ions
129	Δ^5-Sterols	[M-side chain-216]$^+$
213	Δ^5-Sterols	[M-side chain-TMSOH (90)-42]$^+$
215	5α- and 5β-Stanols	[M-side chain-TMSOH (90)-42]$^+$
286	5β-Pregnanol (Internal Standard)	[M-TMSOH (90)]$^+$
368	Cholesterol (C$_{27}$)	[M-TMSOH (90)]$^+$
370	5α- and 5β-Cholestanols (C$_{27}$)	[M-TMSOH (90)]$^+$
382	Campesterol (C$_{28}$)	[M-TMSOH (90)]$^+$
384	5α- and 5β-Campestanols (C$_{28}$)	[M-TMSOH (90)]$^+$
396	Sitosterol (C$_{29}$)	[M-TMSOH (90)]$^+$
398	5α- and 5β-Stigmastanols (C$_{29}$)	[M-TMSOH (90)]$^+$

Molecular Markers of Human Activity

Coprostanol (5β-cholestan-3β-ol; **1**) is a metabolic product formed by microbial action in the mammalian gut (*19*; the usual product of cholesterol reduction outside the gut, in mammalian tissues and sediments is 5α-cholestan-3β-ol). Coprostanol is the major sterol in human feces, and has routinely been studied as a marker of (modern) sewage pollution in marine and lacustrine sediments (*20*). This has led to the search for coprostanol in archaeological soils, in order to detect the presence of fecal material.

As shown in Figure 2, the analyses proceeded via solvent extraction of the soil lipids followed by fractionation and analysis using combined GC/MS with selected ion monitoring (SIM) to detect and quantify specific compounds (Table I). Samples were analyzed from a range of sources, including modern latrine deposits, a 17th century garderobe, Late Saxon/early medieval garderobes and suspected Roman cesspits. Coprostanol and its homologs were detected not only in the modern aged cess samples, but also in the control samples, suggesting their ubiquitous occurrence in the environment, albeit at a low concentration. However, by measuring the relative abundances and ratios of the 5β-stanols, a chemical signature distinctive of fecal material was established, independent of the simple occurrence of coprostanol in the soil. It was found that coprostanol, and its homologs produced by the same microbial mechanism in the gut, were reliable markers of the presence of feces in soils when found in the appropriate relative abundances.

A method of analyzing very small quantities of specific marker compounds preserved in soils has thus been applied to archaeological materials, enabling a particular organic residue to be identified where conventional physical methods of analysis might not be successful. A full account of these analyses can be found in reference 17. An example of the type of data obtained from the GC/MS SIM analysis of the soil recovered from a Late Saxon/early medieval garderobe revealed during excavations at West Cotton, Northants, U.K, is shown in Figure 6. The pattern of 5β-stanols is consistent with that seen in fresh human stool, which is entirely consistent with the archaeological interpretation of the function of the excavated feature.

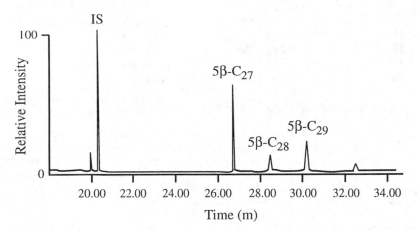

Figure 6. Partial GC/MS SIM (*m/z* 215) chromatograms of an extract of a soil horizon from a Late Saxon/Medieval garderobe from West Cotton, Northants, U.K. IS is an internal standard, 5β-pregnanol, which is added at the extraction stage (see Figure 2) for quantification purposes.

Molecular Markers of Manuring

The identification of evidence of manuring practices from archaeological sites has great potential. Such evidence could enable identification of site limits, field systems, development of early agriculture systems, settlement/farmland relationships, etc. The analysis of animal feces and manure for reliable, characteristic molecular biomarkers in the soil, and the subsequent measurement of the abundances of those biomarkers in the soil, formed the basis of this phase of our investigations. This approach was chosen as certain stable molecules are known to be more resistant to decomposition over long periods of time, and are more likely to be retained in the soil than other chemical evidence of manuring. Initial alterations in the soil organic matter content are not thought to survive long beyond the end of the manuring episode, and other criteria, such as phosphorus enhancement, are very dependent on soil conditions, and have rarely been subject to rigorous analysis in the past. The analyses performed here were carried out on modern experimental material from the Butser Hill Farm site in Hampshire, U.K., where a plot under wheat cultivation had been subject to manuring over half its area for 13 consecutive years; the abundances of various biomarkers (specifically sterols and 5β-stanols) were determined in the manured and non-manured areas. The analyses were again based upon extraction of the lipids from the soil samples followed by fractionation by thin layer chromatography and analysis by GC and GC/MS. A distinct enhancement of specific biological molecules characteristic of cattle feces or manure, predominantly 5β-stigmastanol (3) was found in the manured area when compared to the non-manured area as shown in Figure 7.

The method thus was found to have great potential for archaeological application. The results from the lipid work were also compared and contrasted with those from magnetic susceptibility measurements, total lipid and elemental (C, H and N) analyses, highlighting the advantages of the biomarker approach. A more detailed discussion of these results is given in reference 18. It is worth noting the near gaussian distribution observed for the 5β-stanols across the field which represents a departure from the theoretical distribution resulting from soil erosion and bioterbation effectively "blurring" the boundary between the manured and non-manured areas of the experimental plot.

Detection of Bile Acids in Archaeological Soils and Sediments

The bile acids found in higher animals are biosynthesized directly from cholesterol by saturation of the double bond, epimerization of the 3β-hydroxyl group, introduction of hydroxyls into the 7α and 12α positions, and oxidation of the side chain to a C-24 carboxylic acid. Bile acids act as detergents in the emulsification of dietary fats. Although the primary bile acids in mammals are cholic and chenodeoxycholic, they are transformed in the intestine by microorganisms into the secondary bile acids, lithocholic (4) and deoxycholic (5) acids. Of significance archaeologically is the fact that the formation of bile acids is the most important pathway for the metabolism and excretion of cholesterol in mammals. The rate of excretion of bile acids in human feces is of the order of 0.5 g/day. The analytical protocol that was adopted for the analysis of bile acids is summarized in Figure 2. Analyses were performed on the soil samples recovered from a modern latrine and the archaeological soils and sediments already discussed above in relation to the study of 5β-stanols. The results we obtained concur with those of earlier studies (11) and confirm that bile acids survive in archaeological soils. Figure 8 shows the total ion chromatogram obtained by GC/MS-SIM for the bile acid fraction of the total lipid extract of soil from a Late Saxon/early medieval garderobe.

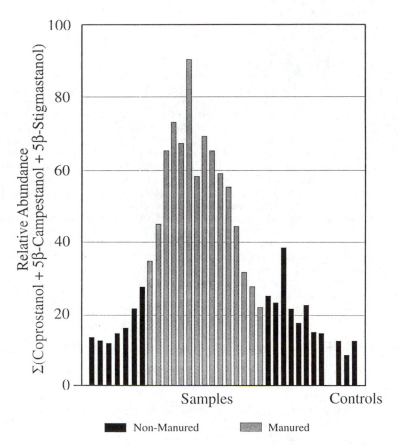

Figure 7. Histogram showing the relative abundance distribution for 5β-stanols in soils samples taken at meter intervals across the cultivated field at Butser Hill Experimental Iron Age Farm compared to control soils taken away from the cultivated area.

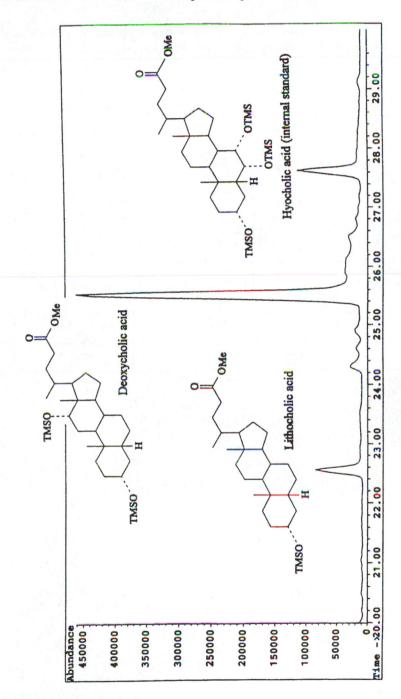

Figure 8. Total ion chromatogram obtained by GC/MS SIM for bile acids in soil taken from a suspected garderobe of Late Saxon/early medieval age.

Threshold 2, Contrast 3, Brightness 9, Halftone Pattern Spiral, Normal Detail 9/21/95 12:17 PM

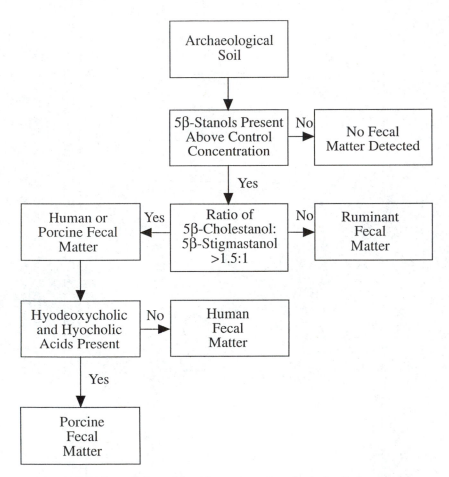

Figure 9. Differentiation of fecal inputs to archaeological soils by a multi-molecular approach based on the combined use of 5β-stanols and bile acids.

The relative proportions of the lithocholic and deoxycholic acid components is consistent with that seen in modern latrine soil. The hyocholic acid (**6**) component at longer retention time is an internal standard added for the purpose of quantification. The results obtained from the range of sample examined confirm the widespread survival of bile acids in archaeological soils and sediments and as such extends the possibilities for using steroid biomarkers as indicators of fecal inputs into archaeological soils. In common with our findings for 5β-stanols in the various control soils examined there are indications that these compounds also survive widely in the environment and as such represent a new class of indicators of fecal matter (pollution) in the ancient and modern environment alike. Measurement of the relative proportions of different sterols and bile acids in individual coprolites has been tentatively proposed as a means of determining the relative contribution of plant and animal foodstuffs to the diet (*7*). The composition of the fecal bile acids also vary with the health or disease state of individuals. However, substantial research is required before such principles can be applied with confidence. The complementary use of 5β-stanols and bile acids as multi-molecular markers in the study of disaggregated fecal matter is discussed further below.

Multi-Molecular Markers for Determining the Sources of Fecal Inputs to Archaeological Soils

It is clear from the results of the various analyses of contemporary reference and archaeological materials summarized above that possibilities exist for distinguishing between the sources of different fecal inputs into soils at archaeological sites on the basis of their steroid biomarker content. For example, a distinction can be drawn between fecal inputs from various animals, e.g., humans (omnivores) vs. ruminants (herbivores), due to the differences in the relative proportions of coprostanol (**1**) and 5β-stigmastan-3β-ol (**3**) present. In the case of humans the relative proportion of coprostanol:5β-stigmastan-3β-ol is approximately 5.5:1, while in the case of ruminants (sheep and cows) the proportions are reversed, i.e., the ratio of coprostanol:5β-stigmastan3β-ol is about 1:4.

However, drawing a distinction between the feces originating from two omnivores is less straightforward since the fecal 5β-stanols of both humans and pigs are dominated by coprostanol. However, a clear distinction can be drawn between human and porcine fecal matter by consideration of the bile acids they produce. As discussed above, lithocholic (**4**) and deoxycholic (**5**) acids are the dominant bile acids in the feces of healthy humans. Although lithocholic acid is present in porcine feces, the high abundance of two bile acids that are insignificant in healthy humans, namely, hyodeoxycholic (**6**) and hyocholic (structure shown in Figure 8) acids, enables a clear distinction to be drawn between the two. We have termed this method of distinguishing between the origins of fecal matter through the combined use of two different classes of steroid biomarkers as a multi-molecular approach and have summarized it schematically in Figure 9.

Acknowledgments

In presenting this paper we should like to acknowledge the contributions of all our collaborators and the experts that have contributed to the success of this research, In particular we should like to note the contributions of the late Dr. Jim Ottaway for his enthusiasm and encouragement in the early stages of this work and for provision of samples. Dr. Peter Reynolds is thanked for provision of samples and Drs. Gillian Campbell and Mark Robinson of Oxford Museum are thanked for their expert advice and assistance with sampling from the Raunds Area Project; Dr. Nick Walsh who is also thanked for assistance with inductively coupled plasma atomic emission

spectroscopy analyses of the soils from the Butser Experimental Iron Age Farm and
Mr. Mark C. Prescott for assistance with the GC/MS analyses.

Literature Cited

1. Shackley, M. *Using Environmental Archaeology*; Batsford: London, 1985.
2. Fry, G. F. In The Analysis of Prehistoric Diets; Gilbert, R. I.; Mielke, J. H., Eds.;
   Academic Press: London, 1985; pp 127-154.
3. Hillman, G. C., In *Lindow Man: The Body in the Bog;* Stead, I. M.; Bourke, J.
   B.; Brothwell, D., Eds.; British Museum: London, 1986; pp 99-115.
4. Heizer, R. F.; Napton, L. K. Science **1969**, *165,* 563-568.
5. Bryant, V. M.; Williams-Dean, G. *Scientific American* **1975**, *232 (1),* 100-109.
6. Wakefield, E. G.; Dellinger, S. C. *Annals of Internal Medicine* **1936**, *9,* 1412-
   1418.
7. Lin, D. S.; Conner, W. E.; Napton, L. K.; Heizer, R. F. *J. Lipid Res.* **1978**, *19,*
   215-221.
8. Napton, L. K.; Heizer, R. F. In *Archaeology and the Prehistoric Great Basin
   Subsistence Regime as Seen from the Lovelock Cave, Nevada;* Heizer, R. F.;
   Napton, L. K., Eds.; Contributions of the University of California Archaeological
   Research Facility No. l0; University of California at Berkeley Press: Berkeley,
   1970.
9. Fry, G. F. In *Miscellaneous Paper No. 23, University of Utah Anthropological
   Papers No. 99 (Miscellaneous Collected Papers 19–24);* University of Utah
   Press: Salt Lake City, 1978.
10. Wales, S.; Evans, J. In *Science and Archaeology – Glasgow, 1987: Proceedings
    of a Conference on the Application of Scientific Methods to Archaeology*;
    Slater, E. A.; Tate, J. A., Eds.; Oxbow: Oxford, 1987; pp 403-412.
11. Knights, B. A.; Dickson, C. A.; Dickson, J. H; Breeze, D. J. *J. Archaeol. Science*
    1983, *10,* 139-152.
12. Pepe, C.; Dizabo, P. *Revue d'Archaeometrie* **1990**, *14,* 23-28.
13. Pepe, C.; Dizabo, P.; Scribe, P.; Dagaux, J.; Fillaux, J.; Saliot, A. *Revue
    d'Archaeometrie* **1989**, *13,* 1-11.
14. Morgan, E. D.; Cornford, C.; Pollack, D. R. J.; Isaacson, P. *Science and
    Archaeology* **1973**, *10,* 9-10.
15. Dormar, J. F.; Beaudoin, A. B. *Geoarchaeology* **1989**, *6,* 85-98.
16. Bethell, P. H.; Evershed, R. P.; Goad, L. J. In *Prehistoric Human Bone:
    Archaeology at the Molecular Level*; Grupe, G.; Lambert, J. B., Eds.; Springer
    Verlag: Berlin, 1993; pp 229-255.
17. Bethell, P. H.; Goad, L. J.; Evershed, R. P.; Ottaway, J. *J. Archaeol. Science*
    1994, *21,* 619-632.
18. Bethell, P. H.; Evershed, R. P.; Reynolds, P. J.; Walsh, P. J. *J. Archaeol.
    Science* **1996**, *23,* in press.
19. Rosenfeld, R. S.; Fukushima, D. K.; Hellman, L.; Gallagher, T. F. *J. Biol. Chem.*
    1954, *211,* 301-311.
20. Readman, J. W.; Preston, M. R.; Mantoura, R. F. C. *Marine Pollution Bulletin*
    (Reports) **1986**, *17,* 298-308.

RECEIVED October 9, 1995

Chapter 14

Investigation of Fiber Mineralization Using Fourier Transform Infrared Microscopy

R. D. Gillard and S. M. Hardman[1]

Department of Chemistry, University of Wales, Cardiff, P.O. Box 912, Cardiff CF1 3TB, United Kingdom

The mineralization of cellulose and protein fibers has been simulated in the laboratory using oxygenated aqueous solutions. The mechanism peculiar to a given solution has been shown to depend on the initial metal-ligand bonds formed and on kinetic factors relating to the mineral product formation. These experiments rationalize both the occurrence and the comparative rarity of mineralized organic fibers from archaeological deposits. FTIR microscopy has revealed that traces of organic component can survive long-term burial and, in appropriate circumstances, permit their identification even in highly mineralized samples. Remnant dye on excavated textile has also been identified using FTIR microscopy on a scale far below that which is possible by extractive techniques. A spectral data library of fibers, dyes and mordants has been built into the computer analysis program.

The mechanisms of fiber decay and mineralization are poorly understood. The relative rarity of mineralized material limits extensive analysis. We have successfully reproduced mineralization in the laboratory for copper and copper alloys. Samples from these experiments have been analyzed and the results compared with those on archaeological material. FTIR microscopy was used extensively for this work. This paper presents results and discusses implications.

Background

Well-preserved textiles are rare except in certain environments, for instance peat bogs (anaerobic waterlogging), desert (virtual desiccation) and permafrost (extreme cold) (1, 2). Much information about textiles is obtained indirectly from sources such as tools of manufacture, impressions on pottery, paintings, drawings, ancient texts and mineralized fibers.

[1]Current address: School of History and Archaeology, University of Wales, Cardiff, P.O. Box 909, Cardiff CF1 3XU, United Kingdom

0097–6156/96/0625–0173$12.00/0

Definition. Mineralization is the combination and/or replacement of the organic matrix with an inorganic matrix. Mineralized textile fragments are normally small and fragile, generally occurring where the textile is in direct contact with a metal artifact, e.g., coins, knives or cloak pins. They are most frequently found with copper alloy and iron objects. Lead and silver corrosion products have also been reported to participate in the mineralization process (3-5). Early silver artifacts, however, often contain considerable quantities of copper. Where mineralization is observed on such objects, copper may instigate mineralization. Calcium carbonate has also been reported to preserve fiber information (6).

Types of Mineralization. Mineralized information occurs in two forms. In positive casts (which generally provide more information) metal ions penetrate the fiber and coordinate with the organic matrix. This mechanism provides sites for nucleation allowing further metal ions to "lock on". The corrosion products then gradually replace the fiber as it decays giving a positive three-dimensional fiber impression composed primarily, if not totally, of metal corrosion products. Such casts can provide information on the fiber type, yarn (e.g., the direction of spinning and evidence of plying) and weave pattern. Negative casts form when corrosion products deposit on the surface of the fiber which then decays, leaving behind a negative impression in the corrosion products.

The extent of mineralization can vary considerably even within the same piece of textile. Variables involved include: (1) The different chemical and physical properties of protein and cellulose fibers; (2) The type of metals/alloys present, their respective corrosion mechanisms and the corrosion products produced; (3) The possible use of dyes, mordants and metal-wrapped yarns; (4) The burial conditions, e.g., soil type, redox potential (Eh), pH and microbial activity.

"Mineralization" Versus "Fossilization". In the past, the term "fossilization" has been wrongly interchanged with "mineralization" to describe textile remains on metal artifacts (7). "Fossilization" a specific geological term, refers to organic material preserved in rock strata. In contrast "mineralization" is a much more general term encompassing all preservation of organic information by an inorganic mineral.

FTIR Microscopy

Infrared analysis was performed using a Nicolet 510 Fourier transform infrared (FTIR) spectrometer in conjunction with a 620 processor and Spectra-Tech FTIR research microscope.

FTIR microscopy enables spectra to be obtained from samples as small as 10μm x 10μm (i.e., a single crystal or fiber). Analysis can be performed using either transmitted or (if necessary) reflected radiation. Transmission microscopy is preferred whenever possible because the energy throughput is greater. By reflection from its surface, non-destructive analysis can be performed on an artifact when samples cannot be taken.

Sample preparation. All results herein were all collected in transmission mode. Each sample was run for 250 scans at a resolution of 4 cm^{-1}. Fibers were selected and prepared using a low power optical microscope (10X - 30X magnification). Using a clean scalpel and dissecting pin, a small sample was removed and placed on a microscope slide. To prevent diffraction of the IR beam, the sample was slightly flattened using a small roller directly applied to its surface, then placed on a small (13mm x 2mm) NaCl plate and transferred to the FTIR microscope stage. After focusing the fiber at 150X magnification, shutters above and below the field of the sample stage were used to select visually the area for analysis (generally an area of 40μm x 100μm was chosen for fibers). The shutters blocked off the IR beam to the rest of the sample and ensure that the spectrum produced is only that of the area of interest. This technique ("Redundant Aperturing and Targeting") also makes possible analysis of the least contaminated area of the sample.

Textile Chemistry

In order to understand fiber mineralization, the chemistry of the fibers themselves must be considered. Protein and cellulose fibers consist of crystalline and amorphous regions. The degree of crystallinity is important to the fibers survival: crystalline regions resist penetration and attack owing to their strong intermolecular bonds and tighter packing. This resistance reduces chemical accessibility and the number of reactive sites available.

Cellulose Fibers. The building block of cellulose comprises two anhydroglucose units joined *via* an ether link to form cellobiose (Figure 1). In the burial environment, cellulose fibers are sensitive to acidic conditions: hydrolysis occurs and the ether links are broken.

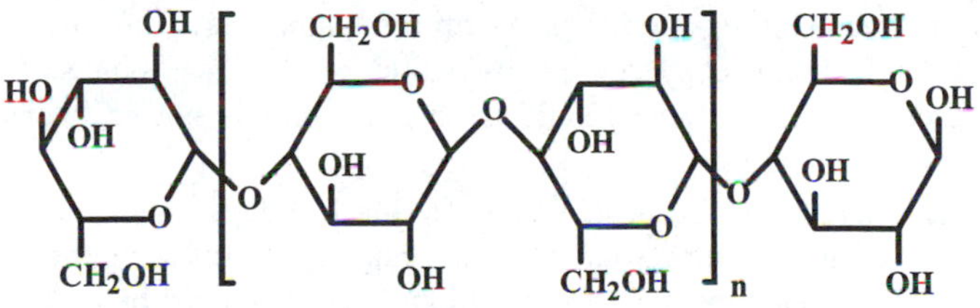

Figure 1. Cellulose consists of repeating units of cellobiose. Degree of polymerization = 2n+2.

Protein fibers. Protein fibers are much more complex than cellulose fibers. Each type is built up of a characteristic sequence of amino acids joined together by peptide linkages. The general structure is shown in Figure 2 where R_1, R_2 and R_3 define the amino acid radicals: -H for glycine, $-CH_3$ for alanine, $-CH_2C_6H_5$ for phenylalanine, etc..

$$\begin{array}{ccc} \overline{}CH.CO.NH.CH.CO.NH.CH.CO.NH\overline{} \\ | & | & | \\ R_1 & R_2 & R_3 \end{array}$$

Figure 2. The general structure of protein.

The polypeptide chains are held together by hydrogen bonds, ion-ion salt linkages between carboxyl and amide end groups and, in the case of wool, disulfide bonds. The degree of crystallinity and packing is determined by the amino acids present since bulky side groups influence the properties and morphology of the fiber (*8*). Thus, silk is a polypeptide of 15 amino acids arranged in a ß-sheet conformation, while the polymer of wool consists of about 19 amino acids (varying with wool type) arranged in an α-helical structure.

The chemistry of wool is dominated by the reactive disulfide bond in the cystine residue which predominates in the amorphous regions of the fiber. The sulfur-sulfur bond can form both within (intra-molecular) and between (inter-molecular) polypeptide chains. The inter-molecular disulfide "bridges" are particularly important as they help to hold the polypeptide chains together, providing some structural support. Wool fibers are particularly sensitive to alkaline environments; the peptide and salt links are hydrolyzed and the disulfide bond cleaves to form cysteine and cysteic acid. Protein fibers are less rapidly degraded in acid. Disulfide bonds resist acids above pH2 and although the fibers weaken, they maintain some structural integrity (*9*).

Experimental Methods and Results

Positive Casting.   The mechanisms of positive and negative casting are determined not only by the metal but also by the anions and cations in solution (*10*). For positive casting to occur the metal ions must initially penetrate the fiber and complex with it. This is readily observed when wool and a piece of sheet bronze are immersed in deionized water at room temperature: the wool becomes green and FTIR microscopy reveals oxidation of disulfide bonds. Silk also becomes green under the same conditions but its color is not so intense. Both solutions remain colorless.

If the experiment is repeated at 50 °C for 100 days, the same change in color is observed in the fabrics but at a faster rate. The solutions initially remain colorless but at 90 days the solution containing wool becomes blue. Reflectance electronic spectroscopy on the wool and silk samples shows that while the color intensity of the silk increases almost linearly with time, the color of the wool reaches a maximum after about 90 days (Figure 3). This observation is thought to represent a change in the reactions occurring as a result of the "saturation" of easily accessible sites in the amorphous regions of the fiber.

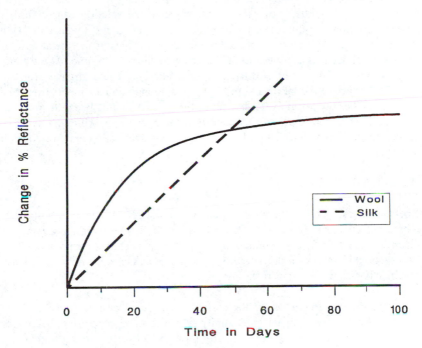

Figure 3. Plot showing rate of change in color intensity with time for silk and wool samples in a water/copper environment at 50 °C.

Negative Casting. Negative casting requires a build-up of metal ions at the surface of the fiber sufficient to exceed the solubility product of a mineral, permitting its deposition. This phenomenon can be reproduced at room temperature by placing a piece of sheet copper (metal or alloy) on top of a piece of textile (surface area of metal/textile = ~1:2) in a 1.5 M solution of NaCl. In contrast to a water/textile/metal environment, the textile does not become green. Instead, a fine blue-green mineral layer is rapidly deposited on the surface of the metal and textile. This deposit has been identified by FTIR microscopy as botallackite (ß-$Cu_2Cl(OH)_3$), consistent with mineralogical observations made elsewhere (*11*). Over 4 months, botallackite "recrystallizes" to atacamite (α-$Cu_2Cl(OH)_3$). Continued exposure of the fabric to the solution causes further "recrystallization" of this metastable mineral to paratacamite (γ-$Cu_2Cl(OH)_3$). Paratacamite is first formed on the fabric around the edges of the metal strip at about 6.5 months and continues to radiate over the fabric with time. At the same time, strands or "tubes" of corrosion product develop. These strands protrude from the surface of the fabric and receive internal longitudinal support from fine strands of fiber (i.e., negative casts). With time, these strands began to form all over the fiber but are most concentrated near the metal strip. Although negative casts are readily reproduced on all textile types, positive casting is more difficult to simulate.

Discussion

Protein Fibers. Wool is the most responsive natural fiber to the uptake of metal ions under the conditions given. This fact is not surprising given the number of reactive sites available in the protein helix. Silk is the next most responsive fiber. Like wool, it binds copper at its carboxyl and amide linkages, however, as it lacks disulfide bonds, it is less reactive.

Cellulose Fibers. In contrast to the protein fibers, cellulose fibers respond very slowly owing to their greater crystallinity which restricts access for the copper ions. Cellulose binds metal ions primarily through its hydroxyl groups, although the nature of the binding is not entirely understood (*12*). Further work is required to determine the conditions necessary to reproduce positive cellulose casts.

Metal Ion Uptake by Fibers

The uptake of metal cations can impart certain properties to textiles. Aspects studied include dye uptake (*13-16*), shrink proofing (*17,18*), wrinkling (*19*), and flame proofing (*20*). The chemical industry has also considered the use of wool to remove toxic metals (in particular, mercury) from contaminated water and industrial effluents (*21-23*).

Others have shown that metal ion uptake depends on many variables including the type of cation, counterion, concentration, temperature and pH. These variables affect not only the rate and amount of metal ion uptake but also the sites at which the copper will complex (*21, 24-30*). Unfortunately, many such results are for conditions that would not occur in the burial environment (e.g., extremes of pH and high temperatures). The test wool is also frequently pre-treated or modified: sometimes only certain chemical fractions or morphological regions are used or analyzed (*24, 31-33*). Consequently, extending such results to untreated wool requires circumspection. Due to the complexity of the α-helix, steric effects must also be considered when applying results from simpler polypeptides to whole wool.

Carboxyl/Copper(II) Complexes.   Despite these difficulties, a number of important conclusions have emerged. In particular, electron spin resonance studies have shown that at pH < 9, the bonding of copper to the wool matrix is predominantly due to green carboxyl/copper(II) complexes (*24, 27-29, 33*). Copper(II) complex formation at disulfide sites is minimal. Copper(II) ions catalyze the oxidation of the disulfide bonds but bind elsewhere, presumably through greater electrostatic attraction. This activity conforms well with the room temperature water/textile/metal experiments described herein.

Structural studies on smaller polypeptides with copper may further suggest information relating to the orientation of the metal ion in the wool helix. The smallest peptide producing these reactions is biuret, $NH_2CONHCONH_2$. In neutral or mildly acidic solutions containing chloride ions, a blue-green bis-biuretcopper(II) dichloride complex is produced. The biuret molecule acts as a bidentate chelating ligand *via* its amide oxygen atoms. The amide oxygen atoms have a square planar arrangement around the copper(II) atom, although the NH_2CONH residues are not co-planar (Figure 4). Electrostatic bonds are also formed between the Cu(II) center and axial chloride ions (*34*).

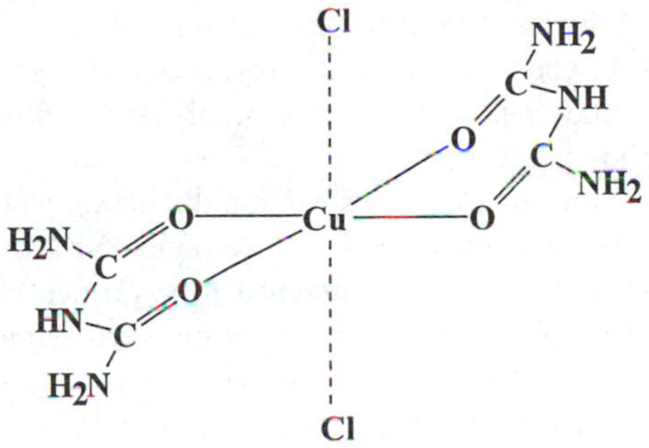

Figure 4. Bis-biuretcopper(II) dichloride.

Infrared Analysis of Copper Complexes.   The formation of copper-organic bonds could have been identified using IR analysis by the appearance of Cu-O, Cu-S or Cu-N absorptions. These occur in the region 500-250cm^{-1} (*35*). At present, this identification is not possible with FTIR microscopy as the lower frequency limit of most microscope mercury cadmium tellurium (MCT) detectors is 750cm^{-1}. This fact prevents the detection of most metal-ligand bonds, although current research is constantly improving this range. Currently, the newest detectors have a lower limit of 600-650cm^{-1}. Although other IR techniques do permit analysis at these low wavenumbers, difficulties with sample preparation, the destructive nature of the techniques and the relatively large sample required make them unsuitable. In contrast, FTIR microscopy enables two or three areas of the same fiber sample to be visually selected and analyzed.

In theory, Cu-O or Cu-N bonds might also be detected through the shifting of the amide I and amide II IR bands to lower frequencies, arising from the copper ion withdrawing electron density from the oxygen and nitrogen, reducing the strength of the C=O and N-H bonds. This shift was not observed in our experiments but was recorded by Jakes and Howard (36).

Mineralization in the Burial Environment. When a textile is buried in a moist environment, in close contact with a metal, initial reactions will occur at the surface of the fiber. The ability of the ions present to penetrate the surface will depend on a number of factors, some of which have been discussed above. Small scale microbial degradation can aid mineralization by opening up the fiber structure. Unfortunately, once microbial deterioration has commenced, it normally continues until the fiber is completely destroyed.

Biocidal metals such as copper help preserve the organic matrix of the fiber by reducing the rate of degradation (37). Thus, for a copper/textile system, the rate and extent of fiber degradation and the mechanisms of mineralization are largely determined by the copper ion concentration. This, in turn, is related to the uptake of the copper ions by the organic matrix. To get positive casting, the copper ion concentration must be high enough for biocidal action, yet sufficiently low to allow the ions to penetrate the fiber. If the concentration is too high, the solubility product of the copper minerals is exceeded before the metal ions can penetrate the fiber. Corrosion products deposit on the fiber surface forming a negative cast. Physical, chemical and microbial action degrade the fiber within this cast, producing the hollow tubes commonly seen in scanning electron microscope (SEM) photographs of mineralized fibers.

High concentrations of metal ions can rapidly build up at the fiber surface, particularly in burial environments which induce rapid corrosion of an associated metal artifact (e.g., a metal cloak pin attached to a garment). The presence of other ions, in particular bulky anions, can hinder the approach and penetration of the metal ions into the fiber. Initial reactions at accessible sites on the surface will impede reactions inside the fiber. As a result, the concentration of ions within the fiber may differ from that surrounding the fiber. Consequently, reactions at the surface may differ from those occurring within the fiber. This differential has been demonstrated when textiles are mordanted with copper, but it does not occur with chromium (38).

Negative Cast Formation. Negative casting is often observed on iron artifacts, while positive casts are found predominantly on copper alloys. Iron, unlike copper, possesses few biocidal properties and therefore cannot protect the fiber from microbial degradation. Consequently, unless the iron ions are able to penetrate and replace the organic matrix rapidly, positive casting is unlikely, particularly in a highly acidic or alkaline burial environment. The relatively rapid rate of iron corrosion, coupled with the insolubility of iron(III) minerals, further restricts positive cast formation. The minerals crystallize on the fiber surface before the metal ions can penetrate throughout the fiber. This has been shown by electron dispersive X-ray analysis (EDXA) of ancient fibers (*39*). In contrast to copper ions which were found distributed throughout the fiber, iron ions had penetrated only the outer surface.

Positive Cast Formation. The process of positive casting is much more complex than negative cast formation and involves two stages. During the first, metal cations (and some anions) penetrate and complex with easily accessible sites in the fiber. This process continues until all available sites are filled. Gradually, ion concentrations in the internal solution will increase until solubility products are exceeded. The second stage of mineralization involves a kinetically controlled deposition of minerals at these sites within the fiber. The rate of this reaction is clearly slow as shown by the build up of visible concentrations of copper(II) ions (through the formation of aquated $Cu(II)$ ions) at 90 days in the bulk solution of the water/wool/metal experiments at 50 °C.

Comparative Mineralization Studies

The mechanisms of mineralization (particularly fossilization) have received considerable attention in geological studies. The most recent interest concerns the processes involved in the mineralization of soft-bodied biotas (*40-43*). The soft parts of an animal, subject to microbial attack, are rapidly lost on burial. In exceptional circumstances, however, they may be preserved in remarkable detail. The results of those studies correlate well with the parameters presented here for fiber mineralization, i.e., (1) Initial preservation and early diagenetic mineral formation is essential for the maximum amount of information to be preserved; (2) anaerobic environments alone are not sufficient for soft part preservation: microbial attack will still occur albeit more slowly than for aerobic conditions; (3) there is a fine balance between release, reaction and precipitation of the mineral and the decay of the organic material.

Crystal Engineering. The "crystal engineering" involved in producing an inorganic positive cast (rather than a mere mass of corrosion product) from an organic matrix must also be considered. This area is of interest in the chemical industry for the design of substrates to generate crystals with precise properties for technological applications such as catalysis and magnetic devices (*44*). Clearly, to form such detailed casts, the organic matrix must be active at the molecular level in controlling crystal nucleation, which is largely achieved through the bonding of metal ions at specific sites in the organic matrix. If the spacing of these metal sites conforms with the space group of the mineral's unit cell, then controlled mineralization replicates or replaces the organic template. Conversely, if the spacing of metal sites does not match the unit cell, then as the mineral's solubility product is exceeded, stable nucleation is achieved through distortion and destruction of the organic template inhibiting positive casting.

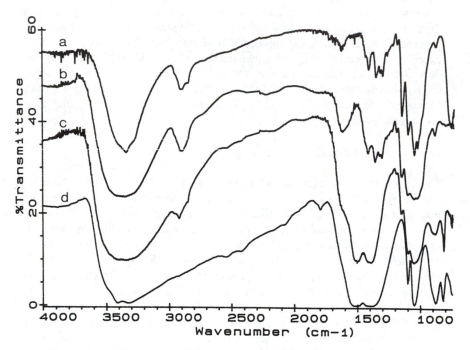

Figure 5. FTIR spectra of (a) an untreated linen fiber, (b-c) two areas from a highly mineralized fiber from a medieval site and (d) malachite, $Cu_2CO_3(OH)_2$.

This inhibition of positive casting is seen in the uptake of gold by wool in aqueous solution. Initially, gold binds with the wool, turning it a bright golden yellow. With time and an increase in metal ion concentration, the wool becomes brittle, then shrinks and collapses, while metallic gold deposits on the wool and in the liquid medium (*23*). In contrast, since copper readily forms accurate positive casts, stabilization of the nuclei must easily be achieved.

Analysis of Archaeological Samples

The results obtained in the laboratory and the conclusions drawn from other fields of study correlate well with results obtained from archaeological samples. Paratacamite and malachite ($Cu_2CO_3(OH)_2$) have been consistently identified using FTIR microscopy on mineralized archaeological fibers.

Despite the processes of mineralization, FTIR microscopy of archaeological samples has often shown measurable quantities of organic material even after long term exposure to ground water. Identification of protein and cellulose fibers has proved possible - even in samples that would have been described previously as totally mineralized, indicating that total replacement is rarely achieved and true pseudomorphs are uncommon. Figure 5 shows FTIR spectra obtained from a highly mineralized fiber from a medieval site. The spectra clearly show the presence both of a cellulose component and malachite.

Given the survival of enough organic matrix for identification by IR, it should be possible to distinguish silk from wool by virtue of their different structures (i.e., β-pleated sheet and α-helix) using polarized IR radiation (*45*). This technique, however, was not available to us at the time of this work.

Dye Detection. Since beginning this work, we have used FTIR microscopy to analyze a wide range of materials including glass, waxes, resins and pigments. In particular, we have shown that the technique can be used to detect dyes on archaeological textiles (avoiding the time-consuming and destructive extraction procedures necessary for the spectrophotometric and chromatographic methods currently used). Figure 6 shows the spectrum obtained from a thread taken from a man's tunic dated 16th century A.D. Although the fiber itself had degraded badly, the presence of indigo is clearly shown. A spectral library of dyes and dyed fibers has been built into the computer analysis program to aid this work and is discussed elsewhere (*46,47*).

Conclusions

Positive cast formation occurs in two stages: (1) Metal ions enter the fiber and bond at certain positions in the organic matrix to form metal-organic complexes; (2) when all available sites are filled, controlled mineral deposition occurs, beginning at the sites occupied by metal ions. In contrast, the formation of negative casts relies upon higher concentrations of metal ions sufficient to exceed the solubility product of the mineral. This process is therefore less controlled than positive cast formation.

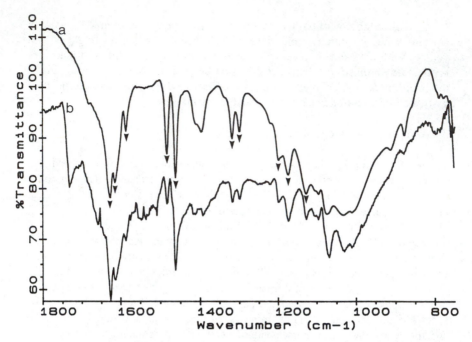

Figure 6. FTIR spectra of a fiber from a man's tunic(b) and an indigo
standard (a), clearly showing the presence of the dye in the fiber.

FTIR microscopy is a valuable analytical tool for archaeological research.
Minimal sample size and preparation make it ideal for the analysis of small,
friable mineralized fibers. The technique has been used to show that mineralized
fibers are rarely totally mineralized and retain sufficient organic component to
permit fiber identification. Its potential for the detection of dyes has also been
shown.

Acknowledgments
We wish to thank the Conservation Unit of the Museums and Galleries
Commission for their support.

References

1. Barber, E. J. W. *Prehistoric Textiles;* Princeton University Press: Princeton,
   NJ,
   1991; pp 126-144.
2. Keller, F. *Lake Dwellings of Switzerland and Other Parts of Europe;*
   Longmans: London, 1866.
3. Harte, N. B.; Ponting, K. G. *Cloth and Clothing in Medieval Europe*;
   Heinemann Educ.: London, 1983.

4. Edwards, G. M. *The Preservation of Textiles in Archaeological Contexts;* University of London Institute of Archaeology, unpublished diploma dissertation, 1974.

5. Stock, S. *An Introduction to the Identification of Animal and Vegetable Fibers Used in Antiquity,* University of London Institute of Archaeology, unpublished dissertation, 1976.

6. Walton, P. *Textiles, Cordage and Raw Fiber from 16-22 Coppergate;* York Archaeological Trust, Council for British Archaeology: London, 1989; Vol. 17: The Small Finds.

7. Carrola, D. L. *American Journal of Archaeology* **1973**, *77*, 334-336.

8. Feller, R. L.; Lee, S. B.; Bogard, J. In *Historic Textiles and Paper Materials Conservation and Characterization*; Needles, H. L.; Zeronian, S. H., Eds.; Advances in Chemistry Series 212; American Chemical Society: Washington, DC, 1986, pp 229-247.

9. Alexander, P.; Hudson, R. F.; Earland, C. *Wool: Its Chemistry and Physics*; Chapman and Hall: London, 1963.

10. Gillard, R. D.; Hardman, S. M.; Thomas, R. G.; Watkinson, D. E. *Studies in Conservation* **1994**, *39*, 132-140.

11. Pollard, A. M.; Thomas, R. G.; Williams, P. A. *Mineralogical Magazine* **1989**, *53*, 557-563.

12. Rattee, I. D. *Textile Research Journal* **1974**, *63*, 728-730.

13. Hartley, F. R. *Australian Journal of Chemistry* **1968**, *21*, 1013-1022.

14. Hartley, F. R. *Australian Journal of Chemistry* **1968**, *21*, 2277-2286.

15. Hartley, F. R. *Australian Journal of Chemistry* **1968**, *21*, 2723-2735.

16. Aspland, J. R. *Textile Chemist and Colorist* **1993**, *25(3)*, 55-59.

17. Barr, T.; Speakman, J. B. *Journal of the Society of Dyers and Colorists* **1944**, *60*, 335-340.

18. McPhee, J. R. *Textile Research Journal* **1965**, *35*, 382-384.

19. Leeder, J. D. *Textile Chemist and Colorist* **1971**, *3*, 193-195.

20. Bensiek, L. *Journal of the Textile Institute* **1974**, *65*, 102-108.

21. Brady, P. R.; Freeland, G. N.; Hine, R. J.; Hoskinson, R. M. *Textile Research Journal* **1974**, *44*, 733-735.

22. Friedman, M.; Harrison, C. S.; Ward, W. H.; Lundgren, H. P. *Journal of Applied Polymer Science* **1973**, *17*, 377-390.

23. Masri, M. S.; Reuter, F. W.; Friedman, M. *Textile Research Journal* **1974**, *44*, 298-300.

24. Guthrie, R. E.; Laurie, S. H. *Australian Journal of Chemistry* **1968**, *21*, 2437-2443.

25. Fukatsu, K.; Mariko, I. *Textile Research Journal* **1986**, *56*, 774-775.

26. Fukatsu, K. *Textile Research Journal* **1988**, *58*, 91-96.

27. Kokot, S.; Feughelman, M.; Golding, R. M. *Textile Research Journal* **1972**, *42*, 704-708.

28. Kokot, S.; Feughelman, M.; Golding, R. M. *Textile Research Journal* **1973**, *43*, 146-153.

29. Kokot, S.; Feughelman, M.; Golding, R. M. *Textile Research Journal* **1974**, *44*, 523-527.
30. Kokot, S. *Textile Research Journal* **1993**, *63*, 159-161.
31. Hinton, E. H., Jr. *Textile Research Journal* **1974**, *44*, 233-292.
32. Kadokura, S.; Miyamoto, T.; Ito, H.; Inagaki, H. *Polymer Journal* **1982**, *14*, 121-126.
33. Kulkarni, V. G. *Textile Research Journal* **1974**, *44*, 724-725.
34. Freeman, H. C. *Advances in Protein Chemistry* **1967**, *22*, 357-424.
34. Nakamoto, K. *Infrared and Raman Spectra of Inorganic and Coordination Compounds*, 4th ed.; Wiley-Interscience: New York, NY, 1986.
35. Jakes, K. A.; Howard, J. H. In *Proceedings of the 24th International Symposium on Archaeometry;* Olin, J.; Blackman, J., Eds.; Smithsonian Institute: Washington, DC, 1986; pp 165-177.
36. Brown, H. J. M. *Environmental Chemistry of the Elements*; Academic Press: London, 1979.
37. Carr, C. M.; Evans, J. C.; Roberts, M. W. *Textile Research Journal* **1987**, *57*, 109-113.
38. Jakes, K. A.; Angel, A. In *Archaeological Chemistry IV*; Allen, R. O., Ed.; Advances in Chemistry Series 220, American Chemical Society: Washington, DC, 1989; pp 451-463.
39. Allison, P. A. *Paleobiology* **1988**, *14*, 139-154.
40. Allison, P. A. *Paleobiology* **1988**, *14*, 331-344.
41. Allison, P. A.; Briggs, D. E. G. In *Taphonomy: Releasing the Data Locked in the Fossil Record*; Allison, P. A.; Briggs, D. E. G., Eds.; Topics in Geobiology 9; Plenum Press: New York, 1991; pp 25-70.
42. Briggs, D. E. G.; Kear, A. J. *Science* **1993**, *259*, 1439-1442.
43. Mann, S.; Heywood, B. R. *Chemistry in Britain* **1989**, *25*, 698-700; 712.
44. Campbell, I. D.; Dwek, R. A. *Biological Spectroscopy;* Benjamin/Cummings: London, UK, 1984; pp 52-55.
45. Gillard, R. D.; Hardman, S. M.; Thomas, R. G.; Watkinson, D. E. *Studies in Conservation* **1994**, *39*, 187-192.
46. Hardman, S. M. *The Mineralization of fibers in Archaeological Environments*; Ph.D. thesis, University of Wales College of Cardiff, 1994; pp 117-137.

RECEIVED October 31, 1995

Chapter 15

X-ray Diffractometric Analyses of Microstructure of Mineralized Plant Fibers

H. L. Chen[1], D. W. Foreman[2], and Kathryn A. Jakes[1]

[1]Department of Textiles and Clothing and [2]Department of Dentistry, Ohio State University, 1787 Neil Avenue, Columbus, OH 43210

Contemporary bast fibers and archaeological mineralized bast fibers were examined by two X-ray diffractometric techniques in order to discern their physical microstructures. Knowledge of this microstructure can lead to an understanding of the mechanism of organic polymer degradation and replacement with inorganic minerals. In comparison to the modern fibers, or sizes of the mineralized fibers sizes based on 101 and $10\bar{1}$ reflections (σ_{101} and $\sigma_{10\bar{1}}$), are larger while that based on the 002 reflection (σ_{002}) is smaller. The unit cell dimensions of the modern and the archaeological fibers are similar and the percent crystalline components of the mineralize samples are significantly lower than the modern ones.

Bast fibers such as flax, hemp, or nettle have been used in the production of textiles, mats, cordage, or baskets for thousands of years (*1*). For fibers to have survived over millennia, they must have been exposed to particular conditions such as extreme dryness, extreme cold, or, as in the case of mineralized fibers, to a moist environment in association with a particular metal for example (*2-6*). Several archaeological textiles, either partially or completely mineralized, have been recovered from burial sites worldwide. These mineralized fibers provide valuable evidence of the textile materials and technology used by ancient people and also provide an opportunity to study the physical and chemical microstructures of ancient fibers. Exploration of these chemical and physical microstructures can lead to an understanding of the changes which occur in fibers in response to specific environmental conditions. The chemical and physical composition of mineralized fibers have been studied only in a few cases (*7, 8*), and their microstructures, that is, the crystallite size, unit cell dimensions, and crystallite orientation, have not been studied prior to the work reported herein.

Study of archaeological mineralized textiles has primarily been limited to macroscopic and microscopic examination of their morphological shapes and the evidence they give for fiber, yarn, and fabric structures (9). In the literature, the term "mineralization" refers to the gross macromolecular replacement of the physical shapes of fibers with minerals, for example, the formation of a negative impression outside the fiber surface or of a positive cast within the fiber (10). In this study, however, the term "mineralization" refers to a phenomenon that occurs at the molecular level, within the fiber microstructure, in which either the atoms of the fiber molecules are replaced with copper atoms, or the spaces between molecular chains are infilled with copper moieties.

Mineralized fibers are formed in the vicinity of metal objects in a moist environment. Jakes and Howard (7) propose that by breaking associative forces in the crystalline regions and decreasing polymer chain length, fiber degradation will aid the penetration into the fiber by ions released in the corrosion of metal nearby. Polymer decomposition allows molecular movement, thus opening the fiber structure, and results in increased absorption. The presence of metal ions within the fiber will inhibit microbiological degradation (11) but other chemical reactions influence further processes of mineralization. As a result of mineralization, the organic components of fibers are assumed to be partially or completely replaced by minerals.

All fibers possess varying amounts of order-disorder in the arrangement of their constituent polymers, ranging from crystalline to amorphous orders. The highly crystalline regions of the cellulosic fiber produce diffraction patterns characteristic of single crystals, which can be detected by the X-ray diffraction (XRD) technique. The monoclinic unit cell crystal system, with three unequal axes ($a \neq b \neq c$) and monoclinic angle β which is not at a right angle ($\alpha = \gamma = 90° \neq \beta$), has been postulated for the native cellulose-I structure (13). The 1,4-β-D- anhydroglucopyranose repeat unit in cellulose is approximately 10.3 Å (13) as measured by X-ray diffraction, and each unit cell contains two cellobiose repeating units (12).

The X-ray method has been used widely in the study of degraded cellulosic materials (10-17). In the case of historical flax fibers recovered from a marine environment (18), it has been suggested that the increase in crystallite size in these materials is due to the rearrangement of associated chains in the amorphous regions and the alignment of the new chains with the crystallites after the absorption of water molecules in the amorphous areas. The same concept of increase in crystallite size due to moisture absorption concomitant with fiber degradation has also been proposed in other studies (*15*, *19*). The percent crystalline component calculated for the same marine flax fibers, however, is less than that of modern flax (*18*), indicating a breakdown in the polymer chain length and a disruption in the overall order in the fiber's internal molecular structure. This increased randomness in polymer chains also results in a reduced orientation of crystallites along the fiber axis of aged cotton fibers (*16*).

Although some compositional studies have been conducted of mineralized fibers (*7*, *8*), no study has been conducted in which the crystallite size and orientation, percent crystalline composition, and other unit cell parameters in archaeological fibers have been determined. These microstructural features can be used to elucidate the processes of mineralization. The objective of the study reported herein is to employ two different X-ray diffractometers to discern the physical microstructures of archaeological mineralized fibers recovered from a particular burial environment and of a contemporary bast fiber employed as a basis for comparison.

Experimental Materials.

Indian hemp (*Apocynum cannabinum*), a typical example of plant fibers used by prehistoric people of Eastern North America, was employed as a contemporary example of bast fiber for comparative purposes. This fiber was chosen due to its microscopic morphological resemblance to those of archaeological mineralized bast fibers used in this study. Plants were collected upon their maturation in summer 1992 in Ohio. The plant stems were hammered to separate the bark from the stem; then the fiber strands were peeled by hand from the bark. The fibers were air dried, and stored in desiccators.

Materials recovered from the James River Site, in James City County, Virginia (ca. 1600 AD) that had been studied (*20*) in the Archaeological and Historic Textiles Materials Laboratory at The Ohio State University included some small mineralized cordage specimens that were subsequently employed in this research. Four mineralized specimens, labelled OSU-16 (from soil block 1F55), OSU-19 (from soil block 1F56), OSU-20 (from soil block 12C04), and OSU-21 (from soil block 12C04), were initially selected. The four specimens weighed 1.0 mg, 4.1 mg, 4.2 mg, and 1.8 mg, respectively. The minute masses of the mineralized samples available gave rise to poor spectral resolution, which reduced the reliability of the data derived,

especially from the samples OSU-16 and OSU-21. Thus discussion of these
two specimens is not included in this manuscript. The soil blocks contained
necklace remnants consisting of decorative arrangements of copper tube
beads threaded with plant fiber cordage. After the blocks were unearthed,
the copper beads were no longer continuously connected with cordage, and
some of the cordage was partially exposed, extending from the inside of the
copper beads. The OSU-19 cordage sample was removed from inside a
rolled flat copper bead and the OSU-20 sample was taken from an eroded
bead in which the sample was partially exposed outside of the copper bead.
The cordage was pulled from the copper beads and no further treatment of
the samples was conducted prior to the instrumental analyses.

X-ray Diffractometry

Two different x-ray diffractometers were used in this study: (a) a powder
diffractometer consisting of a Philips Electronics PW 1316/90 wide range
goniometer with an XRG 3100 X-ray generator, a DMS-41 measuring
system, a theta compensating slit, a graphite monochrometer, and a copper
target operated at 35 kV and 20 mA to obtain the spectral data pairs (2θ and
intensity counts), and (b) a Rigaku RU 300H single crystal diffractometer
with a two dimensional detector, nickel-filtered copper K_α radiation operated
under conditions of 40 kV and 100 Ma to generate an X-ray photographic
image.

X-ray diffractograms of the modern and archaeological mineralized samples
were obtained from the powder diffractometer for the measurement of
crystallite size and percent crystalline component. X-ray photographic films
were obtained from the single crystal diffractometer for the measurement of
crystallite orientation within the fibers.

Data Analysis

The data pairs (spectral angles and intensities) of the individual peak profiles
(Bragg reflection) from these diffractograms were fitted to a Gaussian
distribution curve to obtain the line breadth defined as the full width at half
maximum (FWHM) intensity. Details of the calculation of FWHM and the
derivation of equations for calculating crystallite size, unit cell parameters,
and percent crystalline component are described in (*21*). Due to the
proximity of the 101 and 10$\bar{1}$ reflections in the OSU-19 and OSU-20
diffractograms, data points were collected for the overlapping reflections by
sampling one side of the peak profile and assuming a mirror image profile
for the other side for the crystallite size measurements.

The percentage of malachite component in each of the mineralized samples was calculated from a second power polynomial working curve constructed from a series of mixtures of known percentages of malachite and cellulose. The crystallite orientations were measured directly from the horizontal length of the registered arc in the X-ray films, and the data were then converted to angles in degrees. The crystallite orientation is determined from the horizontal length of the arc; if the crystallites are randomly oriented, the diffraction pattern registers as continuous circle, if the crystallites are preferentially oriented in similar angles with respect to the long fiber axis, the length of diffraction arc narrows. To evaluate the orientation angle of the crystallites from the photographic films, an image analyzer was employed to determine the 0° position at the center of the arc (at 100% intensity) and the end point of the arc (at 2% intensity). The angle between these two points is the crystallite orientation angle.

Results and Discussion

Crystallite Size. The X-ray diffractograms of Indian hemp and the OSU-19 and OSU-20 samples are shown in Figures 1, 2, and 3. The five crystallite reflections observed in the fiber spectra in order of increasing °2θ are: 101, $10\bar{1}$, 002, 040, and 004. In the spectra of each of the mineralized fibers, the peaks of the 101 and $10\bar{1}$ reflections show extensive overlap and the reflections of 040 and 004 are difficult to resolve from the amorphous and background crests.

There are three additional peaks observed in the spectrum of OSU-19 (at 2θ = 14.71°, 26.31°, and 31.06°) and one in OSU-20 (2θ = 26.5°) which are not due to cellulose crystallites. The 2θ and d-values (Å) of these extraneous reflections match those of standard malachite (*22*).

Table I lists the crystallite sizes σ_{101}, $\sigma_{10\bar{1}}$, and σ_{002} reflections for Indian hemp and these two mineralized fibers. The σ_{002} of Indian hemp (65 Å) is much smaller than that reported for flax fibers (170 Å) (*21*).

Table I. Crystallite Sizes σ (Å)

hkl	Indian Hemp	OSU-19	OSU-20
101	46±3	124±1	90±1
$10\bar{1}$	88±1	128±2	87±1
$10\bar{1}$	65±2	45±2	58±1

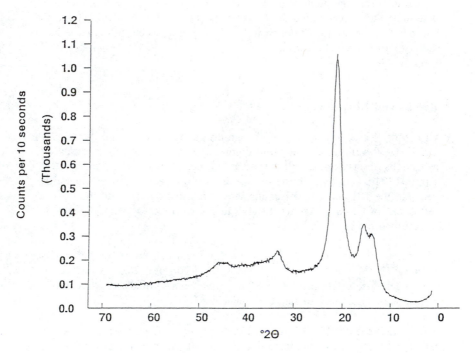

Figure 1. X-ray diffractogram of Indian hemp fiber.
101 reflection=14.86 °2θ, 10$\bar{1}$ reflection=16.41 °2θ,
and 002 reflection=22.56 °2θ.

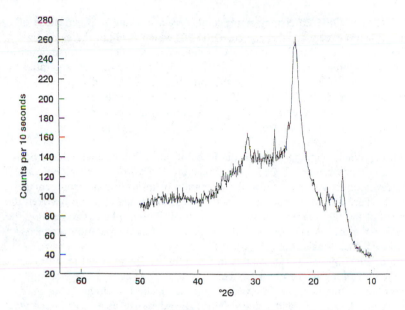

Figure 2. X-ray diffractogram of OSU-19 sample.
101 reflection=16.41 °2θ, 10$\bar{1}$ reflection=17.41 °2θ,
and 002 reflection=22.51 °2θ.

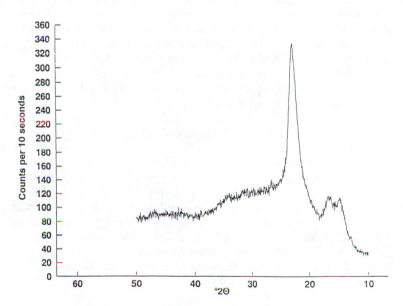

Figure 3. X-ray diffractogram of OSU-20 sample.
101 reflection=14.80 °2θ, 10$\bar{1}$ reflection=16.45 °2θ,
and 002 reflection=22.60 °2θ.

OSU-19 and OSU-20 display significantly larger σ_{101} and $\sigma_{10\bar{1}}$ than those of Indian hemp except for the $\sigma_{10\bar{1}}$ for OSU-20, which is about the same size. The values of σ_{002} for the mineralized samples are smaller in comparison with that of the Indian hemp. The 002 reflection in the fiber crystal is parallel to the b-axis and the fiber surface or the long fiber axis. The decrease in crystallite size derived from the 002 reflections indicates a decrease in the c-axial direction of the crystal. The increase in σ_{101} and $\sigma_{10\bar{1}}$ indicates an increase in the crystal dimensions along the a-axial direction.

Van der Waals bonding between the faces of the cellulose chains have been noted in many reports (*12, 23, 24*). Since all the hydroxyl groups on each of the repeat units are involved in the formation of intra- and intermolecular bonds, there are no primary or secondary valence bonds possible between the layers of the 002 planes. Hence, the layers of the 002 planes are held by the weaker Van der Waals forces, and it is likely that as degradation proceeds, these Van der Waals forces between the layers of 1,4-β-D-anhydroglucopyranose repeat units in the cellulose fibers could readily be broken, resulting in the observed decrease in σ_{002}. Other factors that might contribute to the decrease in σ_{002} during mineralization include a change in crystallite distribution and induced lattice disorder such as a large size crystal defect, each of which could also result from degradation of the fiber.

The increase in the crystal size in the a-axial direction might be due to the fact that in the rearrangement of polymer chains in the amorphous regions resulting from the absorption of water molecules, some less ordered regions near the edges of this a-axial direction of the crystallite could align with the original crystalline areas, thus contributing to the increase in the overall crystallite size in this particular direction. The same concept of increase in crystallite size due to moisture absorption and the rearrangement of molecular chains in the less ordered regions concomitant with fiber degradation has also been proposed by other studies (*15, 19*).

The differences in crystallite sizes between OSU-19 and OSU-20 can be due to the fact that the samples came from different soil blocks and were exposed to different environmental conditions. As we mentioned previously, the OSU-19 sample was removed from inside a copper bead and the OSU-20 sample was taken from an eroded bead in which the sample was partially exposed outside of the copper bead. During the course of mineralization, it is likely that OSU-19 was in contact with a higher copper ion concentration than the OSU-20 sample because the former was "sealed" inside the copper bead while the latter was exposed to other environmental conditions in addition to the copper. As a result, it can be expected that the physical microstructure of OSU-19 would differ from that of the OSU-20 sample and each would differ from the microstructures of the modern sample to differing degrees. This conclusion is confirmed by the crystallite size measurements; the OSU-20 fiber shows a smaller increase in σ_{101} and $\sigma_{10\bar{1}}$ and a smaller decrease in σ_{002} than the same parameters of OSU-19 relative to the modern bast fiber example.

Unit Cell Dimensions. The d spacings calculated based on the 101 (d_{101}) and $10\bar{1}$ ($d_{10\bar{1}}$) of the two mineralized samples are slightly larger than that of Indian hemp, while the d_{002} of these three fibers are similar in magnitude (Table II). The other unit cell dimension measurements are summarized in Table III. The larger unit cell in the *a*-axial dimension in these mineralized fibers is consistent with the finding derived from the crystallite sizes that the crystallite sizes in the a-axial direction (101 and $10\bar{1}$) are also larger in the mineralized samples. Hence, this larger a-axial unit cell dimension may also contribute to the larger crystallite sizes in the same direction. There is no significant difference in unit cell dimensions between OSU-19 and OSU-20.

Table II. Interplanar Spacing (Å)

hkl	Indian Hemp	OSU-19	OSU-20
101	5.91±0.5	6.04±0.7	5.98±1.1
$10\bar{1}$	5.38±1.0	5.41±0.8	5.33±1.0
002	3.95±0.8	3.95±0.5	3.92±0.2

Table III. Unit Cell Dimensions (Å) and Monoclinic Angle β

	Indian Hemp	OSU-19	OSU-20
a-axis	8.04±0.05	8.28±0.10	8.12±0.09
b-axis	10.46±0.08	10.51±0.07	10.47±0.04
c-axis	7.94±0.02	7.94±0.01	7.91±0.02
β	84.68±0.50	83.67±0.90	83.46±0.50

Percent Crystalline Components. The percent crystalline components of cellulose and malachite are reported in Table IV. The Indian hemp cellulose crystalline component is comparable to that reported for flax (78%) (*18*). It was expected that the mineralized fibers would display a lower cellulosic crystalline component than that of Indian hemp. A similar decrease in

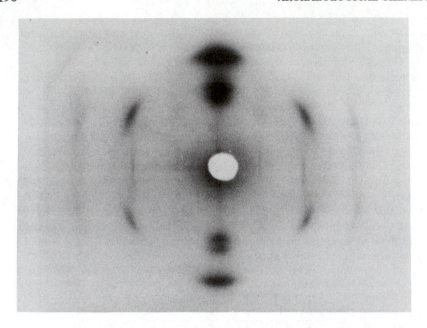

Figure 4. X-ray photographic image of Indian hemp fiber.

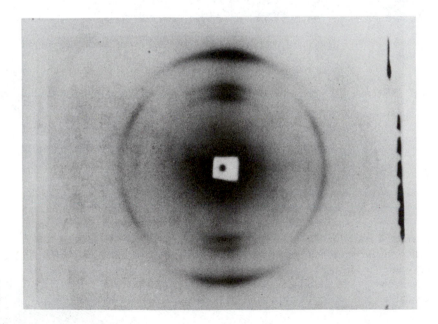

Figure 5. X-ray photographic image of OSU-19 sample.

percent cellulosic crystalline component was found in flax recovered from the marine environment (*18*).

Table IV. Percent Crystalline Components of Cellulose and Malachite

Sample	Cellulose	Malachite
Indian Hemp	76.6±1.47	0
OSU-19	42.5±0.35	7.7±0.50
OSU-20	43.2±0.24	4.6±0.30

With respect to the mineralized fibers, we asked if the loss of cellulosic composition was equal to the amount of infiltration or replacement with copper moieties. We found that the calculated percent mass of malachite in sample OSU-19 (7.7%) is higher than that of OSU-20 (4.6%). The sum of the percent malachite component and the cellulose crystalline component in these mineralized fibers does not yield a net percent crystalline composition comparable to that of Indian hemp. The amorphous and crystalline intensities measured in the mineralized samples are smaller than those of the Indian hemp. It appears, then, that not only does the percent cellulose crystalline component decrease, but also there is decomposition and loss of organic fiber content.

Crystallite Orientations. The results of the X-ray photographic images are shown in Figures 4, 5, and 6. The most intense diffraction arc on the equator line registered farthest from the center of the circle on the X-ray photographic films is due to the 002 reflection, and this equator line is perpendicular to the fiber axis on the film. Each single arc registered on the film is a summation of reflections from planes in different crystallites that are distributed in a similar orientation angle around the fiber axis.

The mean crystallite orientation angle for Indian hemp is 13.2°. In comparison to cotton fiber, which has a range of orientation angle from 20.15° to 24.92° after correction for the effect of convolutions (*25*), Indian hemp displays a higher crystallite orientation. Both of the mineralized samples display larger crystallite orientation angles than Indian hemp, with OSU-19 exhibiting an orientation angle somewhat larger than OSU-20, as might be expected in a material with a greater disturbance of its physical microstructure (Table V).

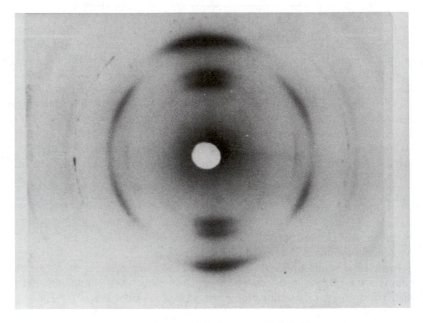

Figure 6. X-ray photographic image of OSU-20 sample.

Table V. Crystallite Orientation Angle

Sample	Angle
Indian Hemp	13.2°
OSU-19	27.1°
OSU-20	20.5°

The larger crystallite orientation angles observed in the archaeological mineralized fibers indicate that crystallite orientation decreases as a consequence of fiber degradation and mineralization. When fibers are buried underground over a long period of time, they deteriorate as a result of interaction with the burial conditions. As the fibers degrade, both intermolecular associations and long molecular chains are broken. The molecules in the crystalline regions of the fiber undergo structural rearrangement and the crystallites become more randomly distributed, losing their orientation with respect to the long fiber axis. In addition to the impact of fiber degradation on crystallite order, the infiltration of fibers with copper ions and the replacement or infilling of the fiber composition with copper moieties must also contribute to the interruption of the orientation of the cellulosic crystallites along the fiber long axis. The orientation of crystallites in cotton fibers (*16*) shows a similar disruption of orientation due to aging.

Conclusion

X-ray diffractometric examination of archaeological mineralized fibers recovered from a specific site has shown that differences are present in the physical microstructures of these fibers in comparison to those of a modern bast fiber. In general, the crystallite sizes σ_{101} and $\sigma_{10\bar{1}}$ of the mineralized samples are larger, but the σ_{002} of each are smaller than those of Indian hemp. The overall unit cell dimensions and monoclinic angles of these mineralized samples are similar in magnitude to those of Indian hemp with the exception of the *a*-axial direction which is slightly larger in the mineralized fibers. As expected, the percent crystalline components in the mineralized fibers are significantly lower than that of the modern fiber sample. In addition, the crystallites in these mineralized fibers are more randomly oriented than those of the Indian hemp.

The presence of malachite reflections in the X-ray diffractograms confirms the infilling or replacement of cellulosic polymers with copper minerals. It is likely that the degradation of the fiber and the process of fiber mineralization aids in the disruption of Van der Waals forces and valence bonds. When both the crystalline and amorphous areas of fibers are damaged, the polymer chain length decreases and the chain mobility increases (*19*). As a result, a

200 ARCHAEOLOGICAL CHEMISTRY

change in the crystallite distribution or some induced lattice disorder might contribute to the decrease in σ_{002} as well as to the decrease in the percent crystalline component. The deterioration of molecular chains in the fiber structure can cause the loss of crystallite orientation, which becomes more randomly distributed along the long fiber axis.

The larger a-axial unit cell dimension in the mineralized fibers may contribute to the larger crystallite sizes in the same direction (σ_{101} and $\sigma_{10\bar{1}}$) in addition to the rearrangement and alignment of additional associated chains in the amorphous regions near the end of the crystallite in the a-axial direction after degraded by water molecules.

The results indicate that copper mineralized fibers show distinctly different fiber microstructures than those of modern bast fibers. The difference in crystallite size and orientation between these two archaeological mineralized samples can be attributed to the different environments to which they were exposed in the burial site. Further analyses of the chemical microstructures of these changes are planned.

Acknowledgments

The authors wish to acknowledge the James River Institute for Archaeology, Inc., Jamestown VA for providing the mineralized fiber cordage, and Dr. Annette Ericksen, Archaeological Data Services, Columbus OH for collecting modern plant fibers. Acknowledgments are also due to Dr. J. Bigham and Mr. F. Jones of Department of Agronomy X-ray Diffractometry Laboratory for the use of the powder diffractometer; and to Dr. M. Caffrey, Department of Chemistry, for the use of the single crystal diffractometer. Support for this research was provided in part by the Ohio Agricultural Research and Development Center.

Literature Cited

1. Barber, E. J. W. *Prehistoric Textiles*; Princeton University Press: Princeton, NJ, **1991**; 9-20.
2. Jakes, K. A.; Sibley, L. R. In *Archaeometry 88*; Maniatis, Y., Ed., Elsevier: New York, **1989**, pp 237-244.
3. Church, F. *Ohio Archaeologist*, **1983**, *33*, 10-16.
4. Jakes, K. A.; Sibley, L. R. *Science and Archaeology*, **1983**, *25*, 31-38.
5. Sylwan, V. *Museum of Far Eastern Antiquities Bulletin*, **1937**, *9*, 119-126.

6. Edwards, M. G. Unpublished Dissertation Thesis, Institute of Archaeology: London, (1974).
7. Jakes, K. A.; Howard III, J. H. In *Conservation and Characterization of Historical Paper and Textile Materials*; Zeronian, H., Needles, H., Eds. Advances in Chemistry No. 212; American Chemical Society: Washington, , DC, **1986**, pp 277-287.
8. Gillard, R. D.; Hardman, S. M.; Thomas, R. G.; Watkinson, D. C. *Studies in Conservation*, **1994**, *39*, 132-140.
9. Sibley, L. R.; Jakes, K. A. *Clothing and Textile Research Journal*, **1982**, *1*, 24-30.
10. Janaway, R. *The Conservator*, **1983**, *7*, 48-52.
11. Hale, F. E. *The Use of Copper Sulphate in Control of Microscopic Organisms*; Phelps Dodge Refining Corp: New York, NY, 1946, pp 1-43.
12. Krässig, H. A. *Cellulose: Structure, Accessibility and Reactivity*; Gordon and Breach Science Publishers: Philadelphia, PA, 1993; pp 12-33.
13. Brown, R. M. *Cellulose and Other Natural Polymer Systems*; Plenum Press: New York, NY, 1982, pp 403-427.
14. Khalifa, B. A.; Abdel-Zaher, N.; Shoukr, F. *Textile Research Journal*, **1991**, *61*, 602-608.
15. Perel, J. *Textile Research Journal*, **1990**, *81*, 241-244.
16. Kalyanaraman, A. R. *Textile Research Journal*, **1984**, *54*, 354-355.
17. Needles, H. L.; Nowak, K. J. In *Historic Textile and Paper Materials II*; Zeronian, S. H.; Needles, H. L., Eds.; ACS Symposium Series 410; American Chemical Society: Washington, DC, 1989, pp 159-167.
18. Foreman, D. W.; Jakes, K. A. *Textile Research Journal*, **1993**, *63*, 455-464.
19. Atalla, R. H. In *Preservation of Paper and Textiles of Historic and Artistic Value II*, Williams, J. C. Ed., Advances in Chemistry Series, No. 193, American Chemical Society: Washington, DC, 1981, pp 69-176.
20. Jakes, K. A.; Sibley, L. R. The Ohio State University, unpublished technical report, 1993.
21. Foreman, D. W.; Jakes, K. A. In *Cellulosics: Pulp, Fiber, and Environmental Aspects*, Kennedy, J. F.; Phillips, G. O.; Williams, P. A., Eds. Ellis Horwood: London, 1991, pp 153-160.
22. *Mineral Powder Diffraction File Databook*, Sets 1-42, Compiled by the International Center for Diffraction Data, Pennsylvania, 1993.
23. Hearle, J. W. S.; Greer, R. *Textile Progress*, **1970**, *2*, 1-187.
24. Blackwell, J.; Kolpak, F. J.; Garkner, K. H. In *Cellulose Chemistry and Technology*; Arthur, J. C., Ed., ACS Symposium Series 48; American Chemical Society: Washington, DC, 1977; pp 42-51.
25. Iyer, P. B.; Iyer, K. R. K.; Patil, N. B. *Journal of Applied Polymer Science*, **1985**, *30*, 435-439.

RECEIVED November 6, 1995

Chapter 16

Clues to the Past: Further Development of the Comparative Plant Fiber Collection

Kathryn A. Jakes

Department of Textiles and Clothing, Ohio State University, 1787 Neil Avenue, Columbus, OH 43210

The Comparative Plant Fiber Collection (CPFC), established to provide comparative plant fiber materials for the identification and characterization of the fibers employed in prehistoric native American textiles, continues to be expanded. This report includes new information concerning morphological distinctions observed in plant fibers observed under scanning electron microscopy, the effect of treatment of the fibers in fiber processing, and the inorganic crystalline inclusions which the fiber products contain. These data may provide further distinctions among fibers within the four categories of plant fibers proposed as the results of previous work. The identification and characterization of fibers employed in textiles helps us reconstruct past methods of textile production and learn how ancient textiles were used. These data can contribute to studies of craft specialization and social differentiation.

Employing plant materials available locally or, perhaps, imported from a distance, prehistoric native Americans produced a wide variety of textile products. Visual examination alone is sufficient to show that these textile products required sophisticated fabrication techniques. While coarsely made bags and mats have been recovered, fine yarns twined in intricate patterns are also seen in textiles from Hopewell (200 BC-500 AD) and Mississippian (1000-1500 AD) sites (1-10). Textiles are material objects; through their

0097–6156/96/0625–0202$12.25/0

study, the materials and methods employed in their manufacture and the patterns of their use may be discerned. Textiles also serve as cultural objects, providing examples of dress and adornment, which can be used to reveal such factors as roles, status, and social differentiation. While the fabrication structure of some textiles recovered from prehistoric native American sites has been studied (*3*), extensive analysis of the fibrous components of these textiles has been hampered by the lack of a comparative collection of fibers representative of those employed in textile production. Identification and characterization of fibers in textiles has been limited, in some cases, to the description of the fibers as "bast" or "vegetal".   The need for a comparative collection was addressed by the establishment of the Comparative Plant Fiber Collection (CPFC) at The Ohio State University. Supported by the National Science Foundation, the collection continues to expand and provides new insights into fiber use in prehistoric textiles.

The Comparative Plant Fiber Collection

The Comparative Plant Fiber Collection was established to provide comparative fibrous material processed from plant stems typical of those employed by prehistoric native Americans of eastern North America. The scope of the collection was limited, at first, to the examination of the fiber products that are very fine, long, and strong and can be used to produce finely twined textiles. Coarse yarns and cords that would result in coarse bags or other textiles and wood splits or entire plant pieces employed in basketry or mats were not originally included in the collection. Dyeplants were not included in the original scope of the CPFC but these are being added. Plant stems were collected from two geographic areas: southern and central Ohio in one area and northern Georgia. One set of the stems were processed within 5 days of collection in the field; the other set was allowed to dry in a desiccator for 6 weeks prior to processing. The fibers were processed from the plant stems in four ways: (1) hammering the stems, then hand peeling the individual fibers or fiber bundles; (2) soaking the stems in water for 2 weeks to simulate the effects of retting, then hand peeling the fibers and fiber bundles from the stems; (3) boiling the stems in demineralized water for 6 hours, allowing them to cool, then hand peeling the fibers from the stems; and (4) boiling the stems in demineralized water with potassium carbonate for 6 hours, and hand peeling the fibers from the stems.

The behavior of the fiber bundles in processing was recorded. The fiber products have been studied by multiple techniques of optical microscopy, scanning electron microscopy, X-ray microanalysis, and infrared microspectroscopy. Details of the establishment of the collection were reported previously (*11*). The results obtained from these first analyses led to the categorization of the plant fibers based on polythetic groups of attributes (*12*). Behavior of the fiber bundles in processing provides information concerning the likelihood of use of certain plant fibers in finely twined textiles; only certain plant stems yield fine, strong, and long bundles which can readily be manipulated into fine yarns. Observation of morphology through the optical microscope also provides more attribute clues which aid classification.

Each of the four groups proposed contains a number of plants whose fibers display generally similar characteristics. Group I plants are those which yield fine, strong, and long fiber bundles which are pliable enough to be twisted or spun into fine yarns. These bundles are easy to process from the plant stems and can be withdrawn in long strands without breaking. The fibers in these bundles are relatively clean-surfaced with periodic dislocations commonly attributed to bast fibers. Fibers of the genus *Apocynum* constitute a subgroup of the Group I fibers because they display characteristic surface folds. Fibers in the other three categories include hard-to-process fibers which do not yield fiber bundles of long lengths or fine diameter. Group II and III fibers display no distinctive surface characteristics. Fibers in Group IV are readily distinguished because of the presence of extensive quantities of crystal inclusions. More detailed descriptions of the group classifications are reported by this author (*12*) elsewhere.

This manuscript reports the results of scanning electron microscopic examination and X-ray microanalysis of some of the CPFC fibers; these analyses revealed features unseen in optical microscopy. Also reported are the results of the study of high temperature "carbonized" and cold plasma ashed fiber products. The findings reported herein will ultimately contribute to the refinement of the Group classifications.

Experimental Methods

A Zeiss Axioplan research microscope was employed in the optical microscopic examination of the fiber products. Scanning electron microscopic (SEM) examination was carried out employing a Jeol JSM-820 scanning electron microscope and X-ray microanalysis was accomplished with a Link Analytical eXL energy dispersive x-ray analyzer. Fibers were mounted on carbon planchettes and carbon coated for SEM-energy dispersive spectrometric (EDS) analysis. Fibers held in open ceramic dishes to allow contact with air and fibers wrapped in foil to eliminate air were "carbonized" in a muffle furnace at 600 °C for 5, 10, and 15 minutes. Fibers also were ashed in a SPI Supplies Plasma-Prep II plasma etching unit with oxygen gas plasma (*13*).

Results and Discussion

Fiber Morphology. Various aspects of fiber morphology were investigated in this continuation of the study of the CPFC. Tables I and II summarize the findings of the SEM and EDS examination of a selection of fibers from Groups I, III, and IV. The possibility of alteration in morphology resulting from the four processing treatments was explored. Butterfly weed (*Asclepias tuberosa*), spreading dogbane (*Apocynum androsaemifolium*), and Indian hemp (*Apocynum cannabinum*) fibers which had been processed from the plant stems by the four treatments were studied with SEM and EDS. The Indian hemp fiber bundles displayed little difference as a consequence of the water soaking or water boiling treatments. In fact, the fiber bundles still possessed attached plant cellular materials. The fibers themselves displayed the surface folds which are characteristic of their genus. The Indian hemp fibers which had been boiled with potassium carbonate appear cleaner than fiber bundles resulting from the other three treatments. The nodes are prominent and the fibrils are apparent within the fiber. There is less attached intercellular material, although particles of material are observed on the fiber surfaces. Some of the particles contain sodium, potassium, and chlorine while others are carbonaceous.

While one would attribute the presence of potassium to the potassium carbonate solution, it should be noted that potassium, chlorine, and sodium were found in the same fibers without the boiling treatment. Butterfly weed fibers also display localized potassium on the surfaces; after boiling with potassium carbonate the fiber surfaces are covered with bumps which have a high potassium and chlorine content. No attempt was made in this exploratory work to quantify the elemental composition of the fibers.

After the water soaking treatment, the spreading dogbane fibers seem somewhat cleaner than those which had been processed by hammering and peeling. In addition, the fibrils are apparent in the fiber surfaces. Water boiled fibers appear clean only in some areas while in others cellular material like parenchyma cells remain. Even after treatment 4, boiling in potassium carbonate, the dogbane fibers still possess areas occluded by agglomerated cells. The evidence obtained indicates that even the boiling treatments are insufficient to remove distinctive surface characteristics such as dislocations or surface folds.

Distinction between Group I and IV categories is based primarily on the presence of the profuse amounts of crystal inclusions found associated with the Group IV fibers. The question then arises whether the fiber morphologies alone would allow their identification if it were possible to eliminate the associated crystals and plant cells by some sort of extensive

Table I. Morphology and Elemental Analyses of Selected Group I Plant Fibers

	Dislocations	Transverse Surface Mark	Surface Folds	Other Features	Elemental Composition
Butterflyweed (OH)*	yes	no	no		
Swamp Milkweed (OH)	yes	no	no		
Common Milkweed (OH)	yes	no	no		
Red Mulberry (OH)	yes	no	no		
Spreading Dogbane (OH)		no	yes	some smooth, some with surface folds	
Spreading Dogbane (GA)			yes	Parenchyma	
Indian Hemp (OH)	yes	no	yes	covering occludes folds, some fibers with folds, some without	C,O
Intermediate Dogbane (OH)	yes	no	no	cambium occludes surface folds	C,O,small Cl,K
Blue Dogbane (OH)	yes	no	no	smooth fibers	C,O,Cl,K small Ca
Stinging Nettle (OH)	yes	yes	no	Parenchyma cell residue	C,O, Small K
Wood Nettle (GA)	no	yes	no	Parenchyma cell residue, logitudinal striations	
False Nettle (OH)	no	yes	yes	longitudinal striations	

*Each fiber common name is followed by the collection location.

Table II. Morphology and Elemental Analyses of Selected Group III and IV Plant Fibers

	Dislocations	Transverse Surface Marks	Surface Folds	Other Features	Elemental. Composition
Red Cedar (GA) *	no			smooth fibers	C,O, small Ca
Paw Paw (OH)	no	no	no	crystals embedded and covered	
Black Walnut (OH)	no	no	no	smooth fibers	
Black Willow (GA)	no	no	no	some surface disruption	
Slippery Elm (OH)	no	no	no	smooth fibers	

*Each fiber common name is followed by the collection location.

retting technique. Further study of the Group IV fibers was conducted to answer this question. While treatment 4 is not sufficient to completely clean the fibers in Group IV, fibers which had been boiled in potassium carbonate did reveal some areas which were free from crystal inclusions and could be examined. Of these, only two of the Group IV fibers, black willow (*Salix nigra*) and slippery elm (*Ulmus fulva*), appear to possess dislocations in any way similar to those seen in Group I fibers when examined with optical microscopy. Other fibers from Group IV which were examined did not appear comparable to Group I fibers. Basswood (*Tilia americana*), black walnut (*Juglans nigra*), and paw paw (*Asminia triloba*) fibers are smooth-surfaced fibers underneath their crystal and plant cell coverings. Further examination of black willow and slippery elm fibers with scanning electron microscopy revealed surfaces which are very distinctly different from those of the Group I fibers. Figure 1 displays the features of a typical Group I fiber, common milkweed; Figure 2 displays the structures observed in black willow fibers. There is some evidence for "ridges" (*14*) in these fibers which could have been misconstrued in optical microscopy as "dislocations".

It has already been noted that certain fibers display surface folds which result in a unique appearance when examined with the optical microscope. The surface folds cause frequent surface disruption. This unique characteristic has been pronounced as a key indicator of a particular genus of plant fibers, *Apocynum* (*11,12*). Further examination of fibers within this group supported this finding. Blue dogbane, though belonging to the Family Apocynaceae, is from the genus *Amsonia*, and does not display the surface folds. Thus, the distinction made between plant fibers is valid at the general level.

Fiber samples from the same genus and species of plant but collected in different locations (Ohio and Georgia) or at different times (1991 and 1993) were compared for morphological differences. Fiber morphology and elemental composition are summarized in Tables I and II; shapes and elemental composition of inorganic inclusions and other features are summarized in Tables III and IV. Fibers from the same genus and species displayed consistent features despite the differences in circumstances of collection. Both examples of common milkweed (*Asclepias syriaca*), and red mulberry (*Morus rubra*) display the surface characteristics of Group I fibers. Calcareous druses, spherical clusters formed by accumulation of crystals, were found in both samples of common milkweed. Both examples of spreading dogbane (*A. androsaemifolium*), and Indian hemp (*A. cannabinum*) exhibit surface folds. Both examples of stinging nettle (*Urtica dioica*) and false nettle (*Boehmeria cylindrica*) display transverse markings. The crystal inclusions observed in basswood (*T. americana*), paw paw (*Asiminia triloba*), and black willow (*Salix nigra*) are consistent for materials collected at

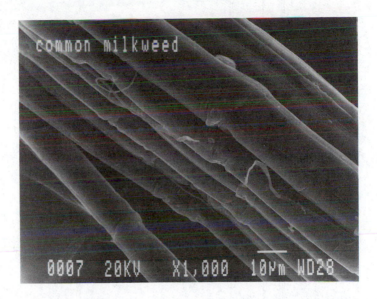

Figure 1. Common milkweed: Typical Group I morphology.

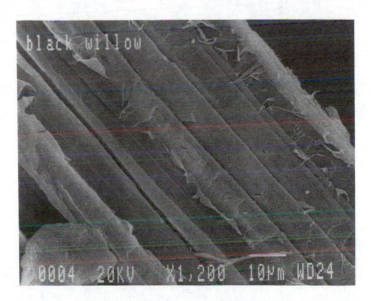

Figure 2. Black willow surface morphology.

Table III. Morphology and Elemental Analyses of Inorganic Inclusions in Group I Plant Fibers

	Large Inclusion			Other Inclusion	
	Shape	Location	Elemental Composition	Description	Elemental Composition
Butterflyweed (OH) *	druse particle/druse	in strings between fibers	Ca,O,C, small S,K K,S, small Ca,Cl,C,O	random particle lumps on surface	Al,Si,S,K,C,O, Si,O
Swamp Milkweed (OH)				particles	Si,O,C
Common Milkweed (OH)	plates, some beginning to form druse	random	Ca,Mg,O,C	particles	Fe, Small K,O,S
Common Milkweed (OH)	druse	random	Ca,O,C	lumps fluffy materials fluffy materials	Si,O,K, small C,Mg,Cl,S Ca,O, small K,S K,Cl,O, small P,Mg,S
Red Mulberry (GA)	druse-like particle druse-like crystals		S,K,C,O, small Si C,O,Mg,Si,S,K, small Ca		
Spreading Dogbane (GA)	lump, not flat faced in lines between fibers	lump, not flat faced in lines between fibers			
Indian Hemp (OH)	plate-like druse-like rectangular	random few random few-infilling parenchyma	Ca,P,O, small C,K Ca,P,O, small C,K Ca,O	fluffy deposits sand globules	Ca,P,K,O, small Al,O,K K,O, small Cl, P

Blue Dogbane (OH)	lumps along fiber surface		
Stinging Nettle (OH)	druse	in string	inorganic coating fibers in some areas small lumps prevalent	Si,O Ca
Stinging Nettle (GA)			particles in parenchyma	Si,O, small Al,C
Wood Nettle (GA)	plate-like particles	in string	particles	K,Ca,S
False Nettle (GA)			random particles / particles within parenchyma	K,O,P,S,Ca small Mg / some Si,O,K, S, small Al,P,C

Si,O, small S,K,Ca,C

* Each fiber common name is followed by the collection location.

Table IV. Morphology and Elemental Analyses of Inorganic Inclusion in Group I Plant Fibers

	Large Inclusion			other Inclusion	
	Shape	Location	Elemental Composition	Description Particles	Elemental Composition
Red Cedar (GA)*	irregular, some particles, some blocks	distributed over surface	Ca,O		
Paw Paw (OH)	long flat blocks with rounded ends druses	fill in parenchyma cells random	Ca, small O		
Paw Paw (GA)	long rectangular blocks,	in lines	Ca,O	small articles	Si
	other flat faced shapes		Ca,O	small particles	Si,Al
Black Walnut (OH)	truncated bipyramids	Ca,O		small particles	Si
	druse	Ca,O			
Black Willow (OH)	truncated bipyramids	within each parenchyma cell	Ca,O	fluffy deposits	Ca,O, Small Si
Basswood (GA)	long double pointed twinned	in lines along fiber	Ca,O	flattened lump	Si,O, small K, Mg,Al
	flat crystals druse	between fiber bundles	Ca,O		
Basswood (OH)	double pointed twinned crystals		Ca,O,C	particles	Ca,S,K,C,O

* Each fiber common name is followed by the collection location.

different locations and different times. The morphological consistencies further justify the categorization system proposed. In fact, the size, shape and chemistry of the formations resulting from calcification processes are "under close cellular control" (*15, 16*) and thus may prove to be class specific indicators. In a similar manner, the process of silicification is specific and the presence of opal phytoliths have been proposed as a route to determining use of plants in prehistoric societies (*17*).

Red cedar (*Juniperus virginiana*) fibers have been placed in Group III; Jakes, Chen, and Sibley (*12*) indicate that no inorganic inclusions were observed in optical microscopic study of red cedar fibers. Subsequent scanning electron microscopic examination of these fibers, however, reveal small calcareous inclusions of irregular shape distributed over the fiber surfaces: The inclusions are not located in a specific region as are crystals noted in other fibers. The inclusions, shown in Figure 3, are 1-3 µm in size, much smaller than the crystals observed in the Group IV fibers which can reach 20 µm in size.

Stinging nettle (*U. dioica*), false nettle (*Boehmeria cylindrica*), and wood nettle (*U. divaricatum*) were examined under the SEM. These nettles display longitudinal striations and periodic transverse surface marks (Figure 4). It is possible that these marks are linked to the presence of residual parenchyma cells. The feature appears to be characteristic of the Urticaceae family but no further distinction according to genera can be determined.

Carbonized fibers. In the original design of the CPFC research, a high temperature carbonizing treatment of fibers was proposed to experimentally replicate the "charred" or "carbonized" fibers found in many prehistoric native American textiles. The terms are used interchangeably in the literature, although technically it is unlikely that the charred fibers have, in fact, been reduced completely to carbon alone. Fiber products produced through the four treatments were subsequently heated at 600 °C in air and without air (nominally) wrapped in foil. Observation of these black materials by reflected light microscopy was attempted. Extensive work in the examination of carbonized Indian hemp fibers has been reported by Srinivasan (*18*). The microscopist can make some observations as the fibers are scanned under the microscope, but the field of view is very irregular in all cases, precluding the collection of useful pictures. Scanning electron microscopic study of these carbonized products is necessary to accurately describe the alterations in structure which occur as a consequence of high temperature carbonization. Srinivasan (*18*) also improved the design of the high temperature experiments. Since the foil wrapping did not eliminate oxygen access entirely but did inhibit fiber motion during heating, she carbonized Indian

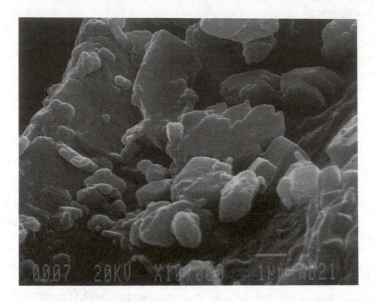

Figure 3. Eastern red cedar inclusions.

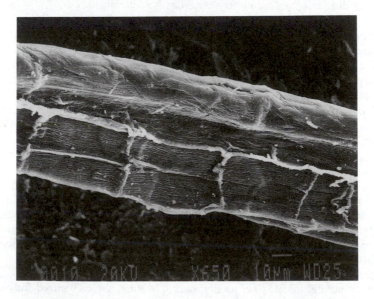

Figure 4. Stinging nettle: Transverse cross markings.

hemp fibers held in open crucibles in an inert argon atmosphere. The fibers coiled and shrank and some residual marks remained which are possibly related to the location of the dislocations. It is apparent from these experiments that much more extensive work is required to adequately describe the change in structure of carbonized fibers which results from high temperature carbonization. Consequent application of such findings to the identification of carbonized fibers in prehistoric textiles also will require extensive research.

Cold Plasma Ashed Fibers. In examination of the carbonized fibers prepared in this research, questions arose concerning the inorganic inclusions revealed through high temperature incineration of the organic matrices in which they are supported. In fact, the carbonization experiments were comparable to some of the methods employed in phytolith research in dietary plants (*19-21*). After further exploration of the methods employed in phytolith preparation and the pitfalls of each, a cold plasma ashing method for phytolith preparation was developed (*13*) which avoided the consequences of high temperature incineration and of strong acid digestion. Cold plasma ashing has been shown to be a promising method for the preparation of phytolith samples from plant material. Fibers can be ashed in place on a carbon planchette, thus allowing the observation of the exposed inorganic inclusions embedded in the skeleton of remaining organic material. The relationship of the location of these crystals and inclusions with other plant cells can be observed. While the larger crystalline inclusions present in Group IV fibers were observed microscopically even prior to ashing, the ashing procedure produces clean-surfaced crystals unobscured by tissue and reveals other less obvious inclusions. An additional advantage of the procedure is that since the ashed fibers are prepared for scanning electron microscopic examination, they can also be analyzed for elemental composition through energy dispersive analysis of X-rays.

The morphological and elemental data on inorganic inclusions obtained through the study of cold plasma ashed fibers from the CPFC provides new clues to enhance the Group classification scheme. The Group IV fibers yield many types of crystal inclusions (Table IV). Basswood (*T. americana*) (both from Ohio and Georgia) possesses both siliceous particles and large (10-20 µm) double-pointed twinned calcareous crystals (Figure 5). The crystals are aligned along the phloem fibers but do not appear to form within the parenchyma cells as do the twinned calcareous crystals observed in paw paw (*Asiminia triloba*) fibers. These flat-faced crystal blocks (Figure 6) are rounded on the ends and appear to have grown to fit within the parenchyma cell. Black willow and black walnut display calcareous truncated bipyramids (Figure 7) as well as some small siliceous particles. The bipyramids can be seen within the walls of the parenchyma cells. The brick-like rectangular

Figure 5. Basswood: Twinned double pointed crystal inclusions.

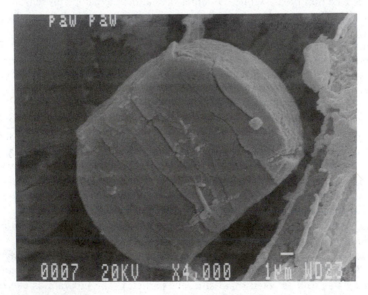

Figure 6. Paw paw: Twinned rounded plate crystal inclusions.

blocks observed in slippery elm (*Ulmus fulva*) also align with the phloem fiber cells, but are not formed inside parenchyma cells. The calcareous crystals of the woody Group IV fibers are so predominant that they obscure the much smaller siliceous particles that are present. While further work is necessary, it appears that the crystal shape and composition of the woody Group IV phytoliths may well prove to be a useful attribute in distinction among members of that Group. Although observed in SEM in fibers which were not ashed, the small inorganic inclusions covering the surface of red cedar fibers are more easily studied in plasma ashed material. Figure 3 displays the particles revealed after ashing the fiber bundles.

While EDS analyses of these inorganic inclusions revealed elemental compositions of calcium and oxygen or silicon and oxygen, the form of these materials was not evaluated. It is likely that the compounds represented are calcium oxalate and silica (*15, 16, 19, 20, 21*).

Optical microscopic examination of Group I fibers reveals that some possess associated inorganic structures. By reducing the organic matrix, plasma ashing of these materials exposes the crystals and other inclusions so that they can be more readily observed. In effect, the inclusions which may be random and infrequent are "concentrated" by the ashing process, as the organic composition of the plant material is reduced and their elemental composition can provide information as well.

The Group I fibers display many structures of both calcareous and siliceous composition (Table III). Druse crystals (Figure 8) are observed in milkweed fibers (*A. tuberosa, A. incarnata, A. syriaca*) and in red mulberry fibers (*M. rubra*), in agreement with the optical microscopic observations. These druses are large and predominantly calcium in composition. Other calcareous plates and particles are observed as well as rounded and irregularly shaped siliceous particles. Common milkweed fibers also display fluffy globular surface structures which, in some areas, have a high potassium content. Swamp milkweed displayed large siliceous faceted inclusions.

Druses were not observed in the dogbanes or Indian hemp fibers. Indian hemp, in one instance, did display an agglomeration of plate-like crystals which could possibly be called a druse. More common in the Indian hemp fibers are plate-like crystals which form within cell walls (Figure 9). Indian hemp also displays fluffy looking globular structures with a high potassium, phosphorous, and calcium content.

Calcareous druses were observed in stinging nettle fibers, while plates were seen in wood nettle fibers. The stinging nettle also displayed a siliceous coating of the fibers in some areas (Figure 10) and small siliceous lumps

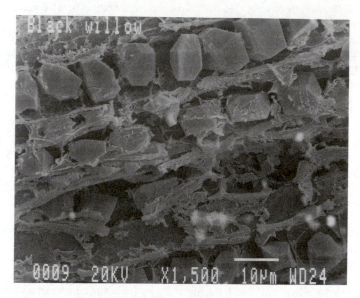

Figure 7. Black willow: Truncated bipyramid crystal inclusions.

Figure 8. Common milkweed: Druse crystal inclusion.

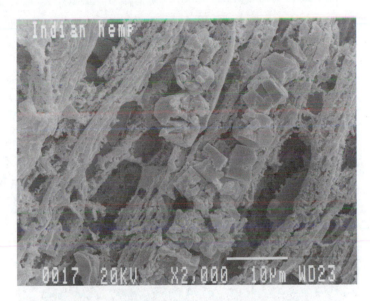

Figure 9. Indian hemp: Block crystal inclusions.

Figure 10. Stinging nettle: Siliceous coating over fibers.

Figure 11. Stinging nettle: Siliceous particle within a cell.

were observed within the cell walls (Figure 11). The number of any of these inclusions is small in comparison to the number seen in Group IV fibers and their presence alone is not considered diagnostic for the plant fiber genera but their presence may provide an additional clue useful in plant fiber indication and group classification of unknown plant fibers. Cold plasma ashing reveals structure previously occluded by plant cellular structures and so is useful in developing an atlas of the plant fiber inorganic inclusions.

Conclusion

Continued work conducted in the study of fibers from the CPFC has provided a variety of insights. The intrinsic morphology of fibers appears to be maintained despite the processing treatment employed. Thus enough of the characteristic features of the plant fibers are maintained in processing that the fibers can be identified by the group classification scheme. The surface features of the Group IV fibers alone are not enough to provide categorization but they are distinct from the characteristic dislocations observed in Group I fibers. Carbonization of fibers still needs much work to provide useful information leading to the identification of charred archaeological fibers. Cold plasma ashing of the plant fibers is a useful method for the preparation and observation of inorganic inclusions in fibers. The Group IV phytoliths and other inclusions are distinct and are multitudinous, and so may aid in subclassification within the group. Group I phytoliths and other inclusions are less prevalent than those observed in Group IV and may only provide a useful clue to identification of fibers.

Acknowledgments

The award of the National Science Foundation Grant BNS-9021275 provided the initial support for this work. Additional salaries and support were provided in part by state and local funds appropriated to the Ohio Agricultural Research and Development Center, The Ohio State University.

Dedication

This paper is dedicated to the memory of Lucy R. Sibley. Just as the lives of the peoples of the past live on through the textiles they left behind, the life of Lucy R. Sibley will continue through her contributions to this field of research.

Literature Cited
1. Willoughby, C. C. *Ohio Arch. and Hist. Quarterly* **1938**, *47*, 273-287.
2. Whitford, A. C. *Anthropological Papers of the Museum of Natural History* **1941**, *38*, 5-21.

3. Church, F. *Midcontinental J. of Archaeology* **1984**, *9*, 1-26.
4. White, E. P. *Excavating in the Field Museum: Survey and analysis of the 1891 Hopewell Mound group excavation.* M.A. Thesis, Sangamon State University: Springfield, IL, U.S.A., **1987**.
5. Drooker, P. B. *Mississippian Village Textiles at Wickliffe*; University of Alabama Press: Tuscaloosa Al, **1992**.
6. Kuttruff, J. T. *Textile attributes and production complexity as indicators of Caddoan status differentiation in the Arkansas Valley and Southern Ozark regions.* Ph.D. Thesis, The Ohio State University: Columbus, OH, U.S.A., **1988**.
7. Kuttruff, J. T. *American Antiquity* **1993**, *58*, 125-145.
8. Sibley, L. R.; Jakes, K. A. In *Conservation and Characterization of Historical Paper and Textile Materials;* Zeronian, H., Needles, H., Eds.; Advances in Chemistry Series 212: American Chemical Society: Washington D.C., **1986**; pp 253-257.
9. Sibley, L. R.; Jakes, K. A. *Clothing and Textiles Res. J.* **1989**, *7*, 37-45.
10. Sibley, L. R.; Jakes, K. A.; Swinker, M. E. *Clothing and Textiles Res. J.* **1992**, *10*, 21-28.
11. Jakes, K. A.; Sibley, L. R.; Yerkes, R. W. *J. Arch. Sci.* **1994**, *21*, 641-650.
12. Jakes, K. A.; Chen, H.-L.; Sibley, L. R. *Ars Textrina.* **1993**, *20*, 157-179.
13. Jakes, K. A.; Mitchell, J. C. *J. Arch.Sci.* in press.
14. Rahman, M. M. M.; Sayed-Esfahani, M. H. *Indian J. Textile Res.* **1979**, *4*, 115-120.
15. Arnott, H.J. In *The Mechanisms of Mineralization in the Invertebrates and Plants*, Watabe,N., Wilbur, K.M., Eds: University of South Carolina Press: Columbia, S.C.**1974**; pp. 55-73.
16. Arnott, H.J.In *Biological Mineralization*, Zipkin, I., Ed.: Wiley Interscience: New York **1973** pp.609-627.
17. Rovner, I. *Quaternary Research* **1971**, 1, 343-359.
18. Srinivasan, R. *Exploration of the effects of carbonization on Apocynum cannabinum fibers.* M.S. Thesis, The Ohio State University: Columbus, OH, U.S.A., **1993**.
19. Pearsall, D. M. *Paleoethnobotany: A Handbook of Procedures* Academic Press Inc.: San Diego, **1989**.
20. Lanning, F. C.; Ponnaiya, W. X.; Crumpton, C. F. *Plant Physiology* **1958**, *33*, 339-343.
21. Lanning, F. C.; Eleuterius, L. N. *Annals of Botany.* **1987**, *60*, 361-375.

RECEIVED October 9, 1995

Chapter 17

Updating Recent Studies on the Shroud of Turin

Alan D. Adler

Department of Chemistry, Western Connecticut State University, Danbury, CT 06810

The Shroud of Turin, a linen cloth alleged to be the burial shroud of Christ, has been precisely radiodated to the 14th century. Nevertheless, its status remains controversial. Is the radiodate accurate? Are the blood images seen on the cloth derived from contact of the cloth with a wounded human body? Is it a painting? If not a painting, what is the mechanism of its formation? Some of the latest research attempting to resolve these matters is presented and reviewed.

The Shroud of Turin can be unequivocally historically traced to the mid-14th century (1). Because it was alleged at that time to be the authentic burial cloth of Christ, it has always been an object of controversy. This 4.3 X 1.1 m linen cloth bears both complete head-to-head, frontal and dorsal, straw colored, "negative" body images of a crucified man with blood colored wounds and scourge marks in accordance with Biblical description of the Crucifixion. The body images are bracketed the entire length of the cloth by parallel burn and scorch marks from fire damage incurred in 1532. Waterstains from extinguishing this fire are also evident, as are patched areas from repairs carried out in 1534 prior to the entire cloth being stitched to a backing cloth to support the damaged original. There is a continuous seam along one side of the cloth producing a "side" strip with rectangular pieces of missing cloth at both ends of this strip. The main body of the cloth adjacent to these missing cloth areas shows selvage edges indicative of repair. There is no historic record of why or when this repair and seam were applied to the original cloth.

In 1978 a group of investigators, Shroud of Turin Research Project (STURP), carried out several on-site investigations of the Shroud at its repository in Turin and also took several sticky tape samples from designated areas of the cloth for further off-site studies. This work and the subsequent research has been summarized in several publications (2-4). STURP's major conclusions were that the Shroud was not a painting, the body image chromophore was an oxidation product of the cellulose of the linen fibers comprising the cloth, and the blood images were blood-derived materials produced from contact of the cloth with a wounded human body. A microscopical investigation of the STURP sticky tape samples by an independent investigator came to the opposite conclusion that the Shroud was a painting with the body images composed of iron oxide in a gelatin protein binder and the blood images composed of the same pigment with the addition of considerable cinnabar (HgS) with traces of

0097–6156/96/0625–0223$12.00/0

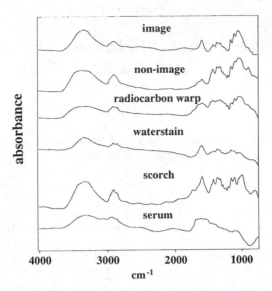

Figure 1. Typical FTIR absorbance patterns of single fiber samples of the Shroud of Turin.

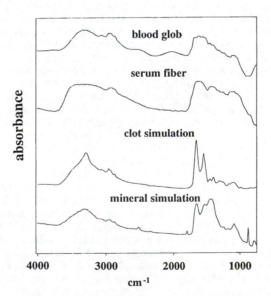

Figure 2. Typical FTIR absorbance patterns of blood samples of the Shroud of Turin compared with blood simulacra.

calcite (5), commonly found components of medieval paint pigments. As it was clear that science could never authenticate the Shroud as the burial cloth of Christ, but could positively disauthenticate it, STURP strongly recommended and supported a radiocarbon dating test (4).

Radiocarbon Dating

Three laboratories in a collaborative study independently radiodated samples from the Shroud of Turin by the Accelerator Mass Spectroscopy (AMS) method and reported a reasonably precise 14th century date in apparent agreement with its unequivocally known historic record (6). Unfortunately, a detailed protocol for sampling the Shroud, assuring both precision and accuracy, recommended by a convened group of consultants (7), was not followed. Only a single sample was taken in the lower corner of the main cloth of the frontal image below the so-called sidestrip from the selvage edge in an obviously waterstained area just a few inches from a burn mark. The selvage edge was trimmed off before portions of the sample were divided among the participating laboratories. Whether such an obviously contaminated sample is truly representative of the rest of the cloth is clearly questionable and the accuracy of the reported date is certainly doubtful.

To assess this question we have carried out further spectroscopic investigations of samples from the STURP sticky tapes (Adler, Selzer, and DeBlase; technical details submitted for publication elsewhere). Nineteen assorted fibers representative of non-image, waterstain, scorch, image, backing cloth, and serum coated fibers were extracted from the tapes and characterized by previously reported methods (2). These were compared with fifteen single fibers taken from three threads from the radiocarbon sample. Similarly, two blood samples (previously designated as globs) were extracted from the tapes and compared against several types of blood controls. The blood controls included two simulacra: a traumatic blood clot exudate (whole blood diluted with bilirubin-enriched human albumin) and mineral simulated blood (iron oxide, cinnabar, and a trace of calcite suspended in gelatin). These samples were all examined by Fourier Transform Infrared (FTIR) microspectrophotometry and the fibers were also studied by scanning electron microprobe. Dried films of the two blood simulacra were also studied by Ultraviolet-Visible (UV-vis) spectrophotometry. Some typical FTIR spectral patterns of these samples from this study are shown in Figures 1 and 2.

The patterns seen in Figure 1 are all distinguishably different from one another clearly indicating differences in their chemical composition. These compositional differences were further confirmed by peak frequency analysis utilizing the computer software that generates the spectral data. In particular the radiocarbon samples are not representative of the non-image samples that comprise the bulk of the cloth. This difference was also supported by the scanning electron microprobe data that showed gross enrichment of the inorganic mineral elements in the radiocarbon samples, even compared to the waterstain fibers taken from the bulk of the cloth. In fact, the radiocarbon fibers appear to be an exaggerated composite of the waterstain and scorch fibers, thus confirming the physical location of the suspect radiosample site and demonstrating that it is not typical of the non-image sections of the main cloth. How much these differences in chemical composition actually affected the accuracy of the radiodate is not clear. However, these data are consistent with a recently proposed mechanism in which it has been experimentally demonstrated that conditions comparable to those suffered by the Shroud in the 1532 fire can produce a large error in radiodating by large kinetic isotope effects (8). Alternatively, considering the presence of the selvage edge, this area may contain newly woven material as a repair.

Some recent image analysis studies comparing the blood marks on the Shroud of Turin with those on the Cloth of Oviedo also cast doubt on the accuracy of the Shroud's radiodate (Whanger, Duke University, personal communication, May 1994). The Cloth of Oviedo, alleged to be the *sudarium* associated with Christ's death, contains blood images similar in appearance to those on the Shroud, and can be historically traced to the 7th century (*9*). In Figure 3 the equally scaled dorsal head wound marks on the two cloths are compared with one another. The similarity of these two complex patterns is evident enough to suggest that these two cloths were in contact with the same wounded body, presumably within the same short time period. Should further research reveal stronger relationships between these two relics, the accuracy of the 14th century date of the Shroud will be clearly doubtful, as the Cloth of Oviedo is considered at least 7th century.

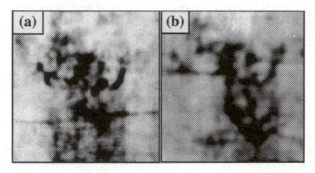

Figure 3. Comparison of dorsal head wound marks on the Shroud of Turin (a) and the Cloth of Oviedo (b).

Blood Images

Analysis of the FTIR data in Figure 2 compared with various controls shows that bilirubin can be spectrally detected in both the blood samples and the yellow serum coated fibers in agreement with the previously reported chemical data (*2*). The pattern match of the simulated clot appears only fair, but spectral analysis utilizing the computer software shows that reducing the protein pattern and increasing the bilirubin pattern makes a very good fit to the blood sample pattern. Conversely, the mineral simulated blood pattern is a complete mismatch except for the presence of protein. Bilirubin is clearly required to obtain a proper spectral match.

The same conclusions are drawn from the near UV-vis spectra of the two simulacra. The clot simulation is in good agreement with the previously reported spectra of Shroud blood specimens and that taken from the on-site examination of the blood images on the whole cloth matching the observed peaks at 420, 450, 520, 580, and 630 nm (*10*). Increasing the amount of bilirubin in this simulacrum will also improve the fit as with the FTIR data. However, the mineral blood simulation showing only two broad low peaks at 470 and 514 nm is again a complete mismatch. This is not surprising as it should be recalled that the two on-site X-ray examinations of the Shroud did not reveal the presence of any mercury compounds, particularly in the blood image areas (*11, 12*). Thus these two new pieces of spectral evidence completely reinforce all the previous chemical, immunological, and forensic work demonstrating that the blood images on the cloth are blood-derived materials produced from contact of the cloth with clotted blood wounds on a human body (*2-4*) and are not composed of an artist's applied mineral pigment mixture (*5*).

Body Images

Peak frequency analysis of the FTIR data also shows that the image fibers contain more conjugated carbonyl absorption than the non-image fibers, consistent with previous identification of the chromophore with a cellulose oxidation product *(2-4)*. Similarly, although the spectral presence of protein in the characteristic amide absorption regions is readily seen on the serum fibers, it is not detectable on the image fibers, as stipulated by the painting hypotesis. This supports the previously published work refuting the painting hypothesis *(2-4)*.

Numerous copies of the Shroud of Turin exist and it has now been thoroughly historically documented that several of these painted copies were "sanctified" by being pressed to the original *(13)*. This process would clearly contaminate the Shroud with artist's materials by contact transfer. Therefore it cannot be maintained that the Shroud is a painting simply on the basis of the microscopical detection of such materials *(5)* in the face of the large corpus of evidence against such a simplified explanation *(2-4)*. The accumulated physical, chemical, and forensic data do not support the contention that the images on the Shroud of Turin are paintings. In particular, the image studies very clearly rule against this supposition *(14)*.

Image Formation Mechanisms

Establishing that the Shroud is not a painting still allows the possibility of its production by some other type of artistic rendition technique. However, many possible formation processes have been tested against the observed properties of this image and have all been found inadequate in some way if they are to be accepted as the explanation of this complex object *(3, 14, 15)*. Image studies have shown that the body image and the blood images are not always in stereometric register *(14, 16)*. As the blood can only have been transferred onto the cloth by contact, this implies that the body images were produced by some type of non-contact mechanism *(14)*. Some image studies *(14)* would suggest some type of radiational or energy transfer type mechanism. However, the nature of this process at this point in time remains a mystery. This should not be interpreted as proof that the image was produced by some supernatural process, but simply reflects the present state of our knowledge of this interesting object. Hopefully, future studies will not only resolve this mystery, but will provide a sound basis for undertaking the preservation and conservation of this cloth and its images *(17)*.

Acknowledgements

I am indebted to several of my colleagues for help, information and advice in preparing this manuscript. Luke Adler of LA Engineering Computer Services prepared the figures and utilized photographic materials supplied by Drs. Gil Lavoie and Alan Whanger.

Literature Cited

1. Wilson, I. *The Mysterious Shroud*; Doubleday & Co., Inc.: Garden City, NY;**1986**; pp 1-156. .

2. Heller, J.; Adler, A. *Can. Soc. Forens. Sci. J.* **1981**, *14*, 81-103.
3. Schwalbe, L.; Rogers, R. *Anal. Chim. Acta* **1982**, *135,* 3-49.
4. Jumper, J.; Adler, A.; Jackson, J.; Pellicori, S.; Heller, J.; Druzic, J. In Archaeological Chemistry-III; Lambert, J. B., Ed.; Advances in Chemistry 205; American Chemical Society: Washington, DC, **1984**; pp 447-476.
5. Mc Crone, W. *Acc. Chem. Res.* **1990**, *23*, 77-83.
6. Damon, P.; et. al. *Nature* **1989**, *337*, 611-615.
7. Harbottle, G.; Heino, W. In Archaeological Chemistry-IV; Allen, R. O.,Ed.;_ Advances in Chemistry 220; American Chemical Society: Washington, DC,. **1989**; pp 313-320.
8. Kouznetsov, D.; Ivanov, A.; Veletsky, V. Presented at the 209th National Meeting of the American Chemical Society, Anaheim, CA, April **1995**; paper HIST 007.
9. Ricci, G. *The Holy Shroud*; Centro Romono Di Sindonologia: Rome, Italy; **1981**; pp 137-143.
10. Heller, J.; Adler, A. *Appl. Opt.* **1980**,*19,* 2742-2744.
11. Mottern, R.; London, R.; Morris,R. *Materials Eval.* **1980**, *38*, 39-44.
12. Morris, R.; Schwalbe, L.; London, J. *X-ray Spectrometry* **1980**, *9*, 40-47.
13. Fossotti, L. *Shroud Spectrum Inter.* **1984**, *13,* 23-39.
14. Jackson, J.; Jumper, E.; Ercoline, W. *Appl. Opt.* **1984**, *23*, 2244-2270.
15. Carter, G. In Archaeological Chemistry-III; Lambert, J. B., Ed.;.Advances in Chemistry 205; American Chemical Society: Washington DC, **1984**; pp 425-446.
16. Lavoie, G.; Lavoie, B.; Adler, A. *Shroud Spectrum Inter.* **1986**, *20*, 3-6.
17. Adler, A.; Schwalbe, L. *Shroud Spectrum Inter.* **1991**, *42*, 7-15.

RECEIVED August 15, 1995

Chapter 18

A Re-evaluation of the Radiocarbon Date of the Shroud of Turin Based on Biofractionation of Carbon Isotopes and a Fire-Simulating Model

D. A. Kouznetsov, A. A. Ivanov, and P. R. Veletsky

E. A. Sedov Biopolymer Research Laboratories, Inc., 4/9 Grafski Pereulok, Moscow 129626, Russia

The inherent uncertainties of radiocarbon dating, particularly with respect to variations in conditions external to the artifact in question, led us to question the accepted radiocarbon date of the Shroud of Turin. In our work, we devised a laboratory model to simulate the fire conditions to which the Shroud was subjected at Chambéry in 1532. Our results showed that radiocarbon ages of experimental textile samples incubated under fire-simulating conditions are subject to significant error due to incorporation of significant amounts of ^{14}C and ^{13}C atoms from external combustion gases into the textile cellulose structure. We also took into account the known phenomenon of biological fractionation of carbon isotopes by living plants which can lead to enrichment of a textile by ^{13}C and ^{14}C isotopes during linen manufacture.

Radiocarbon dating has played a significant role in archaeology since its introduction over four decades ago. However, the inherent uncertainties of this method (1) complicate efforts at chronological resolution and control. For example, it was recognized early on that variation through time of atmospheric concentrations of ^{14}C complicates dating efforts, particularly when the calibration curves used for this purpose have very steep or very shallow slopes during time periods critical to the artifact in question. Moreover, accurate and precise radiocarbon analysis usually dates natural, not cultural, events. It is only the association of those events with cultural practices that can date cultural events of interest by establishing an association through archaeological observation and critical judgment rather than by physical analysis. Radiocarbon results deliver probability, not certitude, and cannot be treated uncritically by ignoring or neglecting the inferential component in the interpretation of a dating series or by failing to consider radiocarbon results in conjunction with independent evidence like chemical structure patterns, stylistic details in works of art, a site's antiquity, etc. Thus, under most circumstances, radiocarbon dating is an indirect way to determine the age of archaeological subjects (1, 2).

The Shroud of Turin is an image-bearing linen textile which many people associate with the crucifixion and death of Jesus Christ. Radiocarbon testing of this relic by a large international team of scientists yielded a calendar age of 1260-1390 A.D. with 95% confidence (3). The historical uniqueness of the Shroud and knowledge about the nature and limitations of the radiocarbon dating approach merited, in our

0097–6156/96/0625–0229$12.00/0

opinion, a re-evaluation of these results. This re-evaluation was carried out taking into account the following phenomena: (1) biofractionation of C-isotopes by living flax, the source for linen textiles (*2, 4 - 6*); (2) possible chemical modification of the textile cellulose in the Shroud ("carbonization") as a result of the fire to which it was subjected at Chambéry in 1532.

Biofractionation of Carbon Isotopes

The conventional radiocarbon dating calculation model of the linen textile includes an assumption according to which at $t_1 = 0$, that is, at the time of manufacture of the textile, the ^{13}C and ^{14}C content values in the flax stems and in the resulting textile were equal to each other (*1, 4, 6, 7*). However, this assumption is open to question because of the known phenomenon of the biological fractionation of carbon isotopes by living plants which leads to significant enrichment of the textile by ^{13}C and ^{14}C isotopes during flax spinning in the manufacture of linen. Thus, polysaccharides from the long-fibered flax stem contain relatively much more ^{13}C and ^{14}C when compared to other classes of biomolecules (nucleic acids, proteins and lipids). It has been shown that not less than 60% of the total amount of ^{14}C atoms in the flax body is concentrated within the cellulose fraction (*6, 8*). This phenomenon is known as intermolecular C-isotope biofractionation (*6*). Since a key process in linen textile manufacturing is flax spinning, a technology leading to cellulose isolation and purification (*9, 10*), it is logical to note that spinning would lead to the enrichment of the resulting textile by heavy (rare) ^{13}C and ^{14}C isotopes as compared to the total amount of flax stem homogenate because of the removal of most of the non-polysaccharide components during this simple technical procedure.

Chemical Modification of Textile Cellulose ("Carbonization")

It is known that when an organic artifact is subjected to extreme heat (> 300 °C) in the presence of an external substrate, isotopic exchange can take place (*11*). In 1532, the Shroud of Turin was housed in a silver reliquary in the Sainte-Chapelle in Chambéry, France. A severe fire broke out in the building and the intense heat melted a corner of the reliquary. The molten silver penetrated one corner of the folded linen inside, producing the now-familiar pattern of burns and scorches (*12*). According to Stevenson and Habermas (*11*), the fire subjected the Shroud to temperatures of at least 960 °C and to superheated steam from the water used to extinguish the fire. They further assert that although carbon-containing molecules from the silver casing, and the case's silk lining and framing materials would have begun to mix with the Shroud's carbon-containing molecules at any temperature over 300 °C, and although the water bath would have caused additional molecular exchange, no testing or measurements were ever published to demonstrate that the fire damage in any way altered the cloth due to isotopic exchange. The work described below is based on the probability that chemical exchange processes involving unscreened OH groups in textile cellulose chains and CO and CO_2 in the presence of water, heat and silver cations as a catalyst may very well have taken place as a result of this fire.

A Laboratory Fire-Simulating Model (FSM)

In the present study, we have devised an experimental model to simulate the major physical and chemical conditions of the 1532 Chambéry fire. Our aim was to throw light on three key questions: (1) Did fire-induced chemical modification of the Shroud's cellulose content take place? (2) What kind of chemical modification was most probable? (3) What impact would chemical modification make on the accepted radiocarbon dates for the Shroud? Both modern and known-age old textile samples (100 B.C. - 100 A.D.) were incubated under these fire-simulating conditions followed by radiocarbon dating with a correction for the fractionation of C-isotopes and cellulose chemical structure analysis. Intact non-incubated textile samples were used for controls. During the development of our fire-simulating laboratory model, we used a detailed description of the 1532 Chambéry fire (*12*) and expert advice from the Moscow Military Fire Defense Academy. The appearance of several controversial publications on the accuracy of the dating of the Shroud of Turin (*11, 13-15*) was an added reason to initiate this research.

Experimental Section

The general scheme of the experiments for the evaluation of the possible impact of the 1532 Chambéry fire on the Shroud of Turin radiocarbon dating results is shown in Figure 1.

Materials and Reagents. Non-dyed linen textile from the long-fibered flax plant, *Linum usitatissimum* (Krasnodar Textile Factory, 1993); Early Roman period linen sample excavated at En Gedi, Israel, radiocarbon dated at 100 B.C. - 100 A.D. (Israel Antiquities Authority, per M. Moroni); cellulase (1,4-[1,3;1,4]-β-D-Glucan 4-glucano hydrolase; E.C. 3.2.1.4 (Sigma); one unit of this enzyme liberates 1.0 mole glucose from cellulose in one hour at pH 5.0 at 37 °C, 2 hr incubation time); Diaflo YM-1 ultrafiltration membranes, 100 dalton exclusion limit (Amicon); Sephasorb SP500 sorbent (Serva-Heidelberg); 70 cm fused silica capillary electrophoresis columns (i.d. = 50 μm, o.d. = 365 μm; Polymicro Technologies).

Cleaning of Textiles. Textile samples were defatted with an ethanol-benzene (1:2, v/v) mixture for 6 hours and air dried. The defatted samples were submerged in an aqueous solution containing 7.7% formaldehyde, 7.7% borax, and 0.5% sodium dodecylsulfate for 3 minutes and then oven dried at 100 °C for 2 hours. All samples were then washed with chromatographically deionized water (Amberlyte), air dried and stored in sealed, dry flasks.

Near-Infrared Spectrometry. A LOMO-450 multichannel computerized IR spectrometer (LOMO Instruments, St. Petersburg, Russia) equipped with a geometric noise filter for removing spectral variations from any position variations of the samples tested was used in our measurements. A sample compartment equipped for nondestructive reflectance analysis of textiles was employed for all near-IR measurements. The BEST/BEAST computer algorithms, which scale spectral vectors in multidimensional hyperspace with a directional probability (*16, 17*) were used for the computerized analysis of the spectra collected. Each textile sample was repositioned and scanned three times in the sample compartment to reduce positioning artifacts in the spectra. The final spectrum retained for each sample was the Euclidean filtered average of these three scans.

Thermal and Gas Treatment of the Textile Samples (FSM Model). The textile samples were incubated for 90 minutes at 200 °C in an artificial atmosphere containing CO_2 (0.03%), CO (60 μg/m^3), and demineralized H_2O (20 g/m^3) in a Medicel-RX200 Thermogas Laboratory Unit (Medtekhnica, Moscow). In this procedure the vaporized demineralized water had been previously incubated with silver metal (40 g/L for 10 days). The resulting concentration of silver cations was 0.80-1.45

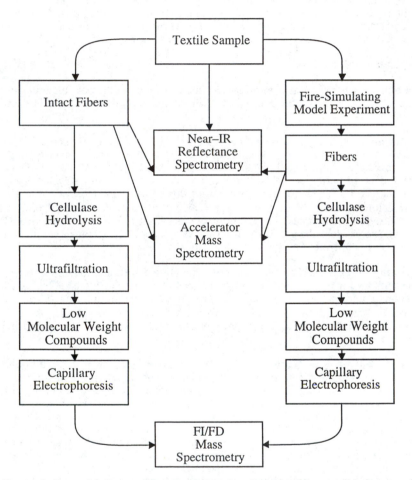

Figure 1. General scheme of the experimental evaluation of the possible impact of the 1532 Chambéry fire on the Shroud of Turin radiocarbon dating results.

μg/L, as determined and monitored with an AAA300 Atomic Absorption Spectrometer (Karl Zeiss, Jena, Germany).

Enzymatic Hydrolysis of the Textile Cellulose. Immediately following incubation of the textile samples by the thermal and gas treatment described above, both experimental (incubated) and control (nonincubated) samples were carefully washed twice with petroleum ether (120 mL/g sample) and then twice with demineralized water (300 mL/g sample) followed by ultrasonic cleaning and oven drying at 80 °C or at room temperature. Samples of approximately 2.0-2.8 g of mechanically disintegrated (crude fibrous material) samples were incubated at 37 °C for 6 hours in 20 mL of 15 mM Tris-HCl (pH 5.0) buffer containing 1.5 mM $MgCl_2$ and 80 units of cellulase per mL. After the incubation, the pool of low molecular weight compounds was separated from the hydrolysate by ultrafiltration through the Diaflo YM-1 membrane in Amicon MMC-10 apparatus (Amicon B.V., Wageningen, The Netherlands). It was then completely desalted by preparative HPLC on a Sephasorb SP-500-1.1 X 15 cm column (2000 psi, 25 °C, 10% water/methanol, v/v) followed by lyophilization of the total monosaccharide fraction (8).

Capillary Zone Electrophoresis. The capillary zone electrophoresis (CZE) results reported here were performed using the capillary columns described above in an Elma 2000 CZE apparatus (NPO Electron Instrument, Zelenograd, Russia) containing a 35 kV high-voltage power supply coupled directly to an MK80 mass spectrometer (NPO Electron Instruments) with on-line coupling (18). The separation parameters were 14.0 kV, 50 A, 7 s injection time at 12 kV, 27 °C thermocooler temperature, 100 mM tetraborate buffer (pH 9.0) separation medium. The most efficient detection and quantification of peaks was achieved using a modified differential refractive index detector (19). In a separate series of determinations, similar but less reproducible results were obtained using a UV-detector at 190 nm. These determinations were made possible by the fact that the tetraborate separation medium induced a red shift in the quartz UV absorption profile of the saccharides (20).

Mass Spectrometry. All electrophoretic fractions were automatically transferred into the pure glycerol matrix-containing copper probe tips (2 μL) inside the MK80 mass spectrometer directly interfaced with the CZE system. The final glucose concentration range in the applied samples was 1.0-30.0 mM depending on the individual CZE fraction which normally corresponds to 1.0-5.0 μL of the post-electrophoretic solution. Both the field ionization (FI) and field desorption (FS) capabilities of the MK80 were used (21). All mass spectra were normalized to the protonated glycerol peak with ensuing computerized interpretation using conventional software. A series of scans was accumulated for each spectrum, the number of which varied from 10 to 100 depending on the injection time.

Determination of Wigley-Muller Correction Parameters for Flax-Dependent Fractionation of C-Isotopes. In a separate series of FSM experiments, the correction parameters proposed by Wigley and Muller (5), $\delta^{13}C$, $d^{14}C$, $D^{14}C$ and $\Delta^{14}C'$, were estimated conventionally (2, 7) for both cleaned pre- and post-incubated textile samples. The $\delta^{13}C$ of the samples were measured with a Nuclide 6-60RMS mass spectrometer (Intertechnique, Rennes, France). For $d^{14}C$ and $\Delta^{14}C'$ determinations the samples were combusted to carbon dioxide using a high-temperature total organic carbon analyzer which was interfaced to a MK2000SE tandem accelerator mass spectrometer (AMS) (NPO-Planeta, Protvino, Russia). The subsequent routine AMS determinations were the basis for further radiocarbon dating calculations (22, 23).

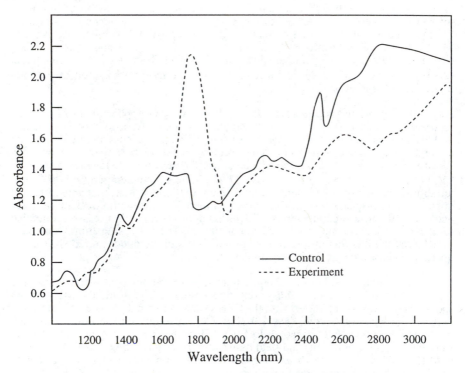

Figure 2. Near infrared reflectance spectra of cleaned modern textile samples incubated (experiment) and non-incubated (control) under the fire simulating model (FSM) conditions. Reproduced from Reference 37 with permission.

These correction parameters are defined as follows:

$$\delta^{13}C = \left[\frac{\left(^{13}C/^{12}C\right)\exp}{\left(^{13}C/^{12}C\right)PDB} - 1 \right] \times 1000\%o \qquad [1]$$

$$d^{14}C = \left[\left(A/A_0 \right) - 1 \right] \times 1000\%o \qquad [2]$$

$$D^{14}C = d^{14}C - 2\left(\delta^{13}C + 25\right)\left(1 + \frac{d^{14}C}{1000} \right)\%o \qquad [3]$$

$$\Delta^{14}C' = \left[\frac{1}{D^{14}C} \times 10^6 \right] \qquad [4]$$

where A = Experimental ^{14}C specific activity (dps/g sample); A_0 = Initial ^{14}C specific activity (dps/g sample), close to the modern atmospheric level; exp = experimentally measured ratio; PDB = conventional paleontological standard of *Belimnitella* fossil (*2, 6, 24*). The relationship of these parameters to the radiocarbon age is discussed in the following section.

Results

Near-IR Reflectance Spectrometry Results. We started our research with near-IR reflectance spectrometry of FSM-treated and control samples of modern linen textiles. Our work was based on findings from other studies on carboxylated Vitamin A derivatives (*25- 27*) and 2-carboxy-D-glucose synthesized from bromoacetylcellulose (*27*) that demonstrated the presence of carboxy-specific peaks in the near-IR spectrum in the 1750 nm region. These studies also showed that the contribution of free unhindered OH groups in the near-IR spectra of various compounds, including monosaccharides, shows significant absorption in the wavelength region between 2600 and 2900 nm. Our spectral results (Figure 2) clearly show a significant decrease in absorbance in the 2600-2900 nm range, indicating that partial dehydroxylation of the cellulose as a result of the FSM treatment of the textile has occurred. This decarboxylation may be associated with the simultaneous carboxylation of glucose residues as indicated by the marked increase of COOH specific signals in the 1700-1900 nm region of the spectrum. The differences between the near-IR spectra obtained indicates that the FSM treatment introduced carboxyl groups into the molecular structure of the fiber. From the observed spectral square ratio, we estimate that our samples underwent approximately 20% carboxylation.

Results from CZE and AMS Determinations. Our initial approach to carrying out comparisons among cellulose sequences from many different textile samples was to compare the CZE patterns obtained as a result of fractionation of low molecular weight compounds (nonmodified and modified β-D-glucose, plus cellobiose) from enzymatically digested cellulose pools isolated from the textiles being studied (*8, 20*). CZE is a high-speed, efficient analytical tool (*19, 28*) for a broad range of soluble substances, and which is also compatible with on-line coupling to a mass spectrometer (*18, 21*). AMS can provide peak identification for the CZE eluates.
    The relative abundance ratios of glucose to carboxyglucose taken from our CZE results (Figure 3) indicate that about 20% of the glucose residues in our FSM-treated

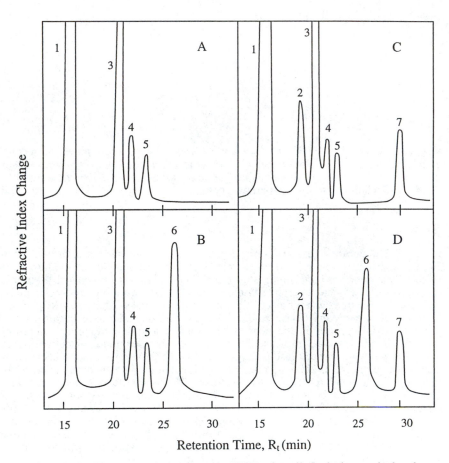

Figure 3. Capillary zone electrophoresis (CZE) of textile hydrolysates isolated from (a) modern intact linen, (B) modern FSM-treated linen, (C) old Palestinian intact linen (2100-1900 yrs BP), (D) old Palestinian FSM-treated linen. Reproduced from Reference 37 with permission.

samples have been carboxylated. The ratios of the relative abundances of the corresponding molecular ions taken from our AMS data (Figure 4) give the same result. These data correlate well with the results of our near-IR spectrometric data discussed above.

The identities of the molecular ions from our AMS results (Figure 4, Patterns II, III, VI and VII) clearly show incorporation of carbon as the carboxylic acid group or as the acetyl group into the cellulose chain on the C-2 of the glucose residues. We propose an overall scheme for such incorporation as shown in Figure 5 for the acetylation of the C-2 carbon in the glucose chain under the conditions of our FSM treatment. Since this process includes the covalent binding of exogenous carbon atoms (from CO and CO_2) by cellulose, such a mechanism would lead to changes in the ^{13}C and ^{14}C contents of the cellulose.

The next step, then, in our research was the direct measurement of the ^{13}C content and ^{14}C activity (AMS) values in FSM-treated and intact old textile samples with subsequent calculation of the $\delta^{13}C$, $d^{14}C$, $D^{14}C$ and $\Delta^{14}C'$ correction parameters. By plotting these parameters against both incubation time and temperature for the FSM treated samples, we can see a significant increase in both ^{14}C (Figures 6 and 7) and ^{13}C(Figure 8) content. The maximum level of this ^{13}C and ^{14}C enrichment occurs during the second hour of incubation at 200 °C. The extent of the ^{13}C enrichment observed after the 1-hour incubation period at 200 °C (Figure 8) correlates well with the carboxylation level reached under the same FSM conditions seen from our near-IR spectrometric data (Figure 2).

The radiocarbon age, t, can be estimated using the Wigley-Muller correction parameters and the following equations:

$$K_0 = \frac{K_{st}}{K_{exp}} \; ; \; K_{st} = \left[\frac{\delta^{13}C}{D^{14}C}\right]_{st} \; ; \; K_{exp} = \left[\frac{\delta^{13}C}{d^{14}C}\right]_{exp} \qquad [5]$$

$$t = \frac{T_{\frac{1}{2}}}{\ln 2} \times \ln\left(\frac{A_0 \times K_0}{A}\right) \qquad [6]$$

where $T_{1/2}$ = ^{14}C half-life (5740 y); exp = experimental data; st = PDB correction standard defined previously. Another equation system was used simultaneously with the computerized treatment of the AMS measurements and the calibration program (*23*). This approach employed the special oxalate ^{14}C standard, A_{ox} (*4, 6*). The pertinent relationships are

$$A_{corr} = A\left[\frac{2\left(25 + \delta^{13}C_{exp}\right)}{100}\right] \qquad [7]$$

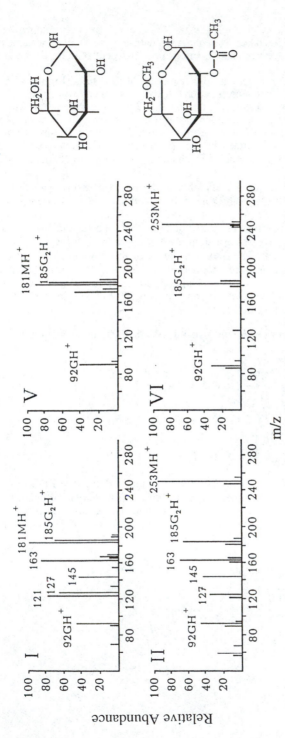

Figure 4. Field ionization (I-IV) and field desorption (V-VIII) mass spectra of the specfic electrophoretic fractions isolated by CZE using the textile hydrolysates as analyzed samples. Two matrix peaks are labeled corresponding to protonated glycerol (m/z 92, GH$^+$) and the protonated dimer (m/z 185, G$_2$H$^+$). The fragment ions at m/z 163, 145 and 127 represent successive losses of H$_2$O from MH$^+$. Reproduced from Reference 37 with permission.

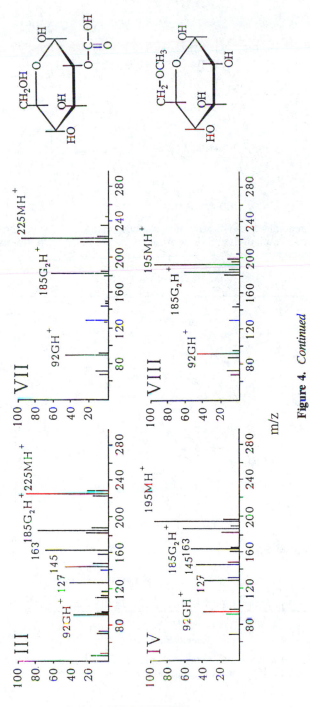

Figure 4. *Continued*

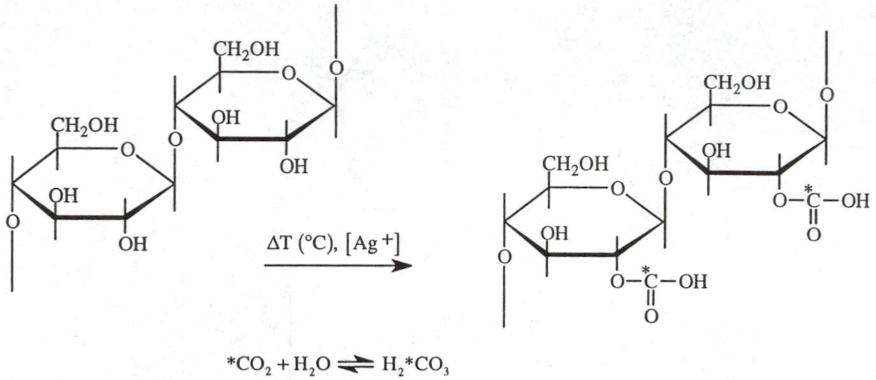

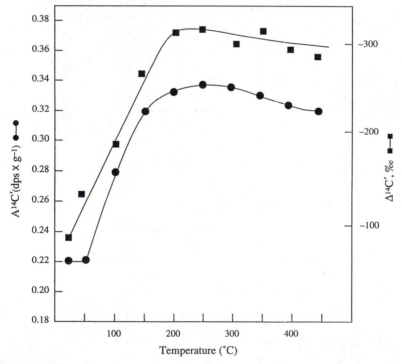

$$*CO_2 + H_2O \rightleftharpoons H_2*CO_3$$

Figure 5. A general scheme illustrating a proposed textile cellulose carboxylation mechanism as a result of the FSM experiments. Reproduced from Reference 37 with permission.

Figure 6. Effect of temperature on the ^{14}C specific activity in linen after incubation of one-hour under FSM conditions. Reproduced from Reference 37 with permission.

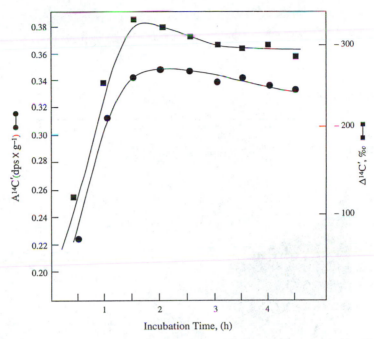

Figure 7. Effect of incubation time on the ^{14}C specific activity in linen at 200 °C under FSM conditions. Reproduced from Reference 37 with permission.

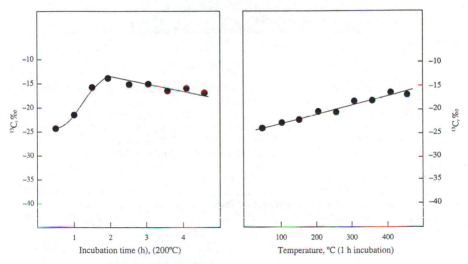

Figure 8. Fire simulation model (FSM): The textile sample ^{13}C-content as a function of temperature and incubation time. Reproduced from Reference 37 with permission.

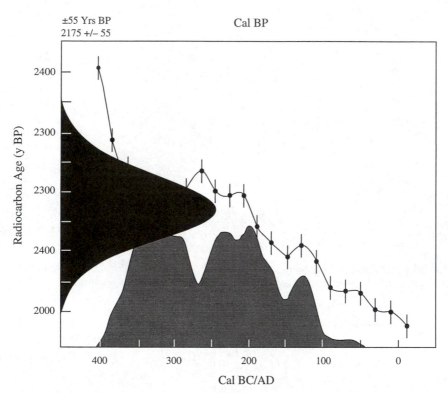

Figure 9. Results of intact old burial linen cloth (En Gedi, Israel) radiocarbon dating reported as a fraction of modern (1950 AD) carbon, as an uncalibrated radiocarbon age in years before 1950, and the calibrated age. $\delta^{13}C = 25.3‰$; radiocarbon age: 2175±55 years BP; calibrated age range: 357-171 B.C. (1 σ, 68% confidence); 386-107 B.C. (2 σ, 95% confidence). The results are reported as a fraction of modern (1950 A.D.) carbon, and as an uncalibrated radiocarbon age in years before 1950 A.D. and the calibrated age. The calibration corrects for variations of the ^{14}C with time, by using the ^{14}C content of known-age tree rings. Values are corrected for $\delta^{13}C$ to the standard value of −25‰ and for $\Delta^{14}C$ taking into account the C-isotopes fractionation index (5, 23). Reproduced from Reference 37 with permission.

$$t = \frac{T_{\frac{1}{2}}}{\ln 2} \ln \frac{0.95A_{ox}\left[1 - \frac{2\left(19 + \delta^{13}C\right)}{1000}\right]}{A_{corr}\left[1 - \frac{2\left(25 + \delta^{13}C\right)}{1000}\right]} \qquad [8]$$

where A_{corr} = ^{14}C activity corrected to the mean C-isotope fractionation index; A_{ox} = oxalate standard activity value proposed by the National Institute of Standards and Technology (NIST) for radiocarbon calculations. Currently, this oxalate standard is an international conventional inter-laboratory standard for radiocarbon dating. Practically, this oxalate standard, obtainable from NIST, should be converted chemically into benzene for ^{14}C scintillation counting or into carbon dioxide for ^{14}C gas counting. In both counting versions, the activity of the A_{ox} should be measured.

Using the above correction parameters, our AMS measurements gave a radiocarbon age of 2175±55 y BP (BP = before 1950) for the intact linen burial cloth obtained from En Gedi, Israel. We measured a ^{13}C of 25.3 ‰, and a calibrated age range of 357-171 BC (1 σ, 68% confidence level) and 386-107 BC (2 sigma, 95% confidence level). The results are reported as a fraction of modern (1950 A.D.) carbon, as an uncalibrated radiocarbon age in years before 1950 A.D. and the calibrated age. The calibration corrects for variations of the ^{14}C with time by using the ^{14}C content of known-age tree rings. Our values are corrected for ^{13}C to the standard value of -25‰ and for ^{14}C taking into account the C-isotopes fractionation index of Wigley and Muller (5, 23). These values correlated very well with previously measured dates for this sample as cited in the introduction above. These results are reported graphically in Figure 9.

On the other hand, when a sample of this identical En Gedi burial cloth underwent FSM treatment, our results using the same methodology outlined above yielded an age of 800±50 y BP (1150-1260 A.D.) with a ^{13}C = 22.0‰, and calibration ranges of 1090-1237 A.D. (1 σ) and 1044-1272 A.D. (2 σ). These results are shown in Figure 10

Discussion

The discussion that follows must allude to the fact that the Shroud of Turin was dated with 95% confidence between 1260-1390 A.D. using the conventional AMS technique (3) on a single piece of cloth isolated from a point very near a water damage stain area on the Shroud (11, 15). Most comments regarding these results added that the Shroud was obviously a forgery (13, 15). On the other hand, even the scientists involved in the dating procedure have made statements that question acceptance of the medieval date. For example, the Oxford laboratory, one of the participating teams, has declared that a major source of error in the dating procedure lies in the pretreatment method (contaminant removal) of the sample (11). Others have declared that it is impossible to trust a single date, or a series of dates on a single feature, to settle an important historical issue; and with respect to radiocarbon dating, it is impossible to claim that all contaminants had been completely removed or that all chemical modifications had been excluded and that the dating range was the actual calendar age of the artifact (1, 2, 6, 7, 24, 29). Taking all of this into account, we must say that problem of the dating of the Shroud of Turin is still unsolved.

The results found in our work provide reason enough to propose that the ^{13}C = -16(-19)‰ as the corrected ^{14}C normalization standard instead of the (-25‰) value which was used by Damon, et al. (3) in their radiocarbon testing of the Shroud of Turin

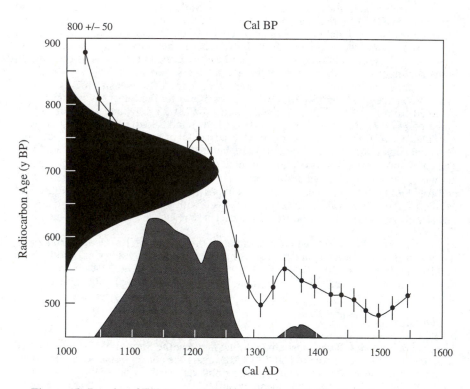

Figure 10. Results of FSM-treated old burial linen cloth (En Gedi, Israel) radiocarbon dating reported as a fraction of modern (1950 AD) carbon, as an uncalibrated radiocarbon age in years before 1950, and as a calibrated age. 1150-1260 A.D. (800±50 years BP). $\delta^{13}C = 22.0‰$; radiocarbon age: 800±50 years BP; calibration age range: 1090-1237 A.D. (1 σ); 1044-1272 A.D. (2σ). Reproduced from Reference 37 with permission.

without taking into account both flax spinning physical and chemical changes and C-isotope flax-dependent biofractionation. In fact, it is impossible to be confident about the ^{14}C level in the Shroud right after its manufacture simply because of the great variation of ^{14}C and ^{13}C concentration levels in flax populations now and in the past. These variations depend on numerous ecological conditions including solar activity and variations in different geographic regions on Earth (*30 - 32*). According to Damon, et al. (*3*), the $^{13}C/^{14}C$ ratio measured in their work using a sample of cloth from the Shroud of Turin was "normal," i.e., equal to -25(-27)‰, which led those authors to conclude that there was no essential enrichment of the Shroud by heavy C-isotopes.

Responding to this conclusion, we would note that if our FSM experiments have given correct information concerning the probability of ^{13}C and ^{14}C carboxylation and related covalent binding with the textile cellulose, it might be possible to conclude that the "normal" values of the $^{13}C/^{14}C$ ratio mentioned above could be a result of the fire-induced incorporation of exogenous carbon into the textile cellulose in the case of relatively low $^{13}C/^{14}C$ content in the original [manufactured textile of the Shroud (pre-fire matter)]. In our view, this statement seems logical in the light of the numerous known data describing the fact that in a number of modern manufactured linen textile samples, the range of varieties of $^{13}C/^{14}C$ indexes is very wide, from -46‰ up to -10‰, depending on such circumstances as the ecological conditions of flax culturing, the geographical and climate context, technological details of manufacture, such as cellulose isolation and purification procedures, etc. (*8*). Moreover, the large 24-37 year cyclic fluctuations of the $^{13}C/^{12}C$ content values in the higher plants including the local Middle East plant populations during the past 800 years have been clearly described as a result of mass-spectrometric analyses of the different "layers" of circular tree rings in old living trees (*32 -35*). This means that even very low as well as very high levels of ^{14}C and ^{13}C in the ancient Middle East long-fibered flax populations should not be excluded. If so, the ^{14}C and ^{13}C content in the original manufactured textile of the Shroud could be significantly greater or less than the so-called "normal" level observed in only a portion of modern linen textile samples ever tested. Thus, the conventional statement about the equality of $^{13}C/^{12}C$ ratios in the original manufactured linen of the Shroud and in recently manufactured textile samples should be labeled an assumption.

Since our data show that the $^{13}C/^{12}C$ equality assumption is incorrect, it is necessary to make appropriate corrections to the radiocarbon dating model. In addition to the ecological effects on flax plant C-isotope content mentioned above (*6, 8 - 10, 32, 35*), it is also necessary to take into account the Craig fractionation asymmetry effect, which if left uncorrected might cause a dating error of 400-500 years for any object of biological origin (*5, 6, 36*).

Conclusion

Despite inherent difficulties in its use, radiocarbon dating can be a useful approach to old textile studies as part of multi-disciplinary research programs which include art-historical dating and chemometric investigations (*20*). Critical judgments by archaeologists based on contextual data and statistical inferences may produce conclusions that an uncritical reading of radiocarbon results would not support. In our view, a more discriminating interpretation of the radiocarbon data may support or refute existing hypotheses concerning old textile chronology and may reveal others that are not immediately apparent.

The Shroud of Turin is one of the most unique old textile relic in existence. It should be dated using a multi-disciplinary approach including radiocarbon dating with special corrections for both biological fractionation of C-isotopes and fire-induced carboxylation with subsequent textile cellulose ^{13}C and ^{14}C enrichment.

As a result of our current experimental study, we have found that biofraction-ation and fire-induced carboxylation corrections modify the conventional radiocarbon methods, and that the use of this modified, corrected method leads us to conclude that the actual calendar age of the Shroud of Turin could be closer to the 1st or 2nd century A.D.

Acknowledgments

This work was supported by a grant from the Guy Berthault Foundation, Meulan, France. The authors are also indebted to the following for their help and advice: M.-C. Van Oosterwyck-Gastuche, M. Moroni, A. D. Adler, A. Volkov, W. Brostow, N. Sazhin, S. Bakhroushin, O. Bakhrushina, S. Berdyshev, J. Teller, and P. C. Maloney.

Literature Cited

1. Shott, M. J. *American Antiquity* **1992**, *57*, 202-230.
2. Taylor, R. E. *Radiocarbon Dating: An Archaeological Perspective*; Academic Press: New York, NY, 1987; pp 99-108.
3. Damon, P. E., et al. *Nature* **1989**, *337*, 611-615.
4. Polach, H. A. In *Proceedings of the Eighth International Radiocarbon Dating Conference, Lower Hutt, New Zealand, 18-25 October, 1972*; Rafter, T. A.; Grant-Taylor, T., Eds.; Royal Society of New Zealand: Wellington, NZ, 1972; pp 98-107.
5. Wigley, T. M. L.; Muller, A. B. *Radiocarbon* **1981**, *23*, 173-190.
6. Kuptsov, V. M. *Absolute Paleogeochronology*; Nauka Publishers: Moscow, 1986; pp 202-208 (in Russian).
7. Van Strydonck, M. J. Y.; Van Der Borg, K.; De Jong, A. F. M. *Radiocarbon* **1992**, *34*, 928-933.
8. Kouznetsov, D. A.; Ivanov, A. A.; Ryabchenko, S. L.; Podobed, O. V. In *Proceedings of the 7th Russian National Meeting on Advanced Methods in Archaeological Science, St. Petersburg, 10-14, December, 1993*; Slokowicz, B. A., Ed.; Sadko Publishers: St. Petersburg, 1993; pp 116-139 (in Russian).
9. Baity, E. C. *Man Is a Weaver*; Viking Press: New York, NY, 1942; pp. 37-55.
10. Lee, J. S. *Elementary Textiles*; Prentice-Hall, Inc.: New York, NY, 1953; pp 47-59.
11. Stevenson, K. E.; Habermas, G. R. *The Shroud and the Controversy*; Thomas Nelson Publishers: Nashville, TN, 1990; pp 46-60.
12. Cardot, V. R. *Savoy Family*; S. and L. Willspey Publishers: Edinburgh, 1934; pp 216-242.
13. Dickman, R. T. In Advances in Science and Technology; Niemer, J. E., Ed.; Arnica Publishers: Prague, 1989; pp 44-57.
14. Harbottle, G.; Heino, W. In *Archaeological Chemistry – IV*; Allen, R. O., Ed.; American Chemical Society: Washington, DC, 1989; pp 313-320.
15. Gove, H. E. *Radiocarbon* **1990**, *32*, 87-92.
16. Carney, J. M., et al. *Analytical Chemistry* **1993**, *65*, 1305-1313.
17. Lodder, R. A.; Hieftje, G. M. *Applied Spectroscopy* **1988**, *42*, 1351-1365.
18. Thompson, T. J.; Foret, F.; Vouros, P.; Karger, B. L. *Analytical Chemistry* **1993**, *65*, 900-906.
19. Bruno, A. E.; Krattiger, B.; Maystre, F.; Widmer, H. M. *Analytical Chemistry* **1991**, *63*, 2689-2697.
20. Kouznetsov, D. A.; Ivanov, A. A.; Veletsky, P. R. *Analytical Chemistry* **1994**, *66*, 4359-4365.

21. Yakimov, S. S.; Zheltkov, R. T. In *Applied Chemistry of Cellulose*; Samarin, L. K.; Bkhurov, V. L., Eds.; Novosibirsk University Press: Novosibirsk, 1990; pp 126-148 (in Russian).
22. Stuiver, M. *Radiocarbon* **1982**, *24*, 1-26.
23. Stuiver, M.; Reimer, P. J. *Radiocarbon* **1986**, *28*, 1022-1030.
24. Stuiver, M.; Pearson, G. W. *Radiocarbon* **1986**, *28*, 805-838.
25. Dmitrovsky, A. A.; Poznyakov, S. P.; Soloviova, N. Y. In *Methods in Biochemistry*; Kretovich, V. L.; Scholtz, K. F., Eds.; Nauka Publishers: Moscow, 1980; pp 125-132 (in Russian).
26. Wetzel, D. L. *Analytical Chemistry* **1983**, *55*, 1165A-1176A.
27. Yablokov, S. L.; Volkogonov, R. T. In *Applied Chemistry and Biochemistry*; Sharov, V. L., Ed.; Novosibirsk University Press: Novosibirsk, 1988; pp 611-639 (in Russian).
28. Colon, L. A.; Dadoo, R.; Zare, R. N. *Analytical Chemistry* **1993**, *65*, 476-481.
29. Schurr, M. R. *American Antiquity* **1992**, *57*, 300-320.
30. Stuiver, M.; Quay, P. D. *Science* **1980**, *207*, 11-19.
31. Stuiver, M.; Braziunas, T. F. *Nature* **1989**, *338*, 405-408.
32. Tieszen, L. L. *J. Archaeol. Science* **1991**, *18*, 227-248.
33. Eddy, J. A. Science **1976**, *192*, 1189-1202
34. Pilcher, J. R.; Baillie, M. G. L.; Schmidt, B.; Becker, B. *Nature* **1984**, *312*, 150-152
35. Tieszen, L. L.; Fagre, T. *J. Archaeol. Science* **1993**, *20*, 25-40.
36. Craig, H. *Journal of Geology* **1954**, *62*, 115-149.
37. Kouznetsov, D. A.; Ivanov, A. A.; Veletsky, P. R. *Journal of Archaeological Science* **1995**, *22*, in press.

RECEIVED December 4, 1995

Chapter 19

Factors That Affect the Apparent Radiocarbon Age of Textiles

A. J. T. Jull, D. J. Donahue, and P. E. Damon

National Science Foundation Arizona Accelerator Mass Spectrometry
Facility, University of Arizona, Tucson, AZ 85721

We comment on a paper in this volume by Kouznetsov et al. (*1*), who report that the ^{14}C age of the textile, which was originally dated as about 2,195±55 yr BP was changed by 1,400 yr to about 800 yr BP, by heating in air containing small amounts of water. These authors also make several claims about the reliability of radiocarbon dating of cellulose and cellulose-containing textiles. They include statements that ^{14}C is not distributed uniformly in flax, and an untested claim that isotopic fractionation of ^{14}C relative to ^{13}C and ^{12}C occurs in some way different from that accounted for by the usual equations employed by radiocarbon laboratories. They use their results to question the validity of all radiocarbon measurements, and specifically to criticize the radiocarbon results on the Shroud of Turin (*2*).

Kouznetsov et al. (*1*) report on some ^{14}C results obtained by accelerator mass spectrometry (AMS) on a linen textile (En Gedi, Israel) which was exposed to a relatively mild procedure of heating to 200 ºC for 90 minutes in air. They report that the ^{14}C age of the textile, which was originally dated as about 2,195±55 yr BP was changed by 1,400 yr to about 800 yr BP, under these conditions. These authors make several claims about the reliability of radiocarbon dating of cellulose and cellu-lose-containing textiles. They include a statement that ^{14}C is not distributed uniformly in flax, and an untested claim that isotopic fractionation of ^{14}C relative to ^{13}C and ^{12}C occurs in some way different from that accounted for by the usual equations employed by radiocarbon laboratories, discussed by Stuiver and Polach (*3*). Kouznetsov et al. (1) have used results described in their paper to question the validity of radiocarbon measurements of textiles in general, and specifically to disparage results on the Shroud of Turin (*2*).

In another result quoted in this paper, Kouznetsov et al. (*1*) state that for textiles subjected to the heat treatment described above, "near IR spectra obtained from whole textile samples, subjected and not subjected to the gas/thermal treatment (FSM) indicates that the treatment introduced carboxyl groups into the molecular structure of the fiber (Figure 2)" and further "it is evident from the relative abundance of ratios of

NOTE: Figures cited in this chapter appear in Chapter 18: Figure 3, page 236; Figure 4, page 238; Figure 6, page 240; Figure 7, page 241; Figure 9, page 242; Figure 10, page 244.

0097–6156/96/0625–0248$12.00/0

glucose to carboxyglucose CZE fractions (Fig. 3) and corresponding molecular ions (Fig. 4), that about 20% of the glucose residues have been carboxylated." These statements are both made in support of the claim that 20% of the glucose residues are carboxylated, without any quantitative information from either method discussed.

We have attempted to reproduce a textile heating effect similar to that reported by Kouznetsov et al. (*1*), without success. Because of their conclusions and also because this work is clearly flawed in several respects, we feel it is important to comment both on their results, discussion and interpretations.

Experimental

In order to check the assertion that heating a textile in an atmosphere containing CO_2 would change its $d^{13}C$ and radiocarbon age significantly, we performed an experiment similar to that of Kouznetsov et al. (*1*) in our laboratory.

A sample of 3.6 mg of the En Gedi textile, from the same piece of material used for the original ^{14}C dating at Arizona, was put into a 9 mm Pyrex tube, and 1.9 cm^3 CO_2 (at 1.0 atm and 25 °C) was cryogenically trapped into the same tube. The amount of CO_2 was determined by measurement of the gas pressure in a known volume, using a capacitance manometer. The amount of carbon in the gas is 0.93 ng carbon and the amount of carbon in the textile is approximately 40% of the mass of the textile, i.e., 1.44 ng. The CO_2 had been previously prepared from combustion of NIST (*3*) oxalic acid standard II (SRM 4990C). The tube was evacuated and sealed using a glass torch and placed in a muffle furnace and heated to 200 °C for 15 1/2 hours. After this time, the tube was placed in a cracking device, and the CO_2 in the tube was recovered. The gas sample was split into two fractions, one to make graphite for AMS analysis, the second for $d^{13}C$ measurements. The textile sample was also recovered. The textile sample was combusted to produce CO_2, and this gas sample was also split into two fractions, one to make graphite for AMS analysis, the second for $d^{13}C$ measurements.

The samples of oxalic II CO_2 and textile CO_2 from before and after the heating experiment were analyzed for $d^{13}C$ on a Fisons Optima stable-isotope mass spectrometer. The remaining samples of oxalic-II and textile CO_2 were converted to graphite, pressed into an accelerator target and measured for ^{14}C by AMS using the University of Arizona instrument.

Results

Results of the study are given in Table I. The radiocarbon measurements are reported as fraction of modern carbon (Fm), where "modern" is taken as 1950 AD carbon, and as a radiocarbon age in years before present (1950 A.D.). As can be seen from Table I, there was a small change in the $d^{13}C$ value of the textile, due to removal of a small amount of organics by charring. The sample was slightly darker after the 200 °C treatment than initially. This effect is in agreement with the pyrolysis treatments of Leavitt et al. (*4*) on white fir cellulose, who observed a fractionation of -0.4‰ on heating this material to 200 °C under vacuum. The change is caused by a small loss of volatile organics during the heating. Leavitt et al. (*4*) noted that the direction of the $d^{13}C$ change in the cellulose was negative, indicating that the volatiles are enriched in ^{13}C by a small amount.

We can also confirm that this mechanism (*4*) is correct, as the CO_2 gas phase after the experiment had $d^{13}C = -17.8‰$ and fraction modern carbon (Fm) = 1.30±0.01, whereas the starting gas was characterized by $d^{13}C = -17.8‰$ and Fm = 1.35 (*3*). A small amount (~80 μg C) of contamination of the CO_2 by the desorbed volatiles (of 2,195 yr BP ^{14}C age) from the textile can account for this effect. The weight of the textile recovered was about 300 μg less than the initial amount.

Table I. Results of ^{14}C and $d^{13}C$ Measurements of En Gedi Textile Exposed to CO_2 Gas at 200 °C

Sample	$d^{13}C$	$Fm^{14}C^a$	^{14}C Age^b
Before heating experiment , 2195±55	−25.3‰	0.7609±0.005	
After heating experiment	−25.9‰	0.7649±0.0042	2153±44
Net change, −42±70	−0.6‰	0.0049±0.0067	

^{a}Fraction of modern ^{14}C, where modern is defined as 95% of the activity of the NIST oxalic acid-I standard.

^{b}Corrected to $d^{13}C$ of −25‰

The results for ^{14}C show that there is no observable change in the ^{14}C age of the En Gedi textile under the conditions of our experiment. Any fractionation in ^{14}C composition is corrected by $d^{13}C$, to −25‰, which is the conventional practice (5). The results also indicate that there is no observable deviation from the assumption that a change in $^{14}C/^{13}C$ is approximately equal to a change in $^{13}C/^{12}C$ (6).

Based on the known amount of CO_2 and our estimate of the volume of the Pyrex tube, the conditions of our experiment were that the experiment was conducted under pCO_2 of about 0.06 atm, compared to Kouznetsov et al.'s use of air, which has pCO_2 of 0.0003 atm. Additionally, we heated our sample for 10.3 times longer than the 90 minute experiment of Kouznetsov et al. If we assume an approximation of a simple first-order process, the rate of our experiment should be 200 times faster than Kouznetsov's experiment. Our results show evidence neither for alteration of the age of the textile, nor for significant isotopic exchange of the textile under these conditions.

Discussion of Isotope Measurements

One may wonder why our radiocarbon measurements on the thermally-treated En Gedi textile differ so markedly from those of Kouznetsov et al. These authors report that they used a tandem accelerator mass spectrometer located at the Russian Academy of Sciences, Protvino. They also indicate the machine is coupled to a total carbon analyzer "interfaced" to the AMS. However, in their paper there are no citations to any published work from the Russian laboratory, or indeed, to any other AMS group. There is no discussion of how the novel gas-sample "interfacing" was achieved. The only traceable radiocarbon date is one done on the original En Gedi textile in our laboratory (AA-12704). The remaining radiocarbon work was all performed at a laboratory which is new and not generally known to Russian scientists or the international AMS community. As this facility is new, and there are no published reports on its performance, the paper presented here by Kouznetsov et al. should contain AMS ^{14}C data on internationally-accepted standards, known-age samples and blank measurements. The operating conditions of the AMS equipment should also be discussed. For example, are data on ^{14}C measurements before the various corrections of the authors available?

Kouznetsov et al. (*1*) include a series of radiocarbon measurements made on textile samples heated in air to various temperatures. In Figures 6 and 7 of this paper, these workers indicate that the ^{14}C activity of the textile, stated in dps/g, increases from 0.22 dps/g carbon to 0.33-0.34 dps/g. We point out that "modern" carbon contains ^{14}C with an activity of 13.5 dpm/g, or 0.225 dps/g (*3*). We also note the following: (1) Kouznetsov et al.'s ^{14}C measurements indicate that the En Gedi textile had an initial ^{14}C age of approximately modern. The sample has been dated previously by our laboratory, and again reported in this paper to be approximately 2,195 yr BP. (2) The samples which had been heated gave ^{14}C activities of up to 0.34 dps/g, equivalent to 150% of the value of modern, pre-bomb ^{14}C. This level cannot be achieved even by complete exchange with contemporary air, which has a ^{14}C level of 110% modern. This indicates that the treated samples were exposed to artificial ^{14}C at a level higher than contemporary carbon and therefore the experiment reported by Kouznetsov et al (*1*) is not as they have reported. Further, it is impossible to derive an age of 700-800 yr BP from the data, using any accepted calculation of ^{14}C ages, or even equation 8 presented by Kouznetsov et al. (*1*) in this article. (3) The results quoted in the captions of Figures 9 and 10 are not consistent. Figure 9 shows that the untreated En Gedi linen has $d^{13}C$ of $-25.6‰$ and a radiocarbon age of $2,175\pm55$ yr BP. Figure 10 indicates that the heat-treated sample has $d^{14}C$ of $-22.0‰$ and a radiocarbon age of 800 years BP. Even assuming that an isotope correction was not applied, a change of 3.6‰ in $d^{13}C$ would result in a change of less than 60 yr BP in radiocarbon age. Either the $d^{13}C$ or the radiocarbon age quoted in Fig. 10 is incorrect. Considering that the measurement of radiocarbon age is completely undocumented, we would assume that it is that measurement which is incorrect. (4) The section of calibration curve shown in Figure 10 does not bear any relation to the curve published by Stuiver and Pearson (*7*). (5) Kouznetsov et al. (*1*) exaggerate the small fractionation effects of stable carbon and ^{14}C. These effects are less than or equal to 9‰ (i.e. 0.9%), accepting the value of $-16‰$ quoted by the authors for flax. Such a change would affect the radiocarbon age by less than 150 yr BP. The authors fail to point out that ^{14}C dates are all normalized to a common $d^{13}C$ value, and that the equations of Stuiver & Polach (*5*) cited compensate for even these effects. (6) The authors discuss some corrections of Wigley and Muller (*6*) for deviations from purely mass-dependent behavior, which they also fail to indicate are very small. These effects cannot cause a change in age of a textile by the anything like the amounts discussed by Kouznetsov et al. The authors also use equations (4 and 8) not derived from the reference of Wigley and Muller (*6*) as is implied. (7) In reference to the comments of Kouznetsov et al. (*1*) on the dating of the Turin Shroud, we point out that if the Shroud sample were heated to 300 °C, it would have charred significantly. We already observe darkening of the En Gedi textile at 200 °C in the experiment reported here. However, the sample of the Shroud dated at Arizona (*2*) showed no evidence of charring. Despite statements made by Kouznetsov et al. (*1*), samples of the Shroud were indeed measured for $d^{13}C$. The quoted values (*2*) were within the usual range for cellulose textiles, and indeed cellulose in general, of about -23 to $-25‰$.

Comments on Kouznetsov et al.'s Chemistry

One of the results of Kouznetsov et al. (*1*) is the report that samples of textile become carboxylated by the heating procedure used by these authors. Unfortunately, there is no quantitation of any of the techniques discussed. Figure 3 of Kouznetsov et al. (*1*) shows a peak in the "heated" samples as opposed to the untreated samples. Neither in their text, nor in their figure legend is any information given about the temperature of heating, or other conditions of this particular experiment. The peak identified as

2-carboxy-β-D-glucose is stated in the text to be 20% of the total sample, yet the peaks for glucose are offscale, so that no quantitative comparison can be made.

For discussion purposes, let us assume that this estimate of 20% 2-carboxy-β-D-glucose is correct, and further that this contamination is recent carbon (from 1994 AD). Then, 20% of the glucose from a textile dated to be 2,195 yr BP (76.1% modern, 0.761 fraction of modern C) are carboxylated at one OH location with a carboxyl group containing C of 110% modern (contemporary) carbon. Glucose contains six carbon atoms. Adding one more as a COOH makes seven carbon atoms. The effect of this addition of one additional carbon to 20% of the molecules on the measured fraction of modern ^{14}C would be:

$$Fm \text{ (heat treated)} = 0.20 \times 1/7 \times F_c$$
$$+ 0.20 \times 6/7 \times (F_{2,195 \text{ yr BP}})$$
$$+ 0.80 \times (F_{2,195 \text{ yr BP}})$$
$$= 0.7707 \ (77.07\% \text{ modern}).$$

In these equations, Fm is the fraction of modern carbon (taken as 1950 A.D.), Fc is 1.10, the fraction of modern carbon for contemporary (present-day) material, and $F_{2195yrBP}$ is 0.7609, the fraction of modern for material of 2,195 yr BP radiocarbon age. The third term represents that portion of the sample unaffected by the treatment. For these values, the fraction of modern of the heat-treated sample would be Fm = 0.7707, and the radiocarbon age would be 2,092 years BP.

Thus, even if the 20% carboxylation of Kouznetsov et al. (1) were correct, a result certainly not demonstrated in this paper, a change in the radiocarbon age of about 100 years would result. It is not possible to generate an age of 800 yr BP even if all glucose molecules became substituted with a carboxyl group of recent age.

In the thermal gas treatment experiment described, the textile is exposed to "an artificial atmosphere containing CO_2 (0.03%), CO (60 μg/m³) and 20g/m³ water." The size of the chamber in the "Thermogas Unit" was not given, but let us assume it was one hundred liters. This would mean there is 300 ppm volume of CO_2, or 30 cm³. This amount of CO_2 contains about 15 mg of carbon at 25 °C. The CO concentration is very small and cannot account for much reaction. Even in the extremely unlikely event that all of the carbon in the gases exchanged at 200 °C in 90 minutes with all of the 2.0-2.8g of textile stated to have been heated, the amount would be about 1.4% of the carbon being modern instead of 0.761 times modern, and this would only change the apparent fraction modern to 0.765, i.e. the apparent age would be 2,150 years instead of 2,195 yr BP. One can perform similar calculations for other volumes.

Conclusion

In conclusion, we believe that the ^{14}C methods described by the authors have not had appropriate control experiments performed. Additionally, the AMS ^{14}C measurements were done on an apparently untested piece of equipment with no reference to normal procedures of reproducibility, standards, control and blank samples. Further, the results obtained cannot be produced by contamination with contemporary carbon, as asserted by the authors.

With a similar experiment, we find no evidence for the gross changes in age proposed by Kouznetsov et al. (1). These authors use a number of procedures on the samples, without any discussion of control samples, blanks or standards run through the same battery of treatments.

Finally, we have shown that even if the carbon displacements proposed by the authors during the heat treatment were correct, no significant change in the measured radiocarbon age of the linen would occur. We must conclude that the challenge by

Kouznetsov and his co-workers on measurements of the radiocarbon age of the Shroud of Turin and on radiocarbon measurements on linen textiles in general are unsubstantiated and incorrect. We further conclude that other aspects of the experiment are unverifiable and irreproducible.

Acknowledgments

We thank the editor of the ACS volume for the opportunity to reply to this paper. We are also grateful to our colleagues at the NSF Arizona AMS Facility, particularly J. W. Beck and G. S. Burr for their comments and suggestions. We are also grateful to G. E. Kocharov, Y. V. Kuzmin and A. Peristykh for their insight into the status of radiocarbon and lack of availability of AMS measurements in Russia. The NSF Arizona AMS Facility is supported in part by grants from the National Science Foundation.

Literature Cited

1. Kouznetsov, D. A.; Ivanov, A. A.; P. R. Veletksy, P. R. In *This Volume*, Chapter 18.
2. Damon, P. E.; Donahue,D. J.; Gore, B. H.; Hatheway, A. L.; Jull, A. J. T.; Linick, T. W.; Sercel, P. J.; Toolin, L. J.; Bronk, C. R.; Hall, E. T.; Hedges, R. E. M.; Law, I. A.; Perry, C.; Bonani, G.; Trumbore, S.; Wölfli, W. *Nature* **1989**, *337*, 611-615.
3. Mann, W. B. *Radiocarbon* **1983**, *25*, 519-527.
4. Leavitt, S. W.; Donahue, D. J.; Long, A. *Radiocarbon* **1982**, *24*, 27-35.
5. Stuiver, M.; Polach, H. A. *Radiocarbon* **1977**, *19*, 355-363.
6. Wigley, T. M. L.; Muller, A. B. *Radiocarbon* **1981**, *23*, 173-190.
7. Stuiver, M.; Pearson, G. W. *Radiocarbon* **1986**, *28*, 805-838.

RECEIVED December 23, 1995

Chapter 20

Analysis of Cellulose Chemical Modification: A Potentially Promising Technique for Characterizing Archaeological Textiles

D. A. Kouznetsov, A. A. Ivanov, and P. R. Veletsky

E. A. Sedov Biopolymer Research Laboratories, Inc., 4/9 Grafski Pereulok, Moscow 129626, Russia

Cellulose chains in many archaeological textile remains contain a significant number of chemically modified β-D-glucose residues. This work, using a capillary zone electrophoresis-mass spectrometric approach, has demonstrated a correlation between cellulose alkylation extent and calendar age of the textile samples tested. The results suggest that if cellulose alkylation is the consequence of microbial activity, this phenomenon could be the basis of a new and efficient dating technique, at least among samples taken from a single site and subjected to a similar environment.

There is currently a great deal of interest in the development of new approaches to chemical investigations of archaeological textile remains (1, 2). These studies may be an important aid to understanding the chemical mechanisms of textile aging and therefore may be useful for improving the accuracy of ancient textile dating procedures. In addition, detailed chemical studies of archaeological textiles could provide important information about ancient technological processes whose procedures have long since been lost (2, 3). Finally, any information on historic textile chemical aging should be important for development of methods for the conservation of textile remains.

Many methods are presently used for the chemical investigation of old textiles. Some of the most useful methods are near infrared (near-IR) and IR reflectance spectrometry, high resolution electron microscopy, atomic absorption spectrometry (AAS), and several chromatographic and electrophoretic techniques (1, 2, 4).

Previous workers have shown that the cellulose chains in many ancient textile remains contain a significant number of β-D-glucose residues that have been chemically modified by functional groups such as the hemiacetal, acetyl, and methyl groups at both the 2- and the 6- positions (5, 6). For this reason, we decided to carry out a comparative study on the possible chemical modification of cellulose from fifteen different archaeological textile remains dated within the range of 1200 B.C. to 1500 A.D. (Table I).

In our opinion, an optimal way to carry out efficient comparison among cellulose sequences of many different museum textile samples such as those we had

0097–6156/96/0625–0254$12.00/0

Table I. Calendar Age and Provenance of Linen Textile Samples Analyzed for Cellulose Enzymatic Hydrolysates by CZE/MS

Museum	Provenance	Calendar Age (y)	Museum Number	Experiment Record Code
Russian National Historical Museum, Moscow	Alexandria, Egypt	3200-3270 BP[a]	34118KL9	A1
Russian National Historical Museum, Moscow	Baguate Site, Egypt (Coptic Linen)	1620-1550 BP[a]	20014AE3	A2
Israel Antiquities Authority, Jerusalem	En Gedi, Palestine	2100-1900 BP		B
Moscow State Institute of Textile Museum	Limerick, Ireland	1650-1600 BP[a]	S411	C1
Moscow State Institute of Textile Museum	Southern England	900-860 BP[a]	R832	C2
Museum of Slavic Applied Art, Vladimir, Russia	Middle Russia	540-480 BP	655E	D
Crimean State Archaeological Museum, Simpheropol, Ukraine	Northern Greece	1140-1070 BP	TK4451	E
West Ukraine Museum of Ethnography and Archaeology, Ternopol	Gniezno, Western Poland	720-670 BP	A8026	F

[a]BP = Before present; these calendar age values were estimated by the radiocarbon method at the Russian Academy of Sciences Radiochemical Center, Moscow; all other samples listed were dated by historical and stylistic evidence only.

in hand is to compare the chromatographic or electrophoretic patterns obtained as a result of fractionation of low molecular weight compounds (nonmodified and modified β-D-glucose and cellobiose) of purified enzymatically digested cellulose hydrolysates isolated from the textiles compared. A combination of capillary electrophoresis and mass spectrometry has been found to be a very efficient method for the analysis of textiles: capillary zone electrophoresis (CZE) provides significant separation efficiency and high analytical speed (*7, 8*) for a broad range of substances in solution, while mass spectrometry provides peak identification. Furthermore, the flow rates from the CZE capillary columns are compatible with on-line coupling to a mass spectrometer (*4, 9*).

Two significant problems have been identified for the analysis of sugars by CZE. (1) Carbohydrates are not charged species under normal conditions; (2) they are difficult to detect because of the low sample volumes necessitated by CZE and low absorbances in the UV. It is possible to overcome both of these difficulties with a single simple process, namely, the addition of borate ions to the sugar medium. This process has the twofold effect of (1) formation of negatively charged sugar-borate complexes that can be separated by CZE (*8*) and (2) a two- to twentyfold enhancement of sensitivity to UV absorbance at 195 nm. Despite this enhancement, however, detectability of sugars is still limited to the nanomole range for UV detectors, which has led to the development of amperometric and differential refractometric methods as alternative means of detection. Using these methods, several workers (2, 7, 8) have achieved very satisfactory detectability limits for monosaccharides and oligosaccharides.

Experimental Section

Materials and Reagents. Nondyed archaeological linen textile samples with known-age certification were purchased from Russian and Ukranian state museums and from one individual. The cellulase (1,4-[1,3;1,4]-β-D-Glucan 4-glucano hydrolase; E.C. 3.2.1.4; was purchased from Sigma; one unit of this enzyme liberates 1.0 mole glucose from cellulose in one hour at pH 5.0 at 37 °C, 2 hr incubation time). Also used were Diaflo YM-1 ultrafiltration membranes, 100 dalton exclusion limit (Amicon); Sephasorb SP500 sorbent (Serva-Heidelberg); 70 cm fused silica capillary electrophoresis columns (i.d. = 50 μm, o.d. = 365 μm; Polymicro Technologies). All chemicals used were Analytical Grade (Serva Heidelberg).

Cleaning of Textiles. Textile samples (each with dimensions of about 2.8 cm^2) were defatted with an ethanol-benzene (1:2, v/v) mixture for 6 hours and air dried. The defatted samples were submerged in an aqueous solution containing 7.7% formaldehyde, 7.7% borax, and 0.5% sodium dodecylsulfate for 3 minutes and then oven dried at 100 °C for 2 hours. All samples were then washed with chromato-graphically deionized water (Amberlyte), air dried and stored in sealed, dry flasks.

Enzymatic Hydrolysis of Textile Cellulose. Approximately 2.0-2.8 g of the mechanically disintegrated textile samples (crude fibrous material) were incubated at 37 °C for 6 h in 20 mL of 15 mM Tris-borate (pH 6.40) buffer containing 4.5-5.0 units of cellulase per mL of solution. At these conditions, 98% cellulose depolymerization is expected (2). Following incubation, the low molecular weight fraction was separated from the hydrolysate by ultrafiltration through the Diaflo YM-1 membranes in an Amicon MMC-10 filtration apparatus (Amicon B.V., The Netherlands) at 2000 psi and then concentrated in a rotor evaporator.

Capillary Zone Electrophoresis (CZE). An Elma 2000 (NPO Electron Instruments) CZE apparatus modified with microcylindrical copper electrodes was used. Fused-silica capillaries pretreated with 0.15 M NaOH and then with 0.10 M tetraborate buffer (pH 9.0) were used. Sample volumes of approximately 50-750 nL were siphon injected and separated at 14.0 kV and 50 A. The on-column laser-based refractive index (RI) detector was designed according to recommendations given by Bruno (7). A general scheme is given in Reference 10.

CZE-Mass Spectrometer (MS) Interface. The CZE system was interfaced with a MK80 Field Ionization (FI)/Field Desorption (FS) mass spectrometer (NPO Electron Instruments) using a approach described by Thompson and Kouznetsov (9, 10).

Mass Spectrometry. All electrophoretic fractions were transferred into the MK80 MS through the CZE-MS interface described above. Each of the MK80 internal cop-per probe tips contained 2 L of the pure glycerol matrix material. For both the FI and FD capabilities of the MS the anode/cathode potential difference was 8.0 kV; a potential gradient of 108 V/cm developed. In both FI and FD versions, ionization occured when a molecule was subjected to a high potential gradient when in the vicinity of the anode. The positive ions are drawn towards the cathode and then into the mass analyzer. In FI, the sample is evaporated and molecules come very closely to or impinge upon the anode (emitter) where they are ionized. In FD, the sample is coated onto the emitter and the ions are desorbed from the solid state (*2, 4, 11*).

All mass spectra were normalized to the protonated glycerol peak followed by computerized interpretation (chemical structure estimate) using a conventional FORTRAN/PAD algorithm. The computer was connected with the databank of the Russian National Center for Ecology Studies in Moscow. A series of scans were accumulated for each sample. The number of scans varied inversely with the injection time. A general scheme of the entire experimental procedure is shown in Figure 1.

Results and Discussion

Our data presented in Figures 2 and 3 show that all eight archaeological textile samples tested (Table I) contain alkylated cellulose chains. This alkylation phenomenon includes formation of such derivatives of glucose residues as 2-acetyl-6-methyl-β-D-glucose and 6-methyl-β-D-glucose (samples A1, A2 and B); 2-carboxy-6-methyl-β-D-glucose (samples C1 and C2); 2-carboxy-β-D-glucose (samples D, E and F); and 2,6-dicarboxy-β-D-glucose (sample E). Furthermore, the presence and abundance of these clearly identified acetyl-, carboxy- and methyl-containing glucose residues in cellulose chains depends on the origins and ages of the original textile samples (Figures 2, 3 and 6; Tables II and III). For example, the content of both 2-acetyl-6-methyl-β-D-glucose and 6-methyl-β-D-glucose residues in textile cellulose increases with the age of the sample (A1>B>A2); Table II). It is also interesting to note that all of these textile samples (A1, A2 and B) were found, and possibly manufactured, in one geographical region, Israel/Egypt. As for the other six textile samples examined (Table I), they have different ages and different geographical origins, and they contain other chemical modifications of cellulose based on permethylation and carboxylation of the latter (Figures 2 and 3). In a group of relatively young Western European textile samples (C1 and C2) and in another group of Eastern European medieval textile samples (D, E and F), the same regularity has been noted: cellulose alkylation extent increases with increase in the calendar age (Figure 2; Table II).

Table II. Abundance of Alkylated Glucose Derivatives in Cellulose Chains from Several Archaeological Textile Samples

Sample (Experiment Record Code)	Content of Minor Glucose Derivatives	Percent Total Cellulose Hydrolysate (M±SEM; n =6)[a]
A1	2-Acetyl-6-methyl-β-D-glucose	5.40±0.08
A1	6-Methyl-β-D-glucose	6.12±0.04
A2	2-Acetyl-6-methyl-β-D-glucose	2.01±0.01
A2	6-Methyl-β-D-glucose	1.99±0.01
B	2-Acetyl-6-methyl-β-D-glucose	4.25±0.02
B	6-Methyl-β-D-glucose	5.12±0.02
C1	2-Carboxy-6-methyl-β-D-glucose	5.98±0.08
C2	2-Carboxy-β-D-glucose	2.74±0.02
D	2-Carboxy-β-D-glucose	2.88±0.03
E	2-Carboxy-β-D-glucose	6.33±0.07
E	2, 6-Dicarboxy-β-D-glucose	3.27±0.02
F	2-Carboxy-β-D-glucose	4.18±0.02

[a]Data obtained using the differential refractive index from CZE.

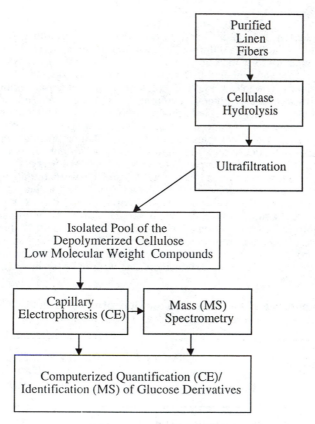

Figure 1. A general scheme illustrating the sequence of steps in the experimental procedure.
(Reproduced with permission from Ref. 10. Copyright 1994 American Chemical Society.)

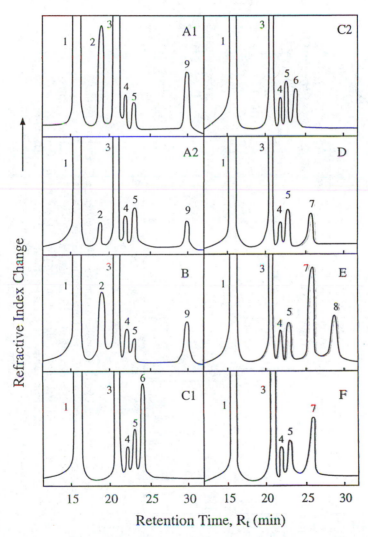

Figure 2. Capillary zone electrophoresis (CZE) separation of enzymatic digestion products form the fibrous cellulose of eight archaeological textile samples. (Reproduced with permission from Ref. 10. Copyright 1994 American Chemical Society.)

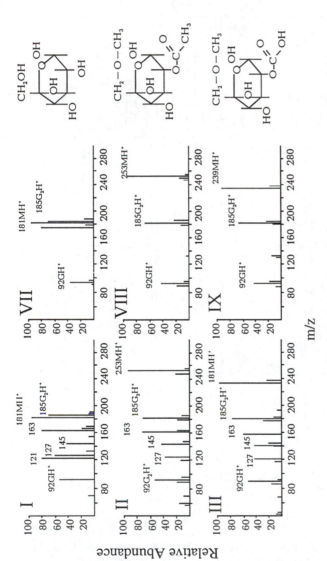

Figure 3. Mass spectra of monosaccharide fractions isolated by CZE of the textile cellulose hydrolysates. I-VI, field ionization technique; VII-XII, field desorption technique. I and VII: nonmodified glucose fraction isolated from all textile samples A1-F; II and VIII: fraction 2 isolated from samples A1, A2 and B; III and IX: fraction 6 isolated from samples C1 and C2; IV and X: fraction 7 isolated from samples D, E and F; V and XI: fraction 8 isolated from sample E; VI and XII: fraction 9 isolated from samples A1, A2 and B. Two matrix peaks are labeled corresponding to protonated glycerol (m/z 92, GH+) and the protonated dimer (m/z 185, G_2H^+). The fragment ions at m/z 163, 145 and 127 represent successive losses of H_2O from MH+.

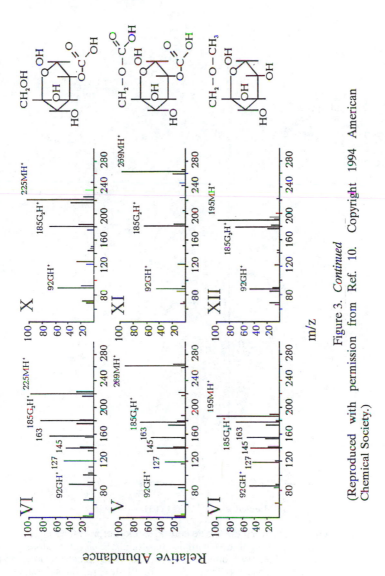

Figure 3. *Continued*

(Reproduced with permission from Ref. 10. Copyright 1994 American Chemical Society.)

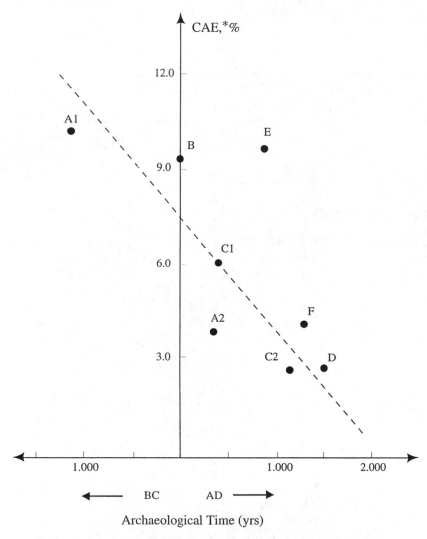

Figure 4. Dependence between the textile cellulose alkylation extent and the calendar age values of the eight linen samples tested. CAE, % = Cellulose alkylation extent or total relative abundance of all alkylated glucose residues in the textile cellulose calculated as a total contribution of all alkyl residues to the textile cellulose pool.

Table III. Key to the Identification of Structures from the Chromatograms of Figure 2

Peak	Retention Time (m)	Identified Structure
1	16	Buffer
2	19	2-Acetyl-6-methyl-β-D-glucose
3	21	β-D-Glucose
4	22	Cellobiose
5	23	Impurity (Non-pyranone/furanose pool)
6	24	2-Carboxy-6-methyl-β-D-glucose
7	26	2-Carboxy-β-D-glucose
8	29	2, 6-Dicarboxy-β-D-glucose
9	30	6-Methyl-β-D-glucose

Cellulose Alkylation Extent (CAE) - Age Correlation. Summarizing the data on cellulose alkylation extent (CAE) listed in Table II, we have found (with the exception of sample E) a crude but clear correlation between the calendar age values of the textile samples tested and their CAE values (Figure 4). Our observation suggested that it would be logical to use advanced statistical methods to clarify the crude CAE/Age dependence suggested by Figure 4.

For this purpose, we have used a bivariate statistics approach which has been specifically developed for a comparison of two or more biopolymer primary structures on the basis of their monomer content (abundance) data obtained by chromatographic or electrophoretic techniques (*12*). Originally, this statistical approach was designed for analysis of proteins and nucleic acids, i.e., for much more heterogeneous types of macromolecules as compared to polysaccharides. Nevertheless, we thought that this approach might be easily generalized for analysis of intramolecular heterogeneity in each class of biopolymers including the alkylated polysaccharides. In general, the correlation coefficient (actually, the product-moment correlation coefficient) is a measure of the "goodness of fit" between a computed (standard) line and a set of experimental points (*12*). A more generalized form for the correlation coefficient is given in vector notation below:

$$R = \frac{\bar{a} \cdot \bar{b}}{\left\{ (\bar{a} \cdot \bar{a})(\bar{b} \cdot \bar{b}) \right\}^{\frac{1}{2}}}$$

The vector $\bar{a}$ is represented by the ordered monomer (pure nonmodified fibrous cellulose) composition of the standard and the vector $\bar{b}$ is represented by the order monomer composition of the test polymer (alkyl-cellulose in archaeological textile samples). For this sample set, we ordered the alkyl-glucose derivative compositions alphabetically for simplicity, although any order would be acceptable so long as the same system is used for both the standard pure nonmodified cellulose assay and the test alkylated cellulose. The correlation coefficient is given as the vector dot product of the monomer composition of the standard and the test polymer divided by the square root of the vector dot product of the standard to itself and the vector dot product of the sample to itself. The dot product of two vectors is the sum of the product of each of the corresponding elements of each vector. If the elements of one

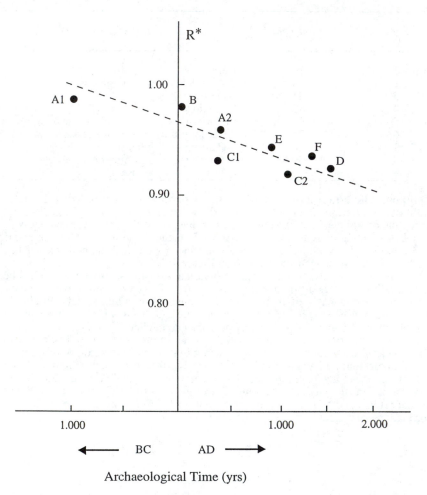

Figure 5. Dependence between the textile calendar age values and the differences in primary structures of the compared homogeneous cellulose chain (standard) and the alkylated textile cellulose chain tested. R* = Regression correlation coefficient characterizing the extent of difference between primary structures of two compared polymers, standard (unmodified cellulose) and tested samples (alkylated cellulose), calculated according to bivariate statistics as the product-moment correlation coefficient (12).

vector are designated p_i, and those corresponding elements of a second vector as q_i then the dot product of the two vectors is $\sum p_i q_i$. The caution is that the vectors may need to be normalized, and that the elements of each vector be ordered in the same fashion. In our work, the monomer assay data are normalized to 100% by virtue of the analysis procedure, and the elements have been ordered by listing the assay results in alphabetical order of the individual alkyl-glucose compounds.

The regression correlation coefficients (R) calculated using the statistical approach described above should be a sensitive criterion indicating the level of difference between the absolutely homogeneous fibrous nonmodified cellulose (standard) primary structure and the primary structure of heterogeneous alkylated textile cellulose. So the greater the difference between the test R value and R = 1.0 (standard), the greater the difference between the primary structures of alkylated cellulose of the textile tested and the pure non-alkylated cellulose (standard). In other words, R is a convenient criterion for the quantitative evaluation of the heterogeneity of alkylated cellulose by comparison of its composition and the composition of 100% homogeneous pure cellulose: the highest level of heterogenity corresponds to the smallest value of R (*12*).

We have calculated the R values for the cellulose pools of each of the eight archaeological textile samples listed in Table I, and the results are presented in Figure 5. These results show a near-linear dependence of the R values on the calendar ages of the textile samples. In our opinion, this type of dependence could be the subject of further investigations leading to improvement in the accuracy of existing dating methods. All modern archaeological textile dating techniques are open to improvement with respect to accuracy and efficiency. The present method of choice, radiocarbon dating, must take into account such additional and sometimes alternative techniques such as the study of stylistic details and microscopic data (*1, 3, 13, 14*). In addition, it is known that there are a number of unique morphological markers for textiles manufactured in several regions during limited historic periods (*3, 13, 14*). Obviously, these markers should be taken into account in each archaeological dating procedure with objects from corresponding sites. From the results of our own work, we propose that it is possible to find not only morphological but also molecular markers for ancient textiles of different geographical origins and calendar ages. We believe that our data (Table II; Figures 2, 4, 5) demonstrate the great significance of this type of research.

Mechanism of Cellulose Alkylation. The carboxylating enzymes of bacterial carboxysomes are active even in bacterial lysates in wet alkali conditions and in the presence of oxygen (*15*). The carboxylization of polyglucose can be promoted easily at pH 9.0 in aerobic conditions by the destroyed (lysated) cells of several environmentally common bacteria of the *Desulfovibrio* genus (*16*). Also, the normal autolysis of air and soil microorganisms may lead to significant carboxylation or methylation of different substrates directed by the released active bacterial enzyles (*15 – 17*). Hence, the alkylation of archaeological textile cellulose could be the consequence of microbial action over a very long time interval, especially since the alkali conditions for bacterial enzyme activity are met in the washing procedures to which textiles are normally subjected. If the cellulose alkylation process is a result of microbial contribution to the chemical modification of textiles, it may be possible to develop a new efficient dating technique based on this process. This dating technique would involve the quantitative evaluation of the textile cellulose alkylation extent estimated in different unknown and known-age archaeological textile samples.

Conclusion

In a separate series of CZE-MS experiments, we have carefully investigated a set of seven known-age burial linen textiles excavated from a single site in Bukhara,

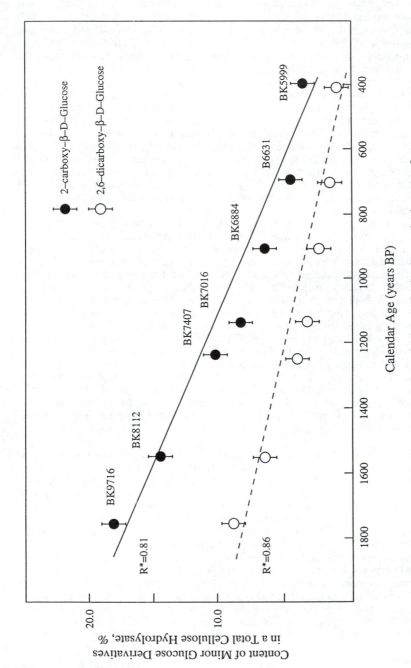

Figure 6. Cellulose carboxylation as a function of calendar age in the case of seven non-dyed known-age burial linen textiles excavated between 1928 and 1987 in Bukhara, Uzbek Republic.

Uzbekistan. Our data show that all of these textile samples contain a carboxylated cellulose, and moreover, the cellulose modification degree is a function of the sample calendar age (Figure 6). Thus, it is logical to assume that textile cellulose chemical modification analysis could be the basis for further development of an additional method for textile dating. Studies of ecological effects on the chemical structure of archaeological objects is still undeveloped but a very important area of research (*18*). We assume that a further search for possible correlations between chemical parameters characterizing both old textiles and biological materials (fossils, grain seeds, wood, etc.) excavated at a single site might lead to interesting conclusions concerning textile structure dependence on environmental factors.

Table IV. Sample Key to the Data Presented in Figure 6

Museum Keeping Code	Estimated Calendar Age	Method of Age Estimation	Age-Dating Laboratory
BK9716	1790-1720 BP	Radiocarbon dating	Soviet Committee on Asian Studies, 1989
BK8112	1580-1510 BP	Radiocarbon dating	Soviet Committee on Asian Studies, 1989
BK7407	1290-1220 BP	Historical evidence and stylistic details	Middle-Asian Museum of Ethnography and Anthropology, Samarkand, UR
BK7016	1140-1060 BP	Historical evidence and stylistic details	Middle-Asian Museum of Ethnography and Anthropology, Samarkand, UR
BK6884	920-830 BP	Radiocarbon dating	USSR Academy of Sciences Institute of Experimental Physics, 1979
BK6631	700-640 BP	Historical evidence and stylistic details	Middle-Asian Museum of Ethnography and Anthropology, Samarkand, UR
BK5999	440-360 BP	Historical evidence and stylistic details	Middle-Asian Museum of Ethnography and Anthropology, Samarkand, UR

In any event, the correlations we have observed among the extent of textile cellulose chemical modification, calendar age values and geographical (regional) origin of the textile (Table IV; Figures 2, 4, 6) seem to merit further investigation. To our knowledge, the present study is a first report on the discovery of alkylated cellulose sequences in archaeological textiles and the relationship of CAE to calendar age. Further studies in this area could help to classify archaeological textile remains using such criteria as cellulose chemical modification types. This classification method could be an additional tool for both dating research in archaeological chemistry and in ancient textile technology studies.

Acknowledgments

We express our gratitude to the following persons for providing known-age samples of archaeological linen textiles: M. Moroni, I. Kappel, O. Krutov, O. Nenasheva, S. Bychkov, I. Tyshko and K.-D. Youldashev. Thanks also to A. Adler, W. Brostow, and A. Volkov for their critical remarks on our preliminary results, and T. Pavelich for his assistance in the preparation of the samples for capillary electrophoresis. This work was supported by the Guy Berthault Foundation, Meulan, France.

Literature Cited

1. Cardamone, J. M.; Brown, P. In *Historic Textile and Paper Materials: Conservation and Characterization*; Needles, H. L.; Zeronian, S. H., Eds.; Advances in Chemistry 212; American Chemical Society: Washington, DC, 1986; pp 41-76.
2. Ageyev, D. T. *Chemical Methods in Archaeology*; Moldova State University Press: Kishinev, Russia, 1992; pp 27-39 (in Russian).
3. Lee, J. S. *Elementary Textiles*; Prentice-Hall: Englewood Cliffs, NJ, 1953; pp 47-59.
4. Yakimov, S. S.; Zheltkov, R. T. In *Applied Chemistry of Cellulose*; Samarin, L. K.; Bakhurov, V. L., Eds.; Novosibirsk University Press: Novosibirsk, Russia, 1990; pp 126-148 (in Russian).
5. Heller, J. H.; Adler, A. D. *J. Canadian Society of Forensic Science* **1981**, *14*, 81-102.
6. Jumper, E. J., et al. In *Archaeological Chemistry – III*; Lambert, J. B., Ed.; ACS Advances in Chemistry Series 205; American Chemical Society: Washington, DC, 1984; pp 447-475.
7. Bruno, A. E., et al. *Anal. Chem.* **1991**, *63*, 2689-2697.
8. Colon, L. A.; Dadoo, R.; Zare, R. N. *Anal. Chem.* **1993**, *65*, 476-481.
9. Thompson, T. J., et al. *Anal. Chem.* **1993**, *65*, 900-906.
10. Kouznetsov, D. A.; Ivanov, A. A.; Veletsky, P. R. *Anal. Chem.* **1994**, *66*, 4359-4365.
11. Carroll, J. A., et al. *Anal. Chem.* **1991**, *63*, 2526-2529.
12. Sokal, R. R.; Rohlf, F. J. *Biochemistry*; W. H. Freeman & Co.: New York, NY, 1981; pp 454-560.
13. Cramer, M. *Das Christlich – Koptische Ägipten Einst und Heute*; Eine Orientierung: Wiesbaden, 1959; pp 46-88.
14. Bresee, R. R.; Chandrashekar, V.; Jones, B. W. In *Historic Textile and Paper Materials: Conservation and Characterization*; Needles, H. L.; Zeronian, S. H., Eds.; American Chemical Society: Washington, DC, 1986; pp 19-40.
15. Price, G. D., et al. *J. Bacteriology* **1993**, *175*, 2871-2879.
16. Hensgens, C. M. H., et al. *J. Bacteriology* **1993**, *175*, 2859-2863.
17. Doyle, R. J.; Koch, A. L. In *Critical Reviews in Microbiology* **1987**, *15*, 169-222.
18. Westbroek, P., et al. *World Archaeology* **1993**, *25*, 122-133.

RECEIVED December 1, 1995

Chapter 21

Historico–Chemical Analysis of Plant Dyestuffs Used in Textiles from Ancient Israel

Zvi C. Koren

The Edelstein Center for the Analysis of Ancient Textiles and Related Artifacts, Shenkar College of Textile Technology and Fashion, 12 Anna Frank Street, Ramat-Gan 52526, Israel

The history, chemical constitutions, and chromatographic properties of the plant dyestuffs that according to Jewish, Greek, and Roman sources have been used in ancient Israel are discussed. This work is the first published critical historical and chemical analysis of the dyestuffs used in ancient textiles from Israel. The archaeological periods included in this investigation span about a thousand years and primarily include the Hellenistic, Roman, and Byzantine periods, from about the 4th century B.C. to about the 7th century A.D. The primary historical sources relied on for the identification of the dye plants that were used in ancient Israel were the Talmud, related rabbinical literature, and the writings of Greek, Roman, and Jewish historians. Chemical sources that are based on chromatographic and spectrometric analyses of archaeological textile dyeings were also used. The plant dyes investigated belong to a variety of chemical groups, which include flavonoids, carotenoids, safflors and aroylmethanes (all yellows), naphthoquinonoids (browns, reds, oranges), tannins (browns); anthraquinonoids and chalconoids (reds); and indigoids (blues). In addition, a reverse–phase gradient elution high–performance liquid chromatographic (HPLC) method is presented for the separation of dye components from a number of the above classes. Both the historical record and the relevant chemical properties of the dyes presented in this study are necessary for the analysis and proper identification of the colorants used in textiles dyed in ancient Israel.

The identification of the colorants used on ancient textiles opens a historical window to the understanding of the processes associated with one of the oldest of chemical technologies – textile dyeing. The analytical inquiries involve chemical detective work to help decipher the identities and sources of the dyestuffs used in antiquity. These investigations of the colorants used by ancient peoples involve a multidisciplinary research that combines history, archaeology, religion, botany, entomology, marine zoology, and microanalytical chemistry. Knowledge of the natural dyes used in different regions in antiquity increases our knowledge of local and international trade and commerce in ancient times. The identification of the textile dyestuffs used can

0097–6156/96/0625–0269$17.50/0

indicate the movement of dyed goods and the transfer of dyeing methods from one geographical area to another. Such an examination of the textiles of past cultures uncovers the development and technological advancement of the "scientific art" of textile dyeing through various archaeological periods. In addition, the study of ancient textiles enhances our understanding of the standard of living, textile art and fashion, and color preferences of ancient peoples. In the religious–historical area, such analyses should help to decipher the mysteries associated with the sources and true colors of the biblical textile dyes – bluish *tekhelet,* purplish *argaman,* and reddish *tola'at shani.* Further, the modern world can also benefit from the instrumental analyses of ancient artifacts. New analytical methods developed for the study of very small archaeological samples may also be applicable to the analyses of modern samples in, for example, industrial quality control, pharmaceutical, environmental, and forensic areas.

The historical record is vital for the determination of whether the dyestuffs identified on ancient textiles excavated in Israel were obtained from local sources or imported into this region. Various classical Greek and Roman authors, such as Aristotle, Dioscorides, Vitruvius, Strabo, and Pliny the Elder, have described various aspects associated with textile dyeing as performed during their time. In addition, a number of articles and books have been written in the past few decades on the subject of natural dyes of historical importance. Some of these books include a historical account of various natural dyes (*1–4*), whi!e others have also included chemical dye constitutions (*5–8*), and a symposium series has devoted itself to this subject (*9*). The publications dealing with the historical documentation of the dyes used in antiquity have been based primarily on Babylonian, Greek, Roman, Egyptian and other geopolitical sources. However, very little has been written about the Jewish historical sources that would elucidate the art of textile dyeing as practiced in ancient Israel – that part of the Middle East that is also known as the 'Land of Israel' (*Eretz Yisrael* in Hebrew) or the 'Holy Land', with its shifting geographical borders over the span of time. A few studies have appeared in Hebrew that discuss some of the historical and botanical aspects of the dyestuffs used in ancient Israel (*10–13*). This article will investigate those relevant sources written in Hebrew and Aramaic that discuss textile dyeing and dyes and that have been unavailable to the interested reader or researcher who is not versed in those languages. (The translations that appear in this article are the author's.) Further, new chemical research in the field of the analytical separation and spectrometric identification of historically important natural dyes has appeared in the literature and this article will include a new chromatographic scheme for the identification of some of these dyestuffs. As such, this work is the first combined critical historical and chemical analysis of the dyestuffs used in ancient textiles from Israel.

The historical document that is undoubtedly the most direct witness to the many facets of life in ancient Israel during the Hellenistic, Roman and Byzantine periods is the Talmud (*14, 15*). The Talmud ("Study" or "Teaching" in Aramaic) formally consists of two parts: the Mishnah (Hebrew for "Repeated Teaching" in an oral manner or "Recitation") and the Gemara ("Teaching" in Aramaic). The Mishnah is a collection of Jewish religious, civil, and agricultural laws that are based on rabbinical interpretations of biblical ordinances and was compiled during the first two centuries of this era. The Gemara consists of rabbinical interpretations of the Mishnah and related commentaries and was compiled between approximately 200 and 500 A.D. Other writings from the talmudic period include the collection of extra–mishnaic texts known as the Tosephta ("Addition" in Aramaic), which was not formally incorporated into the Mishnah but is contemporaneous with that body of work. While the Mishnah was compiled in Israel and was primarily written in Hebrew, the Gemara is mostly in Aramaic. The current printed format of the Talmud is essentially the same as the first printed version published by Daniel Bomberg in 1520–1523 in Venice. In this arrangement, the Gemara follows the relevant Mishnah. The Talmud contains numerous Greek words (written in Hebrew script) that were used by the rabbis ("teachers") of that period. This shows the strong post–Hellenistic Greek influence on the language – especially technical terms – of the Jewish population in Israel during those periods. The Talmud consists of two geographical versions: the Jerusalem Talmud (JT), which is also known as the 'Israeli Talmud' and was produced by the rabbis living in Israel, and the Babylonian Talmud (BT), which is generally the more authoritative and voluminous of the two works and is a product of the rabbis living in Babylonia.

The textile craft, in general, and dyeing, in particular, are discussed in various tractates (or volumes) of the Talmud. As is apparent from history, ancient Israel was at the crossroads of various empires and was conquered by them on numerous occasions. This region thus did not live in a vacuum and various types of goods, including textiles, flowed into and out of this strategic region. Hence, the relevant talmudic passages not only reflect on the state of textile dyeing in Israel of the time but also on the level of that craft in neighboring regions. It will, in fact, be observed that many of these talmudic citations closely parallel those of the classical authors that describe the textile art in their regions.

The periods to which the term 'ancient Israel' will refer are the Hellenistic (332 B.C. – 63 B.C.), Roman (63 B.C. – 324 A.D.), and Byzantine (324 A.D. – 641 A.D.) Periods. Many dyed textile fragments, which have been dated to these archaeological periods, have been excavated in Israel, and some have undergone chemical analyses (16–24). The quality and diversity of dyeings reached a very high level during the Roman era and analyses of these dyeings have elucidated the techniques used by the ancient dyers of this region.

This article will discuss those plants that were described as dyestuffs in the Talmud and were, therefore, cultivated and in use in ancient Israel. Other dye plants that were available in nearby regions, which could have also been imported into this land, are also discussed. Even during talmudic times, products originating from India and from as far away as China found their way to Israel. Some of the more 'exotic' items cited in the Talmud include *hinduyin* (or, in the variant forms, *hindevin, hindoin, hindovin, hindon, hindevon*) (25) and *dartzona* (or *dartzina*) (26a). *Hinduyin* are white clothes composed of linen from India (27a) that were worn by the High Priest in the Temple during the afternoon ritual on the Day of Atonement; the first part of the name – 'hindu' – indicates the geographical origin of the garments. *Dartzina* is the cinnamon tree and is etymologically derived from the Persian *dar–i–cin* ("tree that comes from China") (27b). Cotton ("wool of the vine" in talmudic language) was already mentioned in the mishnaic period (28a, 29a) and later (26b, 30a), and silk, which was imported from India (or China), was also known during those early centuries (28b, 31). The three forms of silk that are mentioned in the Talmud are *siriqon* (26c, 31, 32a, 33a, 34a, 35), *shira* (26c,d, 28b, 32a, 34a, 36a, 37, 38, 39a,b, 40a), and *kalakh* (26c, 28b, 29a, 32a), and related etymological variants. Thus, one should not be surprised if one day a dyestuff that was once only available in India or China is found on an archaeological textile excavated in Israel.

The dyeing colors obtained from the plant dyestuffs and their known chemical constitutions, including functional group and subgroup affiliations, are also discussed.

Finally, a gradient elution method for the separation and identification of dye components from various plant dyestuffs by means of high–performance liquid chromatography (HPLC) is presented.

Textile Dyeing – Chemical Considerations

In this article, the term 'dyestuff' will be used to describe the raw or crude source of the dye, i.e., the plant or part of the plant, and the term 'dye' will refer to the specific coloring substance or component of the crude dyestuff. For example, the yellow 'dyestuff' weld contains the luteolin and apigenin 'dyes'.

A textile may be colored by various means (*41*). Examples include using different naturally colored fibers, painting a textile with an organic or inorganic pigment, or impregnating the fibers with a soluble form of an inorganic salt or oxide. However, true dyeing with organic dyes yielded the most dramatic and widest range of colors. The process of textile dyeing involves the fixing of a colorant into the textile by physical and/or chemical means so as to produce a direct or indirect bond between the dye and the fiber. As a necessary prelude to the dyeing stage itself, the dye must be present in the aqueous bath in a dissolved form in order to allow the dye molecules to diffuse towards the fibers and penetrate into them. It may have been necessary to control the pH and temperature to maximize the dyeability of the fibers. The wide variety, stability, and beauty of the colors of two thousand year-old dyeings that have survived in the dry climates of Israel attest to the fact that the skillful ancient dyer mastered the necessary dyeing parameters.

Natural dyestuffs can be classified according to the following categories: (1) color, (2) source, and (3) chemistry. In the first category, the color of natural dyeings produced were generally red, orange, yellow, brown, and blue. However, many more hues were obtainable by overdyeing with two or more dyestuffs and/or with the formation of a metal–dye complex within the fibers by typically pre–treating the textile fibers with a solution of a metal salt or oxide. The second classification refers to the flora or fauna sources of the dyestuff. Most dyestuffs were vegetal in origin, although certain scale insects and sea snails were also used as dyestuffs (*41*). The last category is based on the chemical structure or functional class of the dyes, which influences the specific dyeing process used for that class of dye. Consequently, the natural dyes can be chemically classified as either direct, mordant, or vat dyes, which are described in the following sections.

The Direct Dyes. These dyes did not require the use of a stabilizing agent in order to fix them into the fibers, though the use of a mordant, as described below, sometimes produced a dyeing that was more fast, i.e., stable.

The Mordant Dyes. Many of the natural dyes did not have a strong chemical affinity for the textile fibers. Hence, a mordant (French: "mordre" to bite) or fixing agent was used in antiquity. Metal salts and tannins (polyphenols) served as bridging agents between the fibers and the dye. Mordanting with a metallic ion was accomplished by first impregnating the woolen fibers with an aqueous solution of a metal salt prior to the dyeing stage and subsequently precipitating it as the hydroxide or oxide, or both, within the fibers. The polyvalent–metal salts that were available were "potassium alum", $K[Al(SO_4)]\cdot 12H_2O$, a source for aluminum ions, and other minerals containing iron, tin, and copper. The resulting precipitated metal–dye complex (*42*), or lake, consisted of a chelate, where the metal ion could be bonded to two or more dye molecules via ionic and coordinate covalent bonds. This complexation is possible due to the close proximity of the carbonyl oxygen to the phenolic oxygen, i.e., the α–hydroxy group. Figure 1 shows the neutral 2:1 complex that 1–hydroxy– and 1,8–dihydroxyanthraquinones can form with divalent metal ions, such as copper or zinc (*43*). Virtually insoluble polymeric complexes can be formed by 1,4– and 1,5–dihydroxyanthraquinones (*43*). The resulting enlarged solid complex that is formed within the fibers, does not easily wash out of the textile and thus the resultant dyeing is relatively fast. The ancient dyers were also undoubtedly aware that the metal mordants served to shade the final color as well. For example, iron chelate complexes with hydroxyanthraquinones dull the colors and can produce chocolate–purple shades with madder dyeings, whereas aluminum complexes produce brighter red shades. These colors were found in the Judean Desert textiles excavated at the Cave of Letters (*17, 18*) and at Masada (*23*), as well as at other archaeological sites in Israel (*22, 24*).

The Vat Dyes. The blue indigotin dye is the most prominent member of this class. This pigment is not soluble in aqueous solutions and in order to solubilize it, a fermentative reduction process was performed in ancient times. First, the insoluble so–called leuco ("white") – or lighter–colored – indigo acid is formed (Figure 2) and, then, this acidic form, in the presence of an alkali from either wood ash, plant ash, lime, or stale urine, reacts to form the soluble salt. Immersion of the textile into the dye solution allows the dissolved dye molecules to penetrate the fibers. Upon removal of the wet dyeing from the bath, air–oxidation will produce the final blue indigotin pigment within the fibers. Successive dyeings will produce darker blue shades and will also yield faster dyeings as the new layer of blue pigment will force the old layer to be driven well within the fibers.

Textile Dyeing – Talmudic Considerations

Most of the talmudic sources that discuss textile dyeing refer to wool and only two mention dyed linen textiles. One source (*44a*) is a general reference to linen dyeings, but the other (*44b*) discusses the dyeing of linen with *heret,* which is shoe-black that is also used for dyeing leather and wool. This minimal discussion runs parallel to the very few ancient dyed linen textiles that have been excavated in Israel. A commentary by the 11th -12th century French sage Rashi (an acronym for Rabbi Shelomo ben Isaac) on a Talmudic passage (*45*) states that "it is more difficult for linen to receive the color than for wool." Other textile materials, such as cotton and silk, which were previously mentioned, and hemp or *qannabus* (*28c,d*), camel hair (*28c, 29a, 32a*), and rabbit hair (*29a, 32a*) were discussed, but not in the context of dyeing. Goat hairs have been mentioned as textile fibers in the Talmud (*32a*). Textiles containing these fibers have been found at a number of Roman– through Byzantine–period archaeological sites in Israel and one sample was found to be dyed (*24*).

Dyed clothes and accessories were described in the Talmud (*26e, 28e, 29b*) as, generally, the attire of women (*45, 46, 47, 48, 49*). An interesting comparison is made between colored garments and fine white linen clothes (*50a*) : "A man is obligated to gladden his children and household on a festival. ... With what does he gladden them? ... Rabbi Judah says: ... (for) men as is fitting for them - with wine; and women, with what? Rabbi Joseph taught: In Babylonia, with colored garments; in Israel, with pressed linen garments." A further parallel is seen from the following two passages: "Thin linen garments from Bet She'an are like colored clothes." (*47*) "Rabbi Hiya taught: One who wants to beautify his wife should dress her in linen garments." (*36b*) Though colored clothing was not the 'proper' accouterment for men (*51a*), nevertheless, the Talmud does mention a ritual mantle or prayer shawl – called a *tallit* – worn by men that was completely dyed with the blue (or violet) *Tekhelet* color (*32b,c*). In fact, the Bible directed that a thread from the fringes on each corner of a man's garment be dyed *Tekhelet* (*52a*) obtained only from a molluskan source (*32d*).

During the talmudic period, and probably prior to it, the dyeing stage of the wool usually preceded the spinning into yarn. The practice of dyeing wool 'in the fleece' can be seen from a number of talmudic passages (*26f, 35, 53a,b, 54, 55*), as, for example, from the following midrashic statement (*54*) : "How much did Adam toil before he wore a single robe until he sheared, washed, carded, dyed, spun, wove, and sewed, and thereafter wore (the garment), and I arise in the morning and find all these (garments) before me." In addition, the following citation discusses the different types of work (or activities) that are not to be performed during the Sabbath – the 'day of rest' – and these acts are listed in the following procession (*53a*): "The main labors

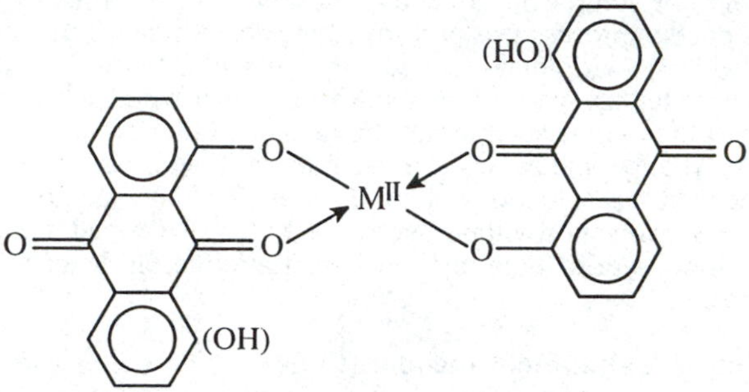

Figure 1. Neutral 2:1 complex of 1-hydroxy- and 1,8-dihydroxyanthraquinones with divalent metals.

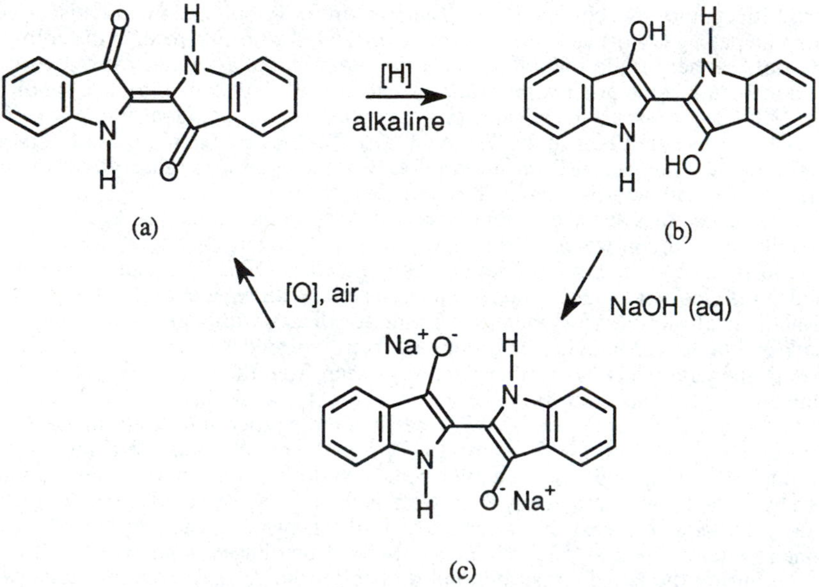

Figure 2. Reduction of indigotin (a) to the soluble leuco salt (c) via the leuco indigo acid intermediate (b) and air-oxidation back to the insoluble pigment.

(prohibited on the Sabbath) are forty less one: ... one who shears the wool, washes it, cards it, dyes it, spins it, ..." That dyeing was performed in the fleece has also been indicated by Pliny (*2a*).

Though dyeing could certainly have been performed for private use in and around the home, textile dyeing was also an industry – as inferred from the talmudic reference to *bet hatzabai'm* ("house of the dyers") (*50b, 56*). Certain regulations regarding the dye works were discussed: "A man may not open a bakery or a dyehouse underneath his fellow's storehouse" (*57-59*) undoubtedly because of the odors, smoke, and heat associated with the operations of these businesses. These two industries also had long–term contracts and credits with numerous clients and, hence, if the landlord wanted to terminate the rental contracts with either business, he had to give the occupants a three–year notice instead of the usual twelve months (*60, 61a, 62a*) Water was an indispensable ingredient in the dye works and the dyer was not permitted to use the water in a public well except for drinking purposes (*61b*).

Dyeing was a specialized profession that was practiced by a *tzaba'* ("dyer"). Two dyers are mentioned in the Talmud by name: 'Amram the Dyer (*30b*) and Menahem ben Signai (or Singai) (*63*). The dyers were probably organized in a crafts guild as can be inferred from the following (*61c*): "The woolers and the dyers are entitled to say: 'any transaction that will come into the city, we will all be partners in it.'" The dyer wore a piece of dyed wool around his ear so that he would be recognized for that profession (*26g, 29c, 64a*). The dyer was thus deemed by the Talmud as a skilled professional practicing his craft like many other craftsmen.

However, despite the skill of the dyer, the dyeing process was an empirical technology and not without accidents (*65*). Financial penalties based on legal judgments are discussed for the mishaps described in the following case studies (*66*): "If one gave wool to a dyer and the vat 'scorched' it, (then the dyer) pays him (the owner) the value of his wool." Dyed wool was always worth more than undyed wool. This was so even if the dyeing yielded an inferior (or "unsightly") or undesirable color: "(If he dyed it badly, and if the value of the improvement (in the wool as a result of its dyeing) exceeded the (dyer's) outlay, (the owner of the wool) only pays him the outlay; and if the outlay were worth more than the improvement, then he pays him the value of the improvement. (If he requested the dyer) to dye it red, but he dyed it black, or black but he dyed it red, ..., (then) Rabbi Judah says that (as in the previous case, the wool owner only pays him the lower of the two values)." If a different color from the one requested is produced due to the presence of a residue from a previous dyeing in the vat, then, it is up to the dyer to recompense the wool owner the value of his wool (*66, 67a*). However, if the dyestuffs that were brought to the dyer by the wool owner caused the damage, then the wool owner pays for the costs of dyeing or the extra worth of the dyed wool, whichever is lower. In one rabbi's opinion, if a different color is produced, then the dyer should pay the wool owner the value of the undyed wool plus the extra worth of the dyed wool because the dyer did not clean his vat before dyeing. However, if this change in color was produced unintentionally, then the dyer should not be fined for the extra worth of the dyed wool (*67a*).

Though gloves were used by at least some dyers (*68a*) their hands were probably nevertheless stained. This is particularly so in the case of blue woad dyeings, as described in the section on woad below.

The pre–dyeing stages described by the Talmud, generally involved grinding the raw dyestuff (*69*) and soaking the plant grounds in water (*26f*). The Talmud discusses dyestuffs that have been processed, i.e., "soaked" and readied for dyeing and the raw dyestuffs that have not yet been processed: "One who takes out (from his private property to the public domain) materials (dyestuffs) that are soaking, (the minimum quantity of this mixture that is necessary for this transport to be considered as a work activity prohibited on the Sabbath is if there is) enough to dye a (woolen) sample to seal a loom shuttle." The dye mixture would be heated in a vat (*yora* or *yura*) (*26h,i, 53c, 66, 68b*) where the dyeing was to be performed and, once completed, the dyed goods would be placed on a reed mat (*68b*). The Talmud recognized that dyeing was not a very short process, as can be inferred from the debate between the School of Shammai and the School of Hillel. This dispute concerned whether the dyeing of wool may be started on the eve of the Sabbath before its start at sundown, which would necessitate leaving the goods in the dye bath through the Sabbath (*53c*).

It should also be mentioned that at least as early as the biblical period, and certainly later, organic substrates other than textiles, such as leather or animal skins, were also colored or dyed. One of the items that was ordained to be used as part of the cover for the biblical Tabernacle that housed the Tablet with the Ten Commandments was red–dyed ram skins (*70*). Leather – shoes, sandals, belts, and saddles – was also dyed a black color in talmudic times (*26j, 30c, 40b*). In addition, the roots of certain plants or trees, such as '*og* (sumac), *agah,* and *vered*, whose identities are uncertain, were used as "colors for animals." This phrasing is somewhat ambiguous as the term "for animals" can indicate that the plant colorant was used in any one or more of the following processes associated with the animal or its product: (1) marking or branding of the livestock for identification purposes, as indicated by Feliks (*12a*); (2) tanning of leather; and (3) coloring of leather. Dioscorides remarks that in Mesopotamia they marked livestock and leather with a sumac dyestuff (*2b*).

Nature's Colors: Plant Dyestuffs of the Talmudic Period and Their Chemical Constitutions

Textile dyes from the vegetable kingdom were obviously more plentiful and widely used than the colorants available from the animal world. Although almost a thousand different dye sources may have been used at one time (*1a*), a much smaller number of dyestuffs found continuous use due to their good wash– and light–fast properties. Different plant parts were used to produce the appropriate dyestuff. In some plants, the leaves were utilized for extracting the colorant, while in others, only the roots were used. The flowers of a small number of plants were also used. Stems, branches, berries or fruits, tree trunks, bark, or seeds of some plants and trees also found use as dyestuffs.

One of the best historical sources for the determination of which dye plants were cultivated in Israel during the talmudic period is the 'botanical chapter' of the JT, chapter 7 of Tractate *Shevi'it.* This tractate includes a discussion of the agricultural laws that were in effect only in Israel. These laws included the sabbatical – or seventh (*shevi'it*) – year for the land and its vegetation. In that year, the land was not to be worked and certain restrictions applied to the commerce in crops collected during that year. Hence, any dye plant discussed in the JT to which at least some of the agricultural laws of the sabbatical year applied, automatically indicates that this plant was definitely cultivated within the geographical borders of the 'Land of Israel' as defined by the rabbis of the talmudic period.

The plant dyestuffs discussed in this study have either been explicitly mentioned in the Talmud as a dye source or cited as such by other historical sources and could have been used in the Middle East. Tables I, II, and III contain relevant botanical, historical, and dyeing information regarding these plants, which yield yellow/brown, red/orange, and blue dyestuffs, respectively. The dye components from these plants are listed in Tables I–III and their chemical structures are presented in Tables IV–VII according to the chemical functional groups of these colorants. The dye components listed are those that will be obtained after hydrolysis of the dye substances. This is due to the nature of the dyeing process itself, which uses elevated temperatures to produce the dyeings, and to the modern analyses of ancient dyeings, which usually contain an acid hydrolysis step for the stripping of the dye from the textile fibers. Thus, the sugar moieties of most of these substances will have dissociated from the dyes following one or both of these procedures. The *Colour Index* (C.I.) (*71*) color classification numbers of the plant dyestuffs catalogued in that index are presented in Tables I–III, and the constitution numbers and chemical structures of the dye components of each dyestuff are given in Tables IV–VII. The parent structures of the colorants in the latter tables are given in Figures 3–11. The colors of wool dyeings produced with these dyestuffs, with or without a complexing mordant that was originally added as a salt of the metal ion, are listed in Tables I–III. The actual final colors are dependent on many factors associated with the dyeing process, such as temperature, age of dyestuff, other additives, pH of soil, season in which plant was picked, etc. Furthermore, the hues listed may be somewhat different for different dyeings with the same dyestuff. The plants are listed in Tables I–III in alphabetical English order according to a common name of the dyestuff, but other common names are also given. In the sections below, the italicized parenthetical name is the transliterated common Hebrew plant name.

Yellow and Brown Dyestuffs. As Table I indicates, most plant dyestuffs produce yellow, beige, and brown colors without the use of a mordant and the most common dyes in these plants are flavonoids and tannins. Flavonoid–based dyes generally yield yellow colors with or without alum, and, by using copper or iron mordants with some plants, greenish dyeings can also be produced. The tannin–rich dyestuffs yield beige or brown dyeings and are virtually unaffected by mordanting with alum. However, iron–mordanted tannin dyeings yield the black iron(III) tannate precipitate within the fibers. In general, all iron–mordanted dyeings are darker than ones obtained with copper, which are darker than aluminum–mordanted dyeings. It seems probable that yellow and brown shades were quite prevalent and cheap, though seasonal factors may have influenced the plant to be used. Unfortunately, many of the yellow dyeings were not fast and, though the historical record describes the use of such dyestuffs, only a few archaeological dyeings excavated in Israel show clear indications of having been dyed with a yellow or brown dyestuff. Further, dyeings containing the iron complexes tend to decompose the textile as the iron oxidizes the organic medium and further acts as a catalyst to erode the textile. The borders around some of the holes found on certain textiles show a thin ring of brown or black color where iron was used.

Other dye components of the yellow–brown family are the anthraquinone emodin, the safflor of Safflower Yellow, the naphthaquinone juglone, the carotenoid crocetin, and the feruloyl (or aroyl) curcumin.

Yellow dye sources were quite abundant. The leaves and stems of many plants produce flavones and flavonols. In addition, leaves and branches of certain trees contain tannins, which produce a brown shade. Wool can be dyed directly with aqueous solutions of flavonoids or tannins without prior mordanting. The yellow–producing flavonoids were used as direct dyes without the use of an intermediary. However, certain flavonoids can be further stabilized in the fiber by complexation with a metal mordant via the adjoining carbonyl oxygen and the negative hydroxyl oxygen, as in the case of the related hydroxyanthraquinones described previously.

Yellow and brown plant dyes belonging to other chemical classes, such as naphthoquinonoids and carotenoids, have also been used throughout history. Saffron, the stigmas of *Crocus sativus* L., whose major colorant is crocetin, has been reported for the orange–yellow colors found on the Cave of Letters textiles (*17, 18*). Flavonoid–containing yellow dyes were undoubtedly used in a number of textiles excavated in Israel; however, the identities of these dyes are still being researched.

The tannins served several purposes, depending on the nature of textile color desired. By themselves, they served as a brown dye source. However, they also served as organic mordants for other dyestuffs. In addition, together with an iron salt, they

Table I (a). Yellow and Brown Plant Dyestuffs of Ancient Israel: Botanical and Historical Information

Botanical Information			Part of Plant Used	Reference to the Plant in Jewish Writings		
Common English Name	Botanical Name of Species	C.I. Natural Color Number		Modern Hebrew Name	Historical Hebraic or Aramaic Names	Uses
Buckthorn berries, Persian berries	Rhamnus palaestinus Boiss.	Yellow 13	unripe berries of shrub	eshhar	(none)	
Galls, gall nuts, nut galls	Quercus infectoria	Brown 6	insect-produced cysts on oak trees	'afatz	afatz, aftza milah	tanning leather, ink-making
Pomegranate	Punica granatum L.	Yellow 7	fruit rinds	rimon	romana, rimona	dyestuff, ink-making, game
Rose	Rosa canina L. or Rosa phoenicia		shrub roots	vered	vered, varda	colorant for animals
Safflower, false saffron, dyer's thistle	Carthamus tinctorius L.	Yellow 5	florets	qurtam	haria', qotzah, qurtami, qurtema, moriqa, dardera	dyestuff, food seasoning and food coloring, medicine for diarrhea
Saffron	Crocus sativus L.	Yellow 6	flower stigmas	karkom	korkema, za'afrana	fragrance, food color,

Table I(a). Continued

Botanical Information		C.I. Natural Color Number	Part of Plant Used	Reference to the Plant in Jewish Writings		Uses
Common English Name	Botanical Name of Species			Modern Hebrew Name	Historical Hebraic or Aramaic Names	
Sumac, dyer's sumac, tanner's sumac	*Rhus coriaria* L.	Brown 6	leaves	*'og*	*sumqa* (or *sumaqa*)	colorant for animals, tanning leather
Turmeric	*Curcuma longa* L.	Yellow 3	rhizomes	*kurkum*	(none)	
Walnut	*Juglans regia* L.	Brown 7	green fruit peels	*egoz*	*amgoza, egoza*	dyestuff, ink-making
Weld, dyer's rocket, dyer's weed	*Reseda luteola* L.	Yellow 2	whole plant and seeds	*rikhpah*	*rikhpah*	dyestuff
Young fustic, Venetian sumac, smoke tree	*Cotinus coggygria* Scop. (old name: *Rhus cotinus* L.)	Brown 1	wood of stem and larger (lower) branches		(none)	

Table I (b). Yellow and Brown Plant Dyestuffs of Ancient Israel: Dyeing Properties

Plant Name	Main Dye Components of Dyestuff after Hydrolysis			Color of dyeing on mordanted wool				
	Flavonoids	Tannins	Others	no mordant	Al^{3+}	Sn^{2+}	Cu^{2+}	Fe^{2+}
Buckthorn	Kaempferol, Kaempferol 7-methyl ether, Quercetin, Rhamnazin, Rhamnetin		Emodin (Anthraquinone)	beige-yellow	dark yellow	orange-brown	greenish yellow	greenish
Galls		Tannin hydrolysates: Ellagic acid, Gallic acid, others		beige				black
Pomegranate		Tannins: Ellagic acid, Gallic acid, others		yellowish	yellowish		greenish brown	purplish dark brown, violet
Rose[a]				beige	beige		greenish yellow	grayish green
Safflower			Safflower yellow (Safflor)	yellow	yellow		greenish yellow	yellowish green
Saffron			Crocetin (Carotenoid)	yellow	yellow - dull orange	dull yellow	green	yellow-brown

Table I(b). Continued

Plant Name	Main Dye Components of Dyestuff after Hydrolysis			Color of dyeing on mordanted wool				
	Flavonoids	Tannins	Others	no mordant	Al^{3+}	Sn^{2+}	Cu^{2+}	Fe^{2+}
Sumac	Fisetin Quercetin	Gallic acid, others						
Turmeric			Curcumin (Feruloyl)	greenish yellow	orange yellow	orange red		brownish black
Walnut			Juglone (Naphthoquinone)	brown	brown		browner than with Al^{3+}	dark brown
Weld	Luteolin Apigenin			yellow	yellow	greenish yellow	yellowish green	olive
Young fustic	Fisetin Dihydrofisetin Myricetin	Tannin hydrolysates			brownish orange	reddish yellow-orange		brownish olive

aColors of dyeings are with the fruits of the rose.

Table II (a). Red and Orange Plant Dyestuffs of Ancient Israel: Botanical and Historical Information

Botanical Information				Reference to the Plant in Jewish Writings		
Common English Name	Botanical Name of Species	C.I. Natural Color Number	Part of Plant Used	Modern Hebrew Name	Historical Hebraic or Aramaic Names	Uses
Alkanet, alkanna, anchusa	Alkanna tinctoria Tausch = Anchusa tinctoria Lam.	Red 20	roots	alkanet, lashon habar, lashon hapar		
Henna, Egyptian privet	Lawsonia alba Lam. = Lawsonia inermis (alba)	Orange 6	leaves	kofer, henna	kufra	fragrance
Lichen, orchil, cudbear	Rocella spp.	Red 28	plant	hazazit	pukh	facial make-up
Madder	Rubia tinctorum L.; other Rubiacaea spp.	Red 8	fresh or dried roots	puah (or fuah)	pua, puva	dyestuff
Safflower, dyer's thistle	Carthamus tinctorius L.	Red 26	orange-red florets	qurtam	haria', qotzah, qurtami, qurtema, moriqa, dardera	dyestuff; food seasoning and food coloring; medicine for diarrhea
Turnsole	Chrozophora tinctoria (L.) A. Juss.		berries and/or leaves	leshishit	shalshushit; gavshushit	dyestuff

Table II (b). Red and Orange Plant Dyestuffs of Ancient Israel: Dyeing Properties

Plant Name	Main Dye Components of Dyestuff after Hydrolysis			Color of dyeing on mordanted wool				
	Anthra-quinonoids	Naphtho-quinonoids	Others	no mordant	Al^{3+}	Sn^{2+}	Cu^{2+}	Fe^{2+}
Alkanet		Alkannin, Alkannan			violet			
Henna		Lawsone		burnt orange	brownish orange		dark orange	brown
Lichen			Orcein purple, litmus blue	bluish red	bluish red		dull purple	brownish violet
Madder	Alizarin, Purpurin, 21 others			dull red to brownish red	orange-red to brick-red (stronger, "livelier", color than unmordanted)	bright orange	purplish	dark chocolate (or purplish) brown
Safflower			Carthamin (Chalconoid)					
Turnsole				bluish or reddish				pink

Table III. Blue Plant Dyestuffs of Ancient Israel

| Botanical Information | | | | Reference to the Plant in Jewish Writings | | | Main Dye |
Common English Name	Botanical Name of Species	C.I. Natural Color Number	Part of Plant Used	Modern Hebrew Name	Historical Hebraic or Aramaic Names	Uses	Components of Dyestuff after Hydrolysis
Indigo	Indigofera tinctoria	Blue 1	leaves	nil	kala-ilan	dyestuff for the imitation of molluskan Tekhelet	Indigotin, Indirubin, Kaempferol (in some species)
Lichen, orchil, cudbear	Rocella spp.	Red 28	plant	hazazit	pukh, kahal, kohal	blue (and/or black) eye make-up	Orcein purple, Litmus blue
Turnsole	Chrozophora tinctoria (L.) A. Juss.		berries, leaves, stems	leshishit	shalshushit, gavshushit	dyestuff	
Woad	Isatis tinctoria L.	Blue 1	fresh leaves, whole plant	isatis	isatis (or is'tis) and variants: satis, satim, satin, is'tin (or isatin)	dyestuff	Indigotin, Indirubin

Table IV. Flavonoid and Isoflavonoid Dyes from Plant Dyestuffs of Ancient Israel

CHEMICAL CLASS Subclass Common name Chemical name[b]	C.I. Constitution Number	Plant Source of Dye: C.I. Natural Color Number of Plant Dyestuff	Substituents[a]				
			Ring A	Ring B		Ring A'	
			R_3	R_5	R_7	$R_{3'}$	$R_{4'}$
FLAVONOIDS							
Flavones							
Apigenin 4',5,7-Trihydroxy	75580	Weld: *Yellow 2*		OH	OH		OH
Luteolin 3',4',5,7-Tetrahydroxy	75590	Weld, Genista: *Yellow 2*		OH	OH	OH	OH
Flavonols (3-Hydroxyflavones)							
Fisetin 3,3',4',7-Tetrahydroxy	75620	Young fustic: *Brown 1*	OH		OH	OH	OH
Kaempferol (Indigo Yellow) 3,4',5,7-Tetrahydroxy	75640	Buckthorn berries: *Yellow 13;* Others	OH	OH	OH		OH
Kaempferol 7-methyl ether 3,4',5-Trihydroxy-7-methoxy	75650	Buckthorn berries: *Yellow 13*	OH	OH	OCH$_3$		OH
Quercetin 3,3',4',5,7-Pentahydroxy	75670	Buckthorn berries: *Yellow 13;* Sumac: *Brown 6*	OH	OH	OH	OH	OH

Continued on next page

Table IV. Continued

CHEMICAL CLASS Subclass — Common name / Chemical name[b]	C.I. Constitution Number	Plant Source of Dye: C.I. Natural Color Number of Plant Dyestuff	Substituents[a] Ring A R_3	Ring B R_5	R_7	Ring A' $R_{3'}$	$R_{4'}$
Rhamnazin 3,4',5-Trihydroxy-3',7-dimethoxy	75700	Buckthorn berries: Yellow 13	OH	OH	OCH$_3$	OCH$_3$	OH
Rhamnetin 3,3',4',5-Tetrahydroxy-7-methoxy	75690	Buckthorn berries: Yellow 13	OH	OH	OCH$_3$	OH	OH
ISOFLAVONOID Isoflavone							
Genistein 4',5,7-Trihydroxy	75610	Genista: Yellow 2		OH	OH		OH

[a]Refer to the parent structure in Figure 3.
[b]Full name of each flavonoid (whether a flavone or a flavonol) and isoflavonoid has the respective "flavone" or "isoflavone" parent molecule name appended; for example, luteolin is 3',4',5,7-tetrahydroxyflavone and genistein is 4',5,7-trihydroxyisoflavone.

Table V. Naphthoquinonoid Dyes from Plant Dyestuffs of Ancient Israel

Common name / Chemical name[b]	C.I. Constitution Number	Plant Source of Dye: C.I. Natural Color Number of Plant Dyestuff	Substituents[a] Ring A R_2	Ring B R_5	Ring B R_8
Alkannan 5,8-Dihydroxy-2-(4-methylpentyl)-	75520	Alkanna, Anchusa, Henna (roots): *Red 20*	$-CH_2CH_2CH_2CH$ with CH_3, CH_3	OH	OH
Alkannin or Anchusin 5,8-Dihydroxy-2-(1-hydroxy-4-methyl-3-pentenyl)-	75530	Alkanna, Anchusa, Henna (roots): *Red 20*	$-CHCH_2C{=}C(CH_3)_2$ with H, OH	OH	OH
Juglone 5-Hydroxy-	75500	Walnut: *Brown 7*		OH	
Lawsone 2-Hydroxy-	75480	Henna: *Orange 6*	OH		

[a]Refer to the parent structure in Figure 4.
[b]Full name of each hydroxynaphthoquinonoid has the parent molecule name appended; for example, juglone is 5-hydroxy-1,4-naphthoquinone.

Table VI. Anthraquinonoid (Madder-Type) Dyes from Plant Dyestuffs of Ancient Israel[a]

Common name Chemical name[c]	C.I. Constitution Number	Ring A				Ring B			
		R_1	R_2	R_3	R_4	R_5	R_6	R_7	R_8
Alizarin 1,2-Dihydroxy	75330	OH	OH						
Anthragallol 1,2,3-Trihydroxy	58200	OH	OH	OH					
Anthragallol 2-methyl ether 1,3-Dihydroxy-2-methoxy		OH	OCH$_3$	OH					
Anthragallol 3-methyl ether 1,2-Dihydroxy-3-methoxy		OH	OH	OCH$_3$					
Christofin 1,4-Dihydroxy-2-ethylhydroxymethyl		OH	C(H)(OH)(C$_2$H$_5$)		OH				
Copareolatin dimethyl ether 6,8-Dihydroxy-4,7-dimethoxy-3-methyl or 4,6-Dihydroxy-7,8-dimethoxy-3-methyl				CH$_3$ CH$_3$	OCH$_3$ OH		OH OH	OCH$_3$ OCH$_3$	OH OCH$_3$
Dammacanthal (nordammacanthal) 1,3-Dihydroxy*-2-aldehyde		OH	CHO	OH					
Emodin[d] 1,3,8-Trihydroxy-6-methyl	75440[e]	OH		OH			CH$_3$		OH
Lucidin 1,3-Dihydroxy-2-hydroxymethyl		OH	CH$_2$OH	OH					

Table VI. *Continued*

Common name / Chemical name[c]	C.I. Constitution Number	Substituents[b]							
		Ring A		Ring B					
		R_1	R_2	R_3	R_4	R_5	R_6	R_7	R_8
Morindadiol 1,5-Dihydroxy-2-methyl	75380	OH				OH			
Morindanigrin 1,3-Dihydroxy-6-methyl		OH		OH			CH$_3$		
Morindone (or morindon) 1,5,6-Trihydroxy-2-methyl	75430	OH	CH$_3$			OH	OH		
Munjistin (munjiston) 1,3-Dihydroxy*-2-carboxylic acid	75370	OH	COOH	OH					
Physcione (parietin) 1,8-Dihydroxy-3-methoxy-6-methyl		OH		OCH$_3$			CH$_3$		OH
Pseudopurpurin 1,2,4-Trihydroxy*-3-carboxylic acid	75420	OH	OH	COOH	OH				
Purpurin 1,2,4-Trihydroxy	75410	OH	OH		OH				
Purpuroxanthin (xanthopurpurin) 1,3-Dihydroxy	75340	OH		OH					
Quinizarin 1,4-Dihydroxy	58050	OH			OH				
Quinizarin-2-carboxylic acid 1,4-Dihydroxy*-2-carboxylic acid		OH	COOH		OH				

Continued on next page

Table VI. *Continued*

Common name Chemical name[c]	C.I. Constitution Number	Ring A		Substituents[b]		Ring B			
		R_1	R_2	R_3	R_4	R_5	R_6	R_7	R_8
(Quinizarin-6-methane)[g] *1,4-Dihydroxy-6-methyl*		OH			OH		CH$_3$		
(Quinizarin-2-methyl alcohol)[g] *1,4-Dihydroxy-2-hydroxymethyl*		OH	CH$_2$OH		OH				
Rubiadin *1,3-Dihydroxy-2-methyl*	75350	OH	CH$_3$	OH					
Rubianin *1,3-Dihydroxy-2-C-glycosyl-*		OH	CH$_2$OH (glycosyl structure: HO, OH, OH)	OH					
Soranjidiol *1,6-Dihydroxy-2-methyl*	75390	OH	CH$_3$				OH		

[a]Madder and related plant roots, as listed by C.I. include C.I. Natural Reds 6 (*Oldenlandia umbellata; Hedyotis umbellata*), 8 (Madder = *Rubia tinctorum*; Wild Madder = *Rubia peregrina*), 14 (*Galium*), 16 (Munjeet), 18 (Morinda = *Morinda citrifolia*; Suranji), and 19 (Mang-kouda = *Morinda umbellata*).

[b]Refer to the parent structure in Figure 5. Numbering of substituents is according to *The Merck Index, 11th Ed.*; Budavari, S., Ed.; Merck & Co., Inc.: Rahway, NJ, 1989.

[c]Full name of each anthraquinonoid has the parent molecule name appended or inserted (at the position with an asterisk); for example, alizarin is *1,2-dihydroxyanthraquinone* and pseudopurpurin is *1,2,4-trihydroxyanthraquinone-3-carboxylic acid.*

[d]Also present in Buckthorn berries, C.I. Natural Yellow 13.

[e]Structure in C.I. has the methyl group misplaced.

[f]Contains an additional methyl group, whose position has not yet been determined, in the substituents column.

[g]Common name not yet coined.

Table VII. **Miscellaneous Dyes from Plant Dyestuffs of Ancient Israel**

Common Name: Chemical class	C.I. Constitution Number	Plant Source of Dye: C.I. Natural Color Number of Plant Dyestuff	Structure
Carthamin (carthamic acid): *Chalconoid*	75140	Safflower: *Red 26*	Figure 6[a]
Crocetin: *Carotenoid (Polyene)*	75100	Saffron: *Yellow 6*	Figure 7
Curcumin: *Feruloylmethane (Aroylmethane)*	75300	Turmeric: *Yellow 3*	Figure 8
Indigotin: *Indigoid*	75780	Indigo, Woad: *Blue 1*	Figure 9
Indirubin: *Indirubinoid*	75790	Indigo, Woad: *Blue 1*	Figure 9
Safflower Yellow (Safflor Yellow A): *Safflor*		Safflower: *Yellow 5*	Figure 10[b]
Tannins: *Polyphenols*		Galls, Sumac, Walnuts: *Brown 6* Pomegranate: *Yellow 7*	
Ellagic acid (hydrolysis product from crude tannins)	75270	(present as the free acid and/or combined in an *ellagitannin*)	Figure 11
Gallic acid *3,4,5-trihydroxybenzoic acid* (hydrolysis product from crude tannins)		(present as the free acid and/or combined in a *gallotannin*)	Figure 11

[a]New structure is as shown in Merck and not in C.I.
[b]An, X.; Li, Y.; Chen, J.; Li, F.; Fang, S.; Chen, Y. *Zhongcaoyao* **1990**, *21*, pp 188-189.

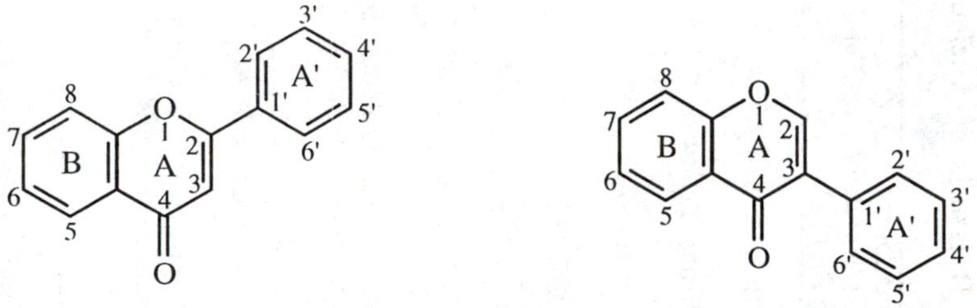

(a)   (b)

Figure 3. Parent structures for flavonoid (a) and isoflavonoid (b) dyes.

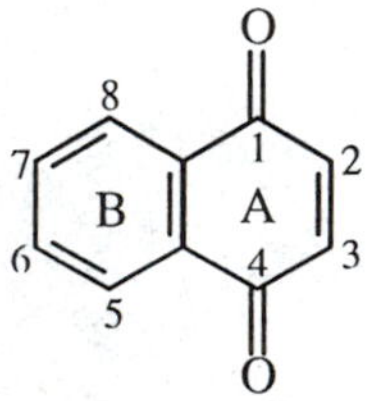

Figure 4. Parent structure for naphthoquinonoid dyes.

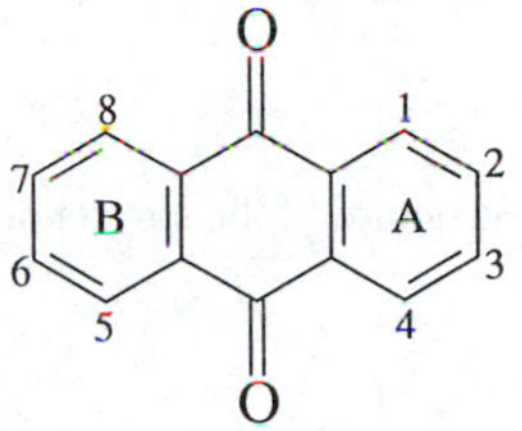

Figure 5. Parent structure for anthraquinonoid dyes.

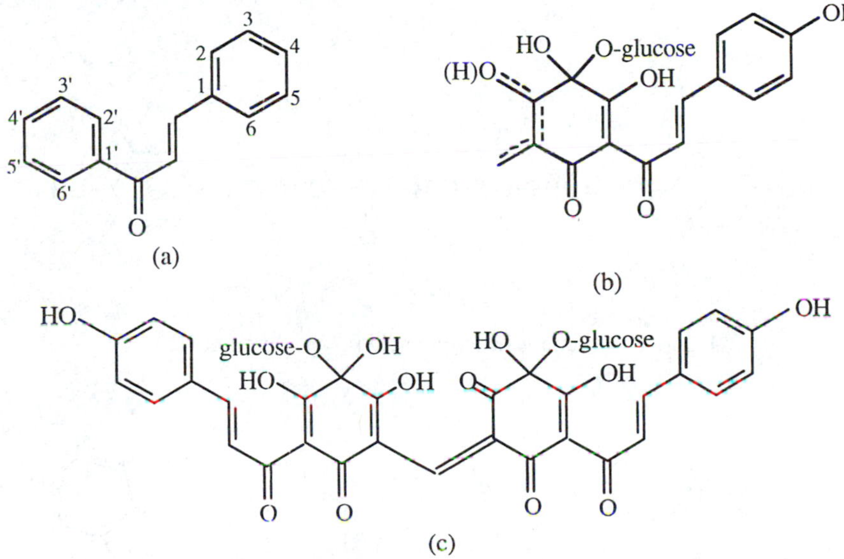

Figure 6. Structures of chalcone (a), chalconoid monomer of carthamin dimer (b), and carthamin (c).

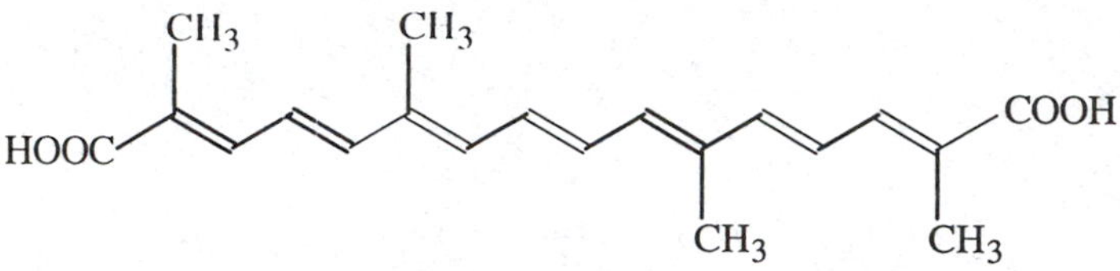

Figure 7. Structure of the carotenoid crocetin.

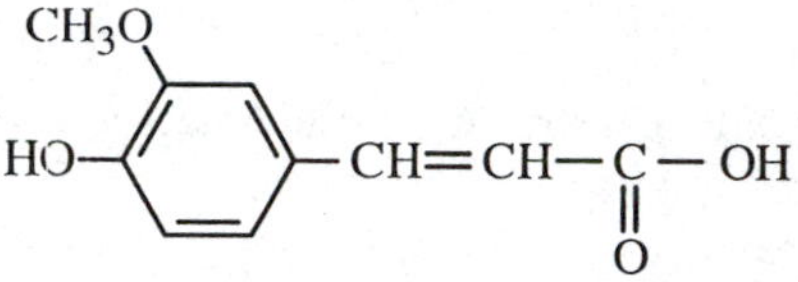

(a)

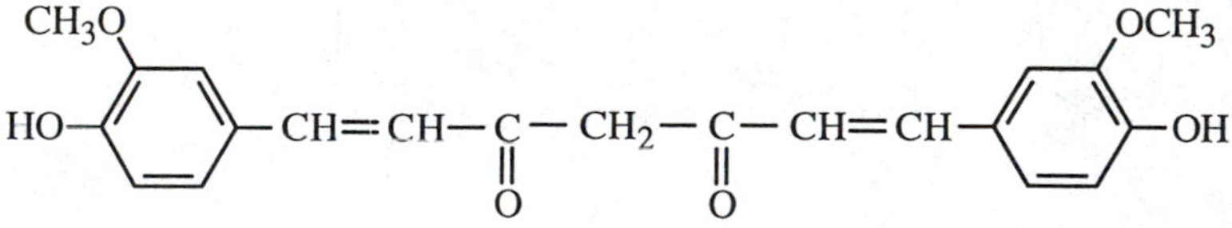

(b)

Figure 8. Structures of ferulic acid (a) and curcumin (b).

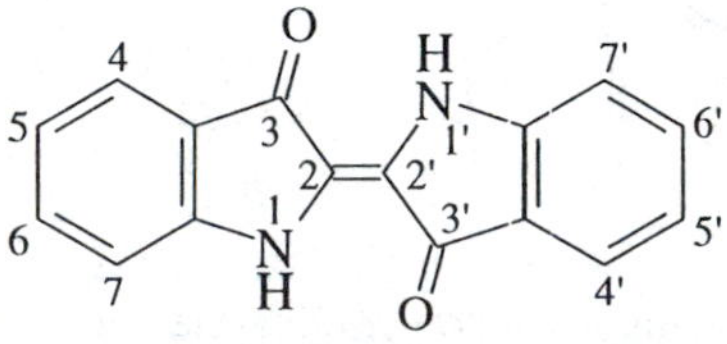

(a) 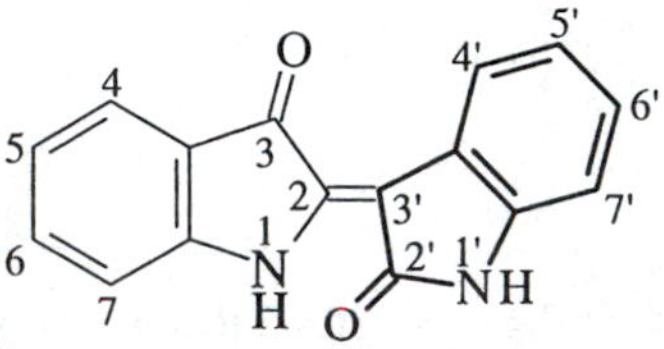   (b)

Figure 9. Structures of indigotin (a) and indirubin (b).

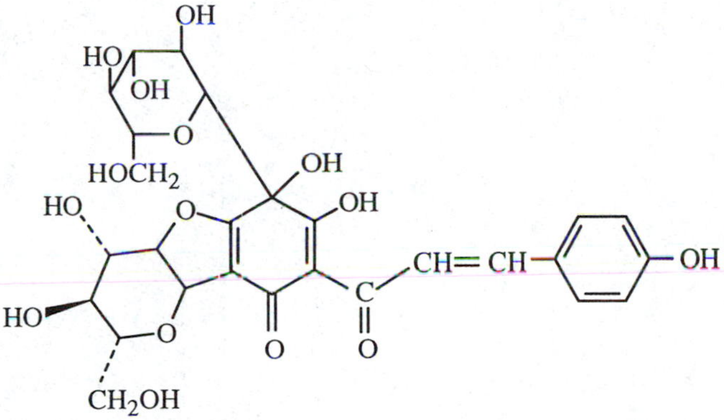

Figure 10. Structure of safflower yellow (Safflor Yellow A).

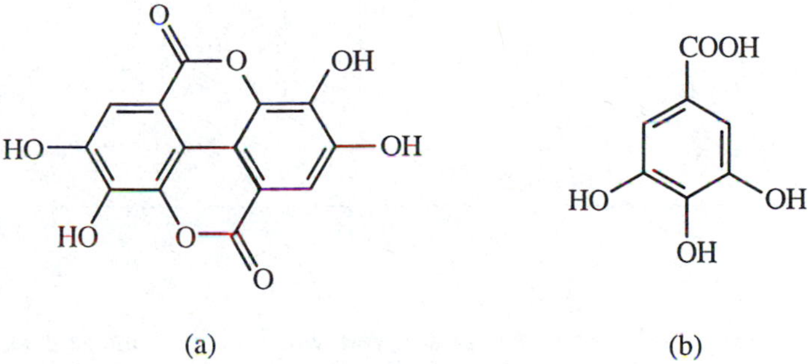

(a)   (b)

Figure 11. Structures of the tannin hydrolysates ellagic acid (a) and gallic acid (b).

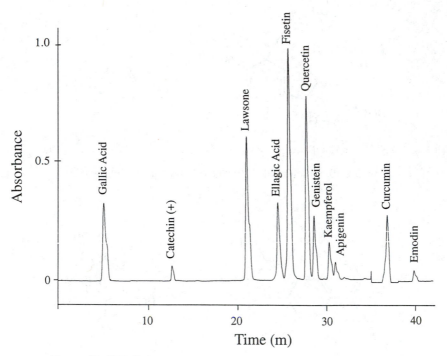

Figure 12. HPLC chromatogram of yellow, brown, and orange dyes.

produced a black lake on the textile. Typical tannin–producing dyestuffs were sumac leaves and gall nuts (also known as nut galls). The galls were round cysts that were produced on the branches of oak trees by insects breeding on these trees and contained within the cysts.

Buckthorn Berries (*Eshhar*). These unripe berries (also known as Persian berries (72) are not mentioned as a dyestuff anywhere in the Jewish sources, though they have been used, at least in other regions, as a source for a yellow dyeing. Pfister (*73a*) mentions that a certain Greek Papyrus No. 10 cites buckthorn berries as a dyestuff and that the best came from Persia and Turkey. It is included in this study, nevertheless, because a species of the tree that provides these berries is native to Israel and, in fact, yellow to green dyeings have been produced from them. The dye components of this plant include flavonols and the anthraquinone emodin.

Gall Nuts (*'Afatz*). There are numerous talmudic references to galls, but never in the context of textile dyeing. The galls are very rich in tannins and are mentioned as an ingredient for tanning leather (*26f*), processing animal hide into parchment for use as a scroll (*40b*), and for the production of ink (*26j, 30c, 74a, 75*). Galls can be found today on certain oak trees in the Upper Galilee region of Israel. Papyrus Holm indicates that galls were used as a dyestuff (*73b,c*) as do Pliny and Theophrastus (*2c*). Dalman (*76*), a historian who studied the crafts of Israel, reports that galls were used in Israel at the end of the 19th century.

This research has found the two tannin hydrolysates gallic and ellagic acids (together with other unidentified compounds) as a result of the acid hydrolysis of the tannins from the raw plant dyestuff and from a tannin–produced textile dyeing. Table VII shows the structures of the two identified components.

Pomegranates (*Rimon*). The pomegranate is cited numerous times in the Bible as, for example(*52b*): "Why have you raised us out of Egypt, to bring us into this evil place? It is no place of seed, or of figs, or of vines, or of pomegranates, or of water to drink." Pomegranate peels are specifically cited in the Talmud as a dyestuff (*26d,f, 53d, 64b, 65, 77a*) and, as such, are often mentioned together with walnut peels. The peels were also used for the production of ink ("water of *nara*") (*29d, 30d, 78*) and, as with walnut peels, for a type of game (*79*). Although the talmudic sources refer only to the pomegranate rinds, Pliny (*2b*) indicates that the flowers are used as a purplish dye and the *Colour Index* states that the flowers serve as a red dyestuff in certain regions in India. Czyzyk (*13a*) supposes that the roots and trunk were also used as a dyestuff.

According to Pliny, the unripe fruits were used for the tanning of leather (*2b*). Pfister (*73b*) cites the Papyrus x and states that pomegranate peels and vinegar were mordants for alkanna dyeings, and that, according to Papyrus Holm 20, pomegranate roots yield tannin and produce a yellow color. Dalman (*76*) reports that pomegranate dyeing was still performed in Israel at the end of the 19th century and that dyeings produced light brown colors and, with iron, black.

According to the *Colour Index*, the constituents of the reddish fruit rinds from this tree are granatonine, an alkaloid that is not considered as a dyeing substance, and hydrolyzable tannins. In the current study, gallic and ellagic acids as well as other unidentified hydrolysis products were found.

Rose (*Vered*). The roots of this plant as described by the JT were used as a "color for animals" (*80a*). The meaning represented by this expression has been previously described. This plant has not been definitively identified although some possibilities are *Rosa canina* and *Rosa phoenicia* (*10a*). It is also mentioned in the

Talmud together with other fragrant plants. Its fruits yield beige, yellowish, and greenish dyeings depending on the mordant used (*10b*), which, by comparison with other dyeings produced, indicates that the fruit dyes are probably mostly flavonoids.

Safflower (*Qurtam*). The origin of the name in Latin, *Carthamus*, is probably derived from the old Hebrew or Semitic root *qrtm*, which, in the verb form, means to cut, nip, truncate as in the process of collecting the flowers of the plant (*81a*). This plant is mentioned as a cultivated dyeing plant in the Talmud (*26k, 29e, 77b*), however, the Jewish sources do not indicate whether a yellow or red dye was intended, or both. The florets were also used for food seasoning and coloring (*82*).

This plant seems to have a significant number of synonymous names in the Talmud: *haria'* (*83, 84a*), *qotzah* (*26k, 77b*), *qurtema* and other variants (*30e, 50c*), *moriqa* (*30e, 84a*), and *dardera* (*30e*). The following passage in the Talmud discusses the uses of safflower as a medication and as an aphrodisiac and incorporates some of these names (*30e*): "The Rabbis taught: If one performs sexual intercourse while standing, then he will contract cramps. If (he does it) while sitting, then he will contract *dalaria* (diarrhea or shortness of breath or impotency). If she is on top and he is on the bottom, then he will (also) contract *dalaria*. ... Rabbi Joshua ben Levi said: The medicine for *dalaria* is *dardera*. What is *dardera*? Abbaye said: *moriqa* (yellow-green) of the thorns. Rabbi Pappa chewed and swallowed the *moriqa,* (while) Rabbi Pappi chewed and spit it out. Abbaye said: Whoever is not fit for sexual intercourse, let him bring three small measures of *kortemi* of the thorns, and he should grind them and cook them in wine and drink it. Rabbi Yohanan said: (That potion was) indeed what restored my youthful (sexual prowess)." A clear identification of these plant names follows (*85*): "And *qotzah* is flower of *qurtemi* and it has a different name in the Holy Language (Hebrew), *haria'* cakes, and in Aramaic *moriqa*, and in Arabic *'uftzur* and with it they dye linen clothes and cotton clothes for the brides for their wedding ceremonies." Also, a commentary by Rabbi Hai Gaon (*83*) states that "*haria'* cakes (are) *moriqa* ... which is *qurtim*." However, there is still some etymological controversies in recent and post-Talmudic Jewish sources regarding the botanical identities of all of these names (*10c*).

Pfister (*73b*), citing Papyrus Holm, states that safflower was added together with alkanna.

The dye substance that has been called "Safflower Yellow" and whose structure has been unknown, has recently been identified as Safflor Yellow A (*86*) and awaits further confirmation.

Saffron (*Karkom*). Saffron is mentioned just once in the Bible and the inference is to a fragrant plant (*87a*): "Nard and saffron, calamus and cinnamon, with all trees of frankincense; myrrh and aloes, with all the chief spices." This crocus, whose Latin name is probably derived from the Semitic 'karkom', is never mentioned as a dyestuff in the Jewish sources. The stigmas are referred to in the Tosephta (*88*) as a food color and it is also mentioned with other fragrant plants. In the Roman empire, the dyers who used (imported) saffron were known as *crocotarii* (*2d*). The main dye of saffron is the carotenoid crocetin.

Sumac (*'Og*). The roots are described as a colorant for animals together with the roots of *vered* (rose) and *egeh* (identity unknown) (*80a*), but there is no mention of it as a dye plant for textiles. According to Dioscorides (*2b*), sumac was used for tanning and dyeing leather. The leaves, berries, and seeds could be used for dyeing, while the roots and stem were used for tanning (*89*). This research has found the flavonol fisetin and the tannin hydrolysate gallic acid, and others, in dyeings produced from sumac leaves.

Turmeric (*Kurkum*). This yellow dye source obtainable from the yellow-colored roots of this plant are not mentioned in the Jewish sources. Though it is native to China, India, and nearby regions, it is listed in this study because it also grows in the Middle East and it has been used historically as a yellow dyestuff source. The main colorant is an aroylmethane (*90*), or, more specifically, a feruloyl, whereby the dye is essentially a dimer of ferulic acid, as indicated in Table VII.

Walnuts (*Egoz*). The walnut tree is already mentioned in the biblical Song of Songs (6:11): "To the walnut garden I went down to see the fruits of the valley, and to see whether the vine had blossomed, whether the pomegranates were in flower." In the Jewish sources, 'walnut peels' are always mentioned together with 'pomegranate peels' and, additionally, sometimes with other dyestuffs. Walnut peels were also used as a mordant (or in a double dyeing process) with the final dyeing or overdyeing produced by pomegranate rinds (*65*).

The walnut tree was cultivated in ancient Israel (*91*) and may have been brought into that area from Persia during the Second Temple period (*11a*). This supposition is partly based on Pliny's statement that the tree was cultivated by the kings of Persia, as implied by its royal name, *Juglans regia*. The Talmud alludes to the high demand of walnuts (*91*) and hence it could certainly have also been imported into Israel at the time even though it was cultivated locally.

The fresh or dried green outer peel (or husk) was chopped and used for producing the colorant for textile dyeing, as was the case with pomegranate peels, as well as for the production of ink (*29d, 78*). It is directly mentioned as a dyestuff for textiles in the Talmud (*53d, 64b, 68c*): "One who takes out (on the Sabbath – from one's private property to the public domain – either) walnut peels, pomegranate peels, woad, or madder, (the minimum quantity of each dyestuff necessary for this transport to be considered as a work activity forbidden to be performed on the Sabbath is if there is) enough of (any one of) them to dye a small article of clothing (such as the forehead band of) a hair–net."

The main dye constituent of walnut peels is the naphthoquinone juglone, which produces brown dyeings on wool and dark brown (or 'black') with iron.

Weld (*Rikhpah*). The modern Hebrew name for this plant, *rikhpah*, is identical to its talmudic appellation. In the Talmud (*77a*), the plant is mentioned together with madder as a dyeing plant that was cultivated in Israel. This is its only citation in the Talmud as a dyestuff. This plant was mentioned by Pliny (*2e*) as being able to be cultivated virtually anywhere, but that in Asia and Syria it grows as a wild plant.

Dalman (*76*) reports that *Rhamnus* berries and weld were used in Israel at the turn of this century to produce yellow dyeings. Masschelein–Kleiner and Maes report on the finding of luteolin, the main dye component of weld, in 5th – 7th century A.D. Egyptian textiles and in some Sudanese textiles (*92*). In that same study, it was found that one textile from Nahal Hever in Israel may have been dyed with weld. The main colorants in weld are flavones, mostly luteolin with some apigenin.

Young Fustic. The *Colour Index* indicates that this shrub or small tree grows in the Levant and thus this dye plant is listed in this study, though the Talmud does not mention it. The wood of this plant contains flavonoids and tannin hydrolysates.

Red and Orange Dyestuffs. Plants producing red and orange dyestuffs are tabulated in Table II and the dye constitutions of these plants are given in Tables V–VII. In the limited number of analyses that have been performed on ancient textiles from Israel up to the present time, no dyeings produced by either alkanna, henna, or safflower have been unambiguously identified. Further investigations on such textiles will be conducted to test for these dyeings. The colors of dyeings that could be produced from these red dyestuffs range from orange, pink, red, brown, and violet, to purple, as indicated in Table II. The red and orange dyestuffs listed in this section belong to three different chemical groups (see Table II).

Alkanet (*Lashon ha–Par*). The roots of alkanna (or alkanet) or anchusa were a source of a red dyestuff of historical importance and as this plant was native to the Middle East, it is listed in Table II. However, this plant is not mentioned in any historical Jewish source. The Greek historian Theophrastus (4th–3rd centuries B.C.) mentions alkanna or anchusa roots and Pliny mentions the blood–red root of alkanet as a dyestuff (*2f*). The main dyes from this dyestuff are naphthoquinonoids.

Henna (*Henna; Kofer*). *Kofer,* the biblical name for henna, is mentioned in the context of a fragrant tree (*87b*): "My beloved is to me a cluster of henna in the vineyards of 'En–Gedi." The henna shrub or small tree is not mentioned as a dyeing plant in Jewish sources but is indirectly referred to as a plant for the production of fragrances (*77c*). However, it is known historically that the leaves of this shrub produced an orange colorant for the coloring of hair and of the body (*2f,g*) and could have been used as a textile dye. This plant has one main colorant, the naphthoquinone lawsone.

Lichen (*Hazazit; Pukh*). It is possible that at some period, a lichen may have been imported and used as a colorant in Israel of the talmudic or even biblical period. Red–producing lichens have been used in Europe and may have been imported into the Middle East in ancient times. The Talmud discusses a red colorant used as a facial makeup (*26l, 27c*). The talmudic word used is a grammatical variant of *fuqos*, which is derived from the Greek φυχοζ, probably a lichen. Hence, the Talmud's *fuqos* was probably a red or purple colorant produced from a lichen. The Greek term, which itself was probably derived from the biblical *pukh*, which is cited in, for example, Kings II 9: 30, has evolved through the Latin form into the modern English 'fucus', meaning a kind of paint for the face or any paint or dye (*93a*). Thus, the biblical *pukh* may have referred to a lichen.

Madder (*Puah*). The most popular dyestuff for the production of red dyeings and related hues was the roots of the madder plant, a *Rubiacaea* species. According to the historian Aharanson (*94*), this plant could still be found early this century in the Trans–Jordan mountains of Gilead, Amon, Moab, and Edom. A single madder dye plant was recently discovered growing on the western bank of the River Jordan, near Kefar Ruppin, Israel (Safrai, Y., personal communication). The madder plant is specifically mentioned as a dye plant together with other dye plants in the 'botanical' chapter of tractate *Shevi'it*, which indicates that it was cultivated in Israel at least during talmudic times. Though the most common species of the madder dyestuff was *Rubia tinctorum* L., *Rubia peregrina* L., which according to Leggett (*1b*), is native to the Middle East, may have also been used. In the latter species, the purpurin component is dominant and alizarin is only a minor constituent (*95*). Other similar species that are native to India and neighboring regions may have also been imported into Israel as direct or indirect trade between India and Israel already existed at that time.

The madder root has been described by various Greek and Roman historians, such as Herodotus, Strabo, Pliny, Dioscorides, Vitruvius, and Theophrastus. Dyeing recipes in Papyrus Holm (*73d*) indicate that pink color can be obtained from madder using alum and vinegar, and purple from madder with indigo. The popularity of the red color, in general, can be attributed to the symbolism associated with this color. Accordingly, Leggett (*1b*) indicates that red symbolized courage.

Essene bones in the Judean Desert caves at Qumran contained madder probably as a result of eating the root as a medicine for the supposed dissolution of kidney stones as well as to prevent their formation (*96*).

The dyes in madder roots consist of hydroxyanthraquinones, some free and others as glucosides, most of which undergo hydrolysis during the dyeing process to yield the free dye and the sugar moiety. A notable exception to this is rubiadin, a C–glycoside that does not undergo hydrolysis. As many as twenty–three dye components have been identified in madder roots (*71, 97, 98*), as shown in Table VI. All of the molecules listed in that table contain the adjacent carbonyl and phenolic oxygens, a structural requirement for the stable chelation with the metal mordant (Figure 1). The major dyes produced from the madder root and from other *Rubiacaea* species are the hydroxyanthraquinones alizarin and purpurin. Alizarin is produced from the hydrolysis of the glycoside rubierythric acid (*71*), and purpurin from the decarboxylation of pseudopurpurin (*97, 99*). The latter phenomenon is especially apparent in aged roots, which may have been used to produce the purpurin–rich madder dyeings found at Masada (*23*).

Most of the red dyeings on archaeological textiles discovered in Israel were dyed with madder. These include the Judean Desert Roman period Masada (*23*) and Cave of Letters (*17, 18*) textiles. The oldest madder dyeing found near Israel was from Kuntillet 'Ajrud in eastern Sinai, dated from 9th–8th centuries B.C. (*21*). Madder dye was also found in ancient textiles from Syria, Israel, Egypt, Sudan, and Cyprus (*10d*).

The ancient dyeing process to produce bright colors with madder roots probably involved the preparation of the dye solution by heating the water mixture containing the roots to well below the boiling temperature. Mild temperatures may have been used to minimize the dissolution of the root tannins into the dye solution, thus avoiding the undesirable dull brown shades produced by the presence of tannins.

Safflower (*Qurtam*). A red dyestuff (carthamin) could also be produced from safflower petals in addition to the yellow colorant already discussed. Normally, the yellow dye component is removed from the dyestuff by repeatedly soaking the florets in water until no more yellow colorant is extracted. The florets are then immersed in an alkaline solution to extract the red dye. The textile is then immersed into the dye solution and then the dye is precipitated into the fibers by means of an acidic solution, such as vinegar. Various chemical structures of carthamin, the main red colorant of safflower, have appeared in the literature, but the most recent consists of a chalconoid dimer, as indicated in Table VII.

Turnsole (*Shalshushit*). The dye plant known as *Chrozophora tinctoria* L. or turnsole in English was the dye plant mentioned only once in the 'botanical chapter' of the JT as *shalshushit* (pronounced "shal–shoo–sheet") (*74b, 80b, 100*). The talmudic name is derived from the triply–fused berries that are characteristic of this plant (*shalosh* means "three" in Hebrew). Pliny (*2i*) was also aware of this triplet characteristic and called the plant *Heliotropum tricoccum.* The modern Hebrew name of this plant has the first consonantal letter removed and is called *leshishit* (pronounced "le–shee–sheet"). As described in the Talmud, the agricultural laws of the sabbatical year pertained to it, which indicate that it was cultivated in Israel.

It is not totally clear as to which part of the plant was used for producing the dyestuff. If one crushes the berries while they are still green, the oozing sap is greenish, which, in a matter of minutes, becomes reddish, and then violet (*10e*). Czyzyk (*13b*) claims that the precursor to the dye is colorless and is found in the stems, leaves, and berries, and after processing with ammonia vapors, the color becomes blue.

However, Löw (*81b*) indicates that the dye is produced from the green parts of the plant, and after processing with ammonia vapors, the green sap eventually turns red. Löw also reports that later sources indicate that only the juice of the berries was used. Wulff (*101*) has reported that traditional dyeing methods employed in modern Iran use the leaves of the plant, which produces violet colors on wool, and with the addition of urine to the dye bath, a turquoise blue dyeing is obtained.

This author has not been successful in obtaining the above reported red and blue dyeings on wool. The solutions of the extracted berry juice did yield bluish and red colors depending on the pH of the solution. However, these colors did not have any affinity whatsoever for the wool. The final colors produced were beige, indicating that the tannins or flavonoids in the berries were the dyes, but, unfortunately, not the other colorants.

Blue Dyestuffs. The two major plants used for producing blue dyeings were woad and indigo, both producing the blue pigment known as indigotin. Turnsole was also mentioned historically as a dyestuff, but it is uncertain as to whether it was used as a blue or red dyestuff or for both, as previously described. It seems probable that a lichen was used for blue eye–makeup, and though lichens were used historically in other regions, their possible use in textile dyeing in Israel of the talmudic period is simply conjecture.

Indigo (*Nil; Kala–ilan*). There have been some uncertainties regarding the identification of the talmudic *kala–ilan* that have been expressed in the modern literature, specifically as to whether it refers to a single plant dyestuff or to a mixture of dyestuffs, as based on the midrashic homily (*102*): "However, one should not say, here I (God) give (you) colorants and *kala–ilan* and they resemble *Tekhelet*." However, based on an investigation of the earliest explanations of this word in historical Jewish sources and on other historical writings and chemical analyses, there is little doubt that *kala–ilan* refers to the indigo plant or to its blue dyestuff.

The first interpretation (or translation) of *kala–ilan* appears during the Geonic period in Babylonia (*Gaon* – literally "Excellency" or "Exalted" – was the head of an academy). This epoch began about 650 A.D., which was about a century and a half following the redaction of the Talmud, and lasted to about the early 11th century. The first written use of the word *nil* in Jewish writings and the first translation of *kala–ilan* that is known to this author is attributed to Rabbi 'Amram Gaon in a Responsum (written response to question) dated from the Hebrew month of Adar, in the Seleucidean year of 169 (857 A.D.) (*103*). (The first written mention of *nil* in Arabic writings from Egypt is only much later – the 11th century A.D. (*4a*). In that response, Rabbi 'Amram clearly states: "*Kala–ilan*: lilang (or *lilagg*) and, in the Arabic language, *nil*." *Lilang* (or *lilak*) and *nilak* are variant Persian forms meaning bluish and have evolved into the English 'lilac' and are derived from the Sanskrit *nila* (or *nili*), which means dark blue or indigo (*93b*). It is well known that the Arabic *nil*, *nila*, or *a–nil* referred to a blue plant dyestuff (*104*). However, since 'isatis' was a well–known word in the Talmud for the woad plant (see later), which itself was a blue dye source, the usage of a different word for blue (*nil*) must refer to a plant different from isatis. As only one scholarly period, the Sevoraic (named after the "explainers" of the Talmud), lay between the sealing of the Talmud and the Geonic period, the identification of *kala–ilan* as *nil*, i.e., indigo, seems to be authentic. Further historical evidence that *kala–ilan* is in fact indigo is from the 10th century Jewish Italian lexicographer Rabbi Nathan who was also known as the 'Arukh', which was the name of his lexicon to the Talmud. In that work, the Arukh clearly translates *kala–ilan* as 'indaco' (*105*). More modern lexicographers, such as Löw and Jastrow, also agree with that translation.

Etymologically, the talmudic name of this dyestuff may be composed of two words, (for example, *kala ilan, kela ilan,* or *k'la ilan*), or as one word. The latter part of the name is pronounced as 'eelan'. However, as all words in the Talmud originally appeared without vowel marks (in modern Hebrew, a system of lines and dots is used as vowel signs), the exact pronunciation of the first part of the name, which may assist in determining its etymological roots, is not certain. Various possible derivations of the name have been given in Rabbi Herzog's doctoral dissertation (*106*).

Kala–ilan was a dyestuff that the Talmud admonished against using as an imitation to a bluish *Tekhelet* color. This *Tekhelet* may only be derived from a sea animal source in order to be ritually fit to color a thread on the fringes from each of the four corners of a garment. The color of dyeings produced by *kala–ilan* and *Tekhelet* were essentially indistinguishable and could only be differentiated by a 'Superior Being' (*62b*). This discernment is indicated by the talmudic passages that interpret the Bible's admonition against fraudulent business practices that would be very difficult for mortal men to discover (*62b*): "The Holy One Blessed Be He said: I am the One who distinguished between a little (seed) of a firstborn and a little (seed) that is not of a firstborn (for the tenth plague of slaying of the firstborn Egyptians prior to the Exodus of the Israelites). I will (therefore be able to distinguish and) 'repay' one who places *kala–ilan* on his fabric and says it is *Tekhelet*." Thus, because of the difficulty in differentiating between these two dyes, the Talmud provides the following caution (*32d*): "*Tekhelet* has no test and should only be obtained from an expert (who knows the difference between these dyes and can be trusted)."

There has been some confusion regarding the blue dye obtainable from the isatis and indigo plants. The 12th century Jewish scholar Maimonides, who wrote his explanations to the Mishnah in Arabic, uses the word *nilag'* (*28f, 77b, 107*) when referring to isatis. This is undoubtedly due to the fact that the Arabic name *a–nil* referred to the blue dyestuff from both isatis and indigo and the plants were known as *nila*. A similar confusion has also entered the area of modern–day identifications of ancient blue–dyed textiles from Egypt and other regions in the Middle East. This is mainly due to the historical name of the blue colorant as simply 'indigo', which, unfortunately, is also the generic name of the plant that generates it. The chemical composition of the blue dye produced from the isatis plant is identical to that produced from the indigo plant. In the past, reports by dye analysts that have indicated the presence of 'indigo' in a dyeing have, unfortunately, led some historians to attribute that indigo dye to the indigo plant. For this reason, the main colorant in both plants is called 'indigotin' and not 'indigo'. Hence, chemical analyses, at this point in time, which cannot distinguish between isatis–dyed or indigo–dyed blue textiles, can only determine the nature of the dye itself and not the plant source.

It is most probable that the dye from the indigo plant began entering the Roman Empire, which included the Middle East, during the first century A.D. This observation is consistent with the statements of Vitruvius and Pliny (*2j*). Vitruvius (1st century A.D.) mentions that it was difficult to obtain indigo in his time, so woad was being used. Pliny indicates that during his time – also the 1st cent. A.D. – the second most important dye after purpura, the 'royal purple' from sea snails, is "indico" from India and that its importation had recently begun. Some time after the 1st century, significant quantities must have entered Israel and it must have posed a religious threat to ritual Tekhelet dyeing. Thus, the injunctions against the use of *kala–ilan* stipulated in the Tosephta (1st–2nd centuries) and later echoed in the Gemara (ca. 200–500 A.D.) are consistent with the entrance of the indigo pigment into Israel in the first or second centuries. This dyestuff probably entered the Middle East as the final blue product in the shape of 'cakes' or lumps, and not as the leafy raw material. In fact, both Pliny and Dioscorides believed that this dyestuff was a mineral, probably because it was sold in the form of mineral–like lumps. Furthermore, *kala–ilan* is not discussed in the Talmud within the context of agricultural laws, which only applied to plants cultivated in Israel. Hence, it can be inferred that *kala–ilan* was not grown in Israel during the talmudic period but was imported.

Lichen (*Hazazit; Pukh*). The Semitic name *pukh* already appears in the Bible as an eye make–up, as in Kings II 9: 30: "When Jehu came to Jezre'el, Jezebel heard of it; And she painted her eyes, and adorned her head, and looked out at the window." As previously discussed in the section on red dyestuffs, the probable etymological development of the word *pukh* led to the Greek 'phykos' and the talmudic 'fuqos' and later into the Latin–English 'fucus', indicating that the biblical *pukh* was a colorant produced from a lichen. In the section on red dyestuffs, it was mentioned that the talmudic 'fuqos' (and maybe also the biblical *pukh*) was a red colorant produced from a lichen. It is also probable that the biblical *pukh* originally referred to a blue colorant that was used for painting the eyes. This can be seen from the close textual associations between *pukh* and another biblical name – *kahal,* which is a bluish eye make–up (*kahol* is "blue" in Hebrew). Evidence for this may be from the discovery of a green–blue powdered pigment that was found in a Byzantine period glass amphora from Israel and analyzed by the author (Koren, Z. C., unpublished.). Solubility tests in various solvents indicated that it was an organic colorant, whose color in the soluble form is extremely fugitive. The colors of this pigment in acidic and basic solutions also indicate that this colorant may have been produced from a lichen. However, even if in fact a lichen was imported into Israel and used as an eye or facial colorant, its use as a textile dye in ancient Israel, lacking further evidence, can only be hypothesized.

Turnsole (*Shalshushit*). This plant dyestuff may have been used to produce bluish dyeings and it was previously discussed in the red and orange dyestuff section.

Woad (*Isatis*). Undoubtedly, the predominant plant that was cultivated in ancient Israel for its blue dyestuff was called *isatis* (or *is'tis*) in the Talmud, which is equivalent to the Greek ισατιζ and the Latin *isatis* names. Since the Talmud was written without vowel marks, the pronunciation of the talmudic name is uncertain. Hence, the Semitic pronunciation of the plant name is either "ee–sa–tees" or "ees–tees". This plant is mentioned numerous times with other dye plants that were growing in Israel during the talmudic period (*26d,f,k,m, 28f, 29f, 51b, 53d, 67b, 77b, 84b, 107*).

According to the Talmud, it may be that in a certain village, most of the labor force was engaged in the production of the dyestuff from woad. The language of the discussion could, however, be only hypothetical. An interesting discussion regarding isatis dyeing in talmudic Israel involves the Priestly Blessing. In this part of the liturgical ceremony, the priests (who were descendants of Aaron) face the congregation, raise their hands towards them and recite the triformed benediction (*52c*). This priestly blessing is termed 'raising the hands' in the Talmud (*40c*): "Mishnah: A priest who has defects on his hands should not 'raise his hands' (i.e., not participate in the ritual). Rabbi Judah says: Even if one (has no defects but) whose hands are colored from isatis or madder, should also not 'raise his hands' because the people look at him (and would thus not be able to concentrate on the blessing itself). ... . Gemara: (However,) one rabbi taught: If the occupation of most of the (working) people of the town is with this (isatis dyeing) – then they are allowed (to participate in the religious ceremony)."

Pfister (*73e*) states that woad was the "indigo" of the Middle East and that isatis was mentioned in Greek papyri (*165*). Its use as a body paint was described by a number of classical historians.

Unidentified Plants

Two unidentified plants, *dan* (or *din*) and *tzad* (or *batzar*), are mentioned together with *isatis* seeds in the Jerusalem Talmud (*74c, 80c*) so that it may be inferred that these were also dyeing plants. In addition, *halavtzin* (or *halavnin*) – or *halavluv* in Hebrew – is also mentioned together with the dye plant *shalshushit* in the JT, which would indicate that it too was a plant dyestuff. The identity of this plant is not certain, though Feliks (*12b*) indicates that it may be the dyer's spurge, *Euphorbia tinctoria* (*E. macroclada*).

The two plants whose roots have been described as being used for "colors for animals," the *vered* (rose) plant and *egeh* (or *ageh, egah,* or *agah*) (*74c, 80a*), have also not been definitely identified. However *egeh* may be *Calycotome villosa* (*12a*) or possibly *Genista fasselata*, a species related to *Genista tinctorum* L. (*12c*). Other possibilities could be *alhagi* species, such as *Alhagi camelorum* (*81a*) or *Alhagi maurorum* (*10f*), and, as indicated in Table I, *vered* may be *Rosa canina* L. or *Rosa phoenicia* (*10a*).

A High–Performance Liquid Chromatographic Analysis for Ancient Dyestuffs

A successful dye analysis must produce a qualitative and quantitative chromatographic fingerprint of each dyestuff. In this way, unambiguous identifications of the dyestuffs used in a particular dyeing may be made. The high–performance liquid chromatographic (HPLC) technique has been shown to be useful for the analysis of natural dyes (*108*). This method has been applied by this author to the analysis of red hydroxyanthraquinonoids, blue indigotin, and red indirubin, which are all obtainable from plant sources, using the same elution scheme for all of these dyes. The results have been published elsewhere (*109*). In this work, another gradient elution method has been used for some of the yellow, brown, and orange dyes investigated in this study.

Prior to the chromatographic analysis, a reproducible sample preparation step is needed for the stripping of the dye from the textile fibers, and this procedure is described below.

Dye Extraction Procedure. The solvent system of choice for the extraction of mordant and direct dyes is the hydrochloric acid–methanol system, which has been widely used. The strong acid breaks the metal mordant–dye lake and also dissolves the metal ion, while the methanol extracts the dye from the fibers (*18, 110*). The advantage of HCl is that the gas is relatively easily driven off from the solution, a factor that is important in the sample preparation process and for the subsequent analysis. The steps involved in stripping the dye are described below.

Mordant– and Direct–Dye Extraction. A dyed thread (ca. 1 cm) or pigment is placed in an open 1.0 or 1.5 mL disposable glass microvial (preferably one with a conical interior or with a glass microvolume insert). The sample is treated with 200 µL each of 3M HCl and methanol at 100 ºC for 5 minutes. The extracted solution is then filtered into another vial by means of a microfilter and the original fibers treated once again as described above. Additional portions of methanol are added to the fibers and these washings are combined with the extract.

The aqueous HCl–methanol solvent system must then be totally evaporated due to: (a) the corrosive nature of the acid to the metalware in the HPLC instrument and in the syringe; (b) highly acidic solutions may irreversibly affect the packing support in the HPLC column; and (c) the pH or the exact quantity of the HCl cannot be accurately and conveniently determined in such a microvolume. Mild evaporation is best accomplished by means of a stream of hot air into the top of the open vial.

The final extraction step involves the addition of 100 µL of methanol to the residue resulting from the solvent evaporation.

Indigoid Extraction. The dyed sample is treated with 50 µL of N,N–dimethylformamide (DMF) at 150 ºC (boiling point = 153 ºC) for 3 minutes in subdued light to avoid the photodegradation of any indigoid. The extracted solution is then transferred into another vial through a microfilter and the original fibers are washed with two 50 µL portions of hot DMF and these washings are combined with the extract.

Prior to chromatographic analysis, filtration of the methanol and DMF extracts is performed by means of a 0.45–µm nylon or polytetrafluoroethylene (PTFE) syringe filter placed in a 4–mm (o.d.) polypropylene housing.

HPLC Analysis. The HPLC gradient elution scheme that has been used for the analysis of the yellow, brown, and orange dyes investigated in this study consisted of a simple linear slope from 0 – 50 min and was composed of 97 – 0 % water; 2 – 0 % H_3PO_4 (5% w/v); 1 – 100 % methanol. The chromatographic system consisted of a Varian Vista 5500 LC instrument with a Merck 150 x 4 mm Lichrosorb 15537 RP–18 standard column (7 µm), 10 µL sample loop and a 1.0 mL/min flow rate. Data processing was performed on a PC using LabCalc software (Galactic Industries). The detector wavelengths used were 254 nm (0 – 35 min), 415 nm (35 – 38 min), and 254 nm (38 – 50 min).

Figure 12 shows the chromatogram of nine different chemical groups: the flavonoids, which include the flavone apigenin, and the flavonols fisetin and kaempferol; the isoflavone genistein; the flavanal catechin; the structurally different tannins ellagic acid and gallic acid; the naphthoquinone lawsone; the anthraquinone emodin; and the aroylmethane curcumin.

This chromatographic method is the first published separation technique that is able to separate as many as nine different yellow, brown, and orange dye groups. Previous HPLC methods have concentrated primarily on flavonoids (*111–113*), tannins (*114, 115*), both of these classes (*116*), and some other classes (*117–119*). Thus, most (if not all) of the dyes on ancient textiles from Israel may be identified using this elution scheme and the one previously published (*109*).

Conclusions

As history has recorded, ancient Israel was at the center of a number of significant historical and religious events. Not a few conquering empires controlled this land and the results were, in part, that goods and technological methods flowed into and out of this region. The analyses of ancient textile dyes can help in understanding the international trade and commerce that evolved in this region in antiquity. For such a study, the botanical history and chemistry of each dyestuff must be known in order to properly identify the dyestuff on an ancient textile. This work is the first critical historico–chemical analysis of the natural dyes used in Israel in antiquity, which should assist in the identification of the colorants in archaeological textiles excavated in this region.

Acknowledgments

The author would like to express his sincere gratitude to the Edelstein Foundation for supporting this research and to the late Dr. Sidney M. Edelstein for his visionary ideas regarding the Edelstein Center at Shenkar. The author is also grateful to Yael Saidian for assistance with the HPLC analyses and to Ruth Precker for many discussions on this topic. Finally, the advice, patience, and collegiality shown to the author by Prof. Mary Virginia Orna during the writing of this article while she was in residence at the Edelstein Center as a Fulbright scholar has been much appreciated.

Literature Cited

Note: The standard format for references to the talmudic and related literature is as follows. For a Mishnah – M. tractate, chapter, Mishnah; for a Tosephta – T. tractate, chapter, law; for a Gemara in the Babylonian Talmud – B. tractate, leaf and side (a=recto, b=verso); for a Gemara in the Jerusalem Talmud – J. tractate, chapter, law, leaf and side.

1. Leggett, W. *Ancient and Medieval Dyes;* Chemical Publishing: Brooklyn, NY, 1944; (a) p vi; (b) pp 87–88.
2. Forbes, R. J. *Studies in Ancient Technology;* E.J. Brill: Leiden, Netherlands, 1964; Vol. 4; (a) p 134; (b) p 123; (c) p 126; (d) p 142; (e) p 124; (f) p 109; (g) p 108; (h) p 107; (i) p 113; (j) pp 110–113.
3. Robinson, S. *A History of Dyed Textiles;* Studio Vista: London, United, 1969.
4. Brunello, F. *The Art of Dyeing in the History of Mankind;* Hickey, B., Transl.; Neri Pozza Editore: Vicenza, Italy, 1973; p 43, note 7.
5. Cardon, D.; *Guide des Teintures Naturelles; Couleurs de la Nature, Nature des Couleurs*; Delachaux et Niestlé: Paris-Lausanne, 1990.
6. Roth, L.; Kormann, K.; Schweppe, H. *Färbepflanzen - Pflanzenfarben: Botanik; Färbemethoden; Analytic; Türkische Teppiche und ihre Motive;* Ecomed Verlagsgesellschaft mbh: Landsberg/Lech, Germany, 1992.
7. Schweppe, H. *Handbuch der Naturfarbstoffe: Vorkommen; Verwendung; Nachweis;* Ecomed Verlagsgesellschaft mbh: Landsberg/Lech, Germany, 1993.
8. Hofmann, R. Ph.D. Thesis, University of Vienna, Austria, 1989.
9. *Dyes in History and Archaeology;* Rogers, P. W., Ed.; Textile Research Associates: York, United Kingdom, 1982– .
10. Precker, R. M.A. Thesis, Bar–Ilan University, Ramat–Gan, Israel, 1992; (a) pp 78–79; (b) pp 80–81; (c) pp 47–51; (d) pp 124–212; (e) p 73; (f) p 82.
11. Feliks, J. *Plant World of the Bible;* 2nd ed.; Masada: Ramat–Gan, Israel, 1968; p 71.
12. *The Jerusalem Talmud, Tractate Shevi'it. Critically Edited. Part Two. Chapters VI – X;* Feliks, Y., Ed.; Rubin Mass: Jerusalem, Israel, 1986; (a) p 108; (b) pp 132–134; (c) p 109.
13. Czyzyk, B. *Otzar ha-Tzemahim (Treasury of Plants);* Author's Pub.: Herzliya, Israel, 1951; (a) p 713; (b) p 331.
14. Steinsaltz, A. *The Essential Talmud;* Galai, C., Transl.; BasicBooks: U.S.A., 1976.
15. *Talmud;* Encyclopaedia Judaica; Keter Pub. House: Jerusalem, Israel, 1973; Vol. 15, pp 750–779.
16. Crowfoot, G. M. In *Qumran Cave I;* Barthelemy, D.; Milk, J. T., Eds.; Discoveries in the Judean Desert I, no. 3.; Oxford University Press: Oxford, United Kingdom, 1955; pp 18–38.
17. Abrahams, D. H.; Edelstein, S. M. In Yadin, Y. *The Finds from the Bar–Kokhba Period in the Cave of Letters;* Judean Desert Studies; Israel Exploration Society: Jerusalem, Israel, 1963; pp 270–279.
18. Abrahams D. H.; Edelstein, S. M. *American Dyestuff Reporter* **1964**, *53*, 19–25.

19. Midgelow, G. W.; In Crowfoot, E. *Textiles.* In *Discoveries in the Wadi ed–Daliyeh;* Lapp, P. W.; Lapp, N. L., Eds.; Annual of the American Schools of Oriental Research 41; ASOR: Cambridge, MA, 1974; p 80.

20. Whiting, M. C.; Sugiura, T. In Crowfoot, E. *Textiles.* In *Discoveries in the Wadi ed–Daliyeh;* Lapp, P. W.; Lapp, N. L., Eds.; Annual of the American Schools of Oriental Research 41; ASOR: Cambridge, MA, 1974; pp 80–81.

21. Sheffer, A.; Tidhar, A. *'Atiqot* **1991,** *20,* 1–26.

22. Masschelein–Kleiner, L.; Maes, L.; Vynckier, J.; Geulette, M. In Sheffer, A.; Tidhar, A. *Textile History* **1991,** *22,* 40 – 41.

23. Koren (Kornblum), Z. C. In *Masada IV;* Aviram, J. ; Foerster, G.; Netzer, E., Eds.; The Yigael Yadin Excavations 1963 – 1965. Final Reports; Israel Exploration Society: Jerusalem, Israel, 1994; pp 257–264.

24. Koren, Z. C. *'Atiqot* **1995,** *26,* 49–53.

25. *M. Yoma* III, 7.

26. *B. Shabbat* (a) 65a; (b) 110b; (c) 20b; (d) 90a; (e) 19a; (f) 79a; (g) 11b; (h) 106a; (i) 102b; (j) 104b; (k) 68a; (l) 64b; (m) 89b.

27. *Babylonian Talmud;* Steinsaltz, A., Ed.; Israel Institute for Talmudic Publications: Jerusalem, Israel, (a) 1982; Vol. 8, Yoma, p 145; (b) 1989; Vol. 2, Shabbat, Part 1, p 272; (c) 1989; Vol. 2, Shabbat, Part 1, p 271.

28. *M. Kilayim* (a) VII, 2; (b) IX, 2; (c) IX, 1; (d) IX, 7; (e) IX, 9; (f) II, 5.

29. *T. Shabbat* (a) X (IX), 3; (b) I, 22; (c) I, 8; (d) XII (XI), 8; (e) IX, 7; (f) X (IX), 7.

30. *B. Gittin* (a) 69b; (b) 52b; (c) 19a; (d) 19b; (e) 70a.

31. *T. Nega'im* V, 4.

32. *B. Menahot* (a) 39b; (b) 39a; (c) 41b; (d) 42b.

33. *B. Hagigah* 16b.

34. *B. Kiddushin* (a) 31a; (b) 32a.

35. *T. Hulin* X, 6.

36. *B. Ketubot* (a) 63b; (b) 59b.

37. *B. Sotah* 48b.

38. *B. Sanhedrin* 67b.

39. *B. Berakhot* (a) 56a; (b) 56b.

40. *B. Megilah* (a) 27b; (b) 19a; (c) 24b.

41. Koren, Z. C. In *Colors from Nature: Natural Colors in Ancient Times;* Sorek, C.; Ayalon, E., Eds.; Eretz–Israel Museum: Tel–Aviv, 1993; pp 15*–31*, 47–65.

42. Orna, M. V.; Kozlowski, A.W.; Baskinger, A.; Adams, T. In *Coordination Chemistry: A Century of Progress;* Kauffman, G. B., Ed.; ACS Symposium Series No. 565; American Chemical Society: Washington, DC, 1994; pp 165–176.

43. Price, R. In *Comprehensive Coordination Chemistry: The Synthesis, Reactions, Properties & Applications of Coordination Compounds*; Wilkinson, G., Ed.; Pergamon Press: New York, NY, 1987; Vol. 6; p 86.

44. *J. Kilayim* (a) IX, 1, 40a; (b) IX, 1, 40b.

45. *B. Nidah* 61b.

46. *T. Ketubot* VII, 8.

47. *J. Ketubot* VII, 7, 45b.

48. *B. 'Avodah Zarah* 20b.

49. *M. Zavim* II, 2.

50. *B. Pesahim* (a) 109a; (b) 55b; (c) 42b.

51. *Midrash Sifrei on Deuteronomy* (a) 226, 5; (b) 105, 95b.

52. *Numbers* (a) 15: 38; (b) 20: 5; (c) 6: 24–26.

53. *M. Shabbat* (a) VII, 2; (b) XIII, 4; (c) I, 6; (d) IX, 5.

54. *T. Berakhot* VII, 5.

55. *T. Baba Qama* XI, 12.
56. *B. Mo'ed Katan* 13b.
57. *M. Baba Batra* II, 3.
58. *T. Baba Batra* I, 4.
59. *B. Baba Batra* 18a.
60. *M. Baba Metzia'* VIII, 6.
61. *T. Baba Metzia'* (a) VIII, 27; (b) XI, 30; (c) XI, 24.
62. *B. Baba Metzia'* (a) 101b; (b) 61b.
63. *M. 'Eduyot* VII, 8.
64. *J. Shabbat* (a) I, 3, 7b; (b) IX, 5, 60b.
65. *J. 'Orlah* III, 1, 17a.
66. *M. Baba Qama* IX, 4.
67. *B. Baba Qama* (a) 95a; (b) 101b.
68. *M. Kelim* (a) XVI, 6; (b) XXIV, 10; (c) XXIX, 10.
69. *T. 'Avodah Zarah* VI, 1.
70. *Exodus* 25: 5; 26: 14; 35: 7, 23; 36: 19; 39: 34.
71. *The Colour Index;* 3rd Ed.; The Society of Dyers and Colourists: London, United Kingdom, 1971; Vols. 3, 4.
72. Robertson, S. M. *Dyes from Plants;* Van Nostrand Reinhold: New York, NY, 1973; p 71.
73. Pfister, R. *Teinture et Alchimie dans l'Orient Helenistique;* Seminarium Kondakovianum Recueil d'Études Archéologie. Histoire de l'Art. Études Byzantines. VII; Kondakov Institute: Prague, Czechoslovakia, 1935; (a) p 16; (b) p 17; (c) p 33; (d) pp 19–20; (e) p 18.
74. *T. Shevi'it* (a) VII, 11; (b) V, 6; (c) V, 7.
75. *J. Gittin* II, 3, 11b.
76. Dalman, G. *Arbeit und Sitte in Palästina;* Georg Olms Verlagsbuchhandlung: Hildesheim, Germany, 1964; pp 74–75.
77. *M. Shevi'it* (a) VII, 3; (b) VII, 2; (c) VII, 7.
78. *T. Gittin* II, 3.
79. *T. Sanhedrin* V, 2.
80. *J. Shevi'it* (a) VII, 2, 18b; (b) VII, 4, 19a; (c) VII, 1, 18b.
81. Löw, I. *Die Flora der Juden;* R. Löwit Verlag: Wien–Leipzig, 1934; Vol. 1; (a) p 401; (b) p 596.
82. *Midrash Tanaim;* Hoffmann, D. Z., Ed.; Books Export Enterprises: Israel; p 78.
83. *M. 'Uqtzin* III, 5.
84. *J. Kilayim* (a) II, 6, 10a–b; (b) II, 3, 7b.
85. *Otzar haGeonim to Tractate Shabbat;* Levine, B. M., Ed.; H. Wagshall: Jerusalem, Israel, 1986; Vol. 2, Part 1, p 67.
86. An, X.; Li, Y.; Chen, J.; Li, F.; Fang, S.; Chen, Y. *Zhongcaoyao* **1990**, *21,* 188 – 189.
87. *Song of Songs* (a) 4: 14; (b) 1: 14
88. *T. Ma'aser Sheni* I, 14.
89. Goodwin, J. *A Dyer's Manual;* Pelham Books: London, United Kingdom, 1982; p 62.
90. Mayer, F.; Cook, A. H. *The Chemistry of Natural Coloring Matters;* American Chemical Society Monograph Series; Reinhold: New York, NY, 1943; pp 93–95.
91. *T. Demai* I, 9.
92. Masschelein–Kleiner, L.; Maes, L. In *ICOM Committee for Conservation, 5th Triennial Meeting;* ICOM: Zagreb, Yugoslavia, 1978; pp 78/9/3/2–78/9/3/10.
93. *Webster's New World Dictionary of the American Language;* Guralnik, D. B., Ed.; Second College Edition; William Collins + World Publishing: Cleveland, OH, 1976; (a) p 563; (b) p 819.

94. Aharanson, A. *Florula Cisiordanica ;* Oppenheimer, H. R.; Evenari, M., Eds.; Société Botanique de Genève: Geneva, Switzerland, 1940; p 304.
95. Wouters, J., *Stud. Conserv.* **1985**, *30,* 119–128.
96. Klein, N. *Teva Va'aretz* **1970–1971**, *13(1)*, 22–23.
97. Schweppe, H. In *Historic Textile and Paper Materials II. Conservation and Characterization;* ACS Symposium Series 410; Zeronian, S. H.; Needles, H. L., Eds.; American Chemical Society: Washington, DC, 1989; pp 188–219.
98. *The Merck Index, 11th Ed.;* Budavari, S., Ed.; Merck: Rahway, NJ, 1989.
99. Hill, R.; Richter, D. *J. Chem. Soc.* **1936**, 1714 ff.
100. *T. Kilayim* III, 12.
101. Wulff, H. E. *The Traditional Crafts of Persia*; MIT Press: Cambridge, MA, 1966; p 192.
102. Midrash *Sifrei on Numbers* 115, 35a.
103. Ginzberg, L. *Genizah Studies;* Geonica II; The Jewish Theological Seminary of America: New York, NY, 1909; pp 309, 333.
104. Balfour–Paul, J. *Hali* **1992**, *61,* 98–105, 140.
105. *The Arukh;* Shlesinger, S. A., Ed.; Yitzu Sifrei Kodesh: Tel-Aviv, Israel; p 493.
106. Herzog, I. In *The Royal Purple and the Biblical Blue: Argaman and Tekhelet. The Study of Chief Rabbi Dr. Isaac Herzog on the Dye Industries in Ancient Israel and Recent Scientific Contributions;* Spanier, E., Ed.; Keter: Jerusalem, Israel, 1987.
107. *M. Megilah* IV, 7.
108. Wouters, J.; Rosario–Chirinos, N. *Journal of the American Institute for Conservation* **1992**, *31,* 237–255.
109. Koren, Z. C. *J. Soc. Dyers and Colour.* **1994**, *110,* 273–277.
110. Walton, P.; Taylor, G. *Chromatography and Analysis* **1991**, *June,* 5–7.
111. Daigle, D. J.; Conkerton, E. J. *J. Chrom.* **1982**, *240,* 202–205.
112. Pietta, P. G.; Mauri, P. L.; Manera, E.; Ceva, P. L.; Rava, A. *Chromatographia* **1989**, *27,* 509–512.
113. Whiting, M. C. *Dyes in History and Archaeology* **1992**, *10,* 7–10.
114. Wouters, J. In *ICOM Committee for Conservation, Working Group 18: Conservation of Leathercraft and Related Objects*, ICOM: Washington, DC, 1993; Vol. 2, pp 669–673.
115. Verzele, M.; Delahaye, P. *J. Chrom.* **1983**, *268,* 469–476.
116. Mueller–Harvey, I.; Reed, J. D.; Hartley, R. D. *J. Sci. Food. Agric.* **1987**, *39,* 1–14.
117. Tonnesen, H. H.; Karlsen, J. *J. Chrom.* **1983**, *259,* 367–371.
118. Fischer, C. H.; Bischof, M.; Rabe, J. G. *J. Liq. Chrom.* **1990**, *13,* 319–331.
119. Rabe, J. G.; Bischof, M.; Fischer, C. H. *Bestauro* **1990**, *96,* 189–194.

RECEIVED October 9, 1995

Chapter 22

Thermoluminescent Analysis of Burned Bone: Assessing the Problems

Patrick T. McCutcheon

Department of Anthropology, University of Washington, Box 353100, Seattle, WA 98195-3100

The mineral phase of bone (hydroxyapatite) is thermoluminescent sensitive. A review of previous research reveals several seemingly insurmountable problems when attempting to date bone material by traditional thermoluminescence (TL). Spurious luminescence (chemical and tribological) recorded during specimen heating masks an underlying TL signal. While organic phase removal and acid-wash pretreatments reduce low-temperature spurious luminescence, this same problem persists in the upper temperature range (275 °C -- 450 °C). In this paper burned bone was subjected to TL analysis, as well as differential thermal (DT) and thermogravimetric (TG) analyses to assess the role of pyrophosphate production during heating. While dating hydroxyapatite has been successful in Electron Spin Resonance (ESR) research with teeth, problems with diagenesis may obfuscate future attempts to TL date bone.

Bone is a common constituent of archaeological deposits. Bone material consists of organic and mineral phases. The mineral phase -- predominantly made up of hydroxyapatite -- is thermoluminescent (TL) sensitive. Thus, TL dating bone is an attractive prospect. The TL sensitivity of hydroxyapatite was first recognized and investigated in the late 1960s (*1, 2*) and has been only sporadically considered since then (*3, 4*). Attempts to TL date the inorganic phase of bone encountered the emission of spurious luminescence (non-radiation induced), which precluded dating the bone. This previous work suggests three sources of spurious or non-radiation induced (NRI) luminescence: (1) chemiluminescence, thermal decomposition of organic phase; (2) triboluminescence, NRI luminescence due to sample preparation technique; and (3) additional NRI luminescence arising from the production of pyrophosphates during heating.

Burning in antiquity, or annealing an experimental specimen, should eliminate both chemiluminescence and the formation of pyrophosphates. Other solutions to triboluminescence problems are currently available (5). Therefore, archaeological burned bone might be a suitable TL subject for dating. The research reported here establishes that bone hydroxyapatite is a suitable material for TL analysis under certain conditions.

Previous Research

Thermoluminescence. Initial investigations into TL analysis of bone did not yield promising results (1, 2) largely because of the presence of organic matter in the bone material. At the time, chemical removal of the organic phase of bone was not attempted because a protocol for separating organic and mineral phases without dissolving the mineral phase had not yet been developed (1). Both preliminary studies encountered spurious luminescence (i.e., triboluminescence and chemiluminescence). When using TL for dating, radiation induced thermoluminescence is the "true signal," and thereby, all non-radiation induced TL is spurious (6). The NRI emission observed during glow-out was attributed to the combustion of the organic phase of bone (chemiluminescence) and grinding of the mineral grains during sample preparation (triboluminescence). "Glow-out" refers to the process of heating the TL sample and measuring its luminescence. The resulting curve is called a glow curve.

In all glow curves produced by Jasinska and Niewiadomski the triboluminescence, which they induced by "a vigorous rubbing of the sample," was noted in the upper temperatures of the glow curve (ca. > 200 °C) (1). This spurious luminescence could be masking a high temperature signal. In order to avoid spurious luminescence in the lower temperatures (i.e., < 200 °C), bone material was incinerated at 400 °C. The authors concluded that all 24 samples tested showed radiothermoluminescence. Their findings suggested that a genuine TL signal could be gleaned from deproteinized bone material (e.g., fossil or incinerated bone). However, most archaeological bone, with the exception of completely calcined or incinerated bone contains organic matter.

Christodoulides and Fremlin (2) attempted to circumvent both triboluminescence (TTL) and chemiluminescence (CL). Two sources of CL were identified: simple oxidization and combustion of the organic phase. To eliminate the first source, glow-out of the TL signal was undertaken in a nitrogen atmosphere. To alleviate the second source of CL the authors used an ethylene diamine extraction procedure, which removed most of the organic phase, reducing the CL below the level of the TTL (2). In addition, Christodoulides and Fremlin suggested further reducing TTL by rejecting the finer powder that has a higher surface to volume ratio. Because TTL is a surface friction phenomena it can be attenuated by using larger-sized grains (2). Following their protocol, the combined effects of CL and TTL were reduced but not completely circumvented. The combined effect of CL and TTL exhibit an equivalent radiation dose on the order of 10 kilorads, which is too high for dating in the 10 - 100 thousand year range (2). Both pioneering studies reach similar conclusions; TL dating bone material is problematic because of spurious luminescence.

In subsequent non-dating investigations (3, 7) CL was circumvented by removing the organic phase of the bone material. In Chapman et al. (3), contemporary rat bones were deproteinized using a hydrazine treatment (8). The authors were able to get a low temperature peak in amorphous calcium phosphate (ACP), synthetic hydroxyapatite (HA), and deproteinized bone mineral. When the specimens were subjected to ultraviolet (UV) light and thermal pretreatment in the range of 400 °C a TL peak was observed at 85 °C (3).

In their study, Chapman et al. (3) adopted a cycling protocol in which a series (total = 16) of 13 minute cycles (each cycle consisting of 10 minute UV exposure and 3 minute readout) were carried out on powdered specimens of bone mineral, synthetic HA, and ACP (3). The results of the cycling technique demonstrated that a logarithmic decay in TL intensity occurs with repeated exposure to UV light. The results suggested that the initial annealing of the sample at 400 °C was responsible for the observed TL peak. A UV stimulated TL peak was observed only when a 10 to 60 minute thermal pretreatment at 400 °C was used. UV irradiation itself was not sufficient to elicit a low-temperature TL peak.

Electron spin resonance (ESR) experiments suggest that the formation of additional and deeper traps occurs during thermal pretreatment (3). Based on the results of comparative analysis on ACP, synthetic HA, and bone mineral, Chapman et al. (3) suggest that the observed TL is not related to crystallite size or impurity content; rather, it is due to an unknown "localized constituent rearrangement" that occurs during thermal pretreatment. The rearrangement is linked to reactions that take place in the formation of a pyrophosphate-type phase. These researchers concluded by suggesting that thermal pretreatment causes the formation of deep traps, which retain electrons that are then subsequently transferred to the 85 °C luminescent centers. However, the mechanism(s) are not known. The cycling program seems to cause a logarithmic decay of TL intensity in the 85 °C TL glow peak by depleting the deeper traps. The relationship between thermal pretreatment and the creation of additional deep traps is not explained, nor is the process by which electrons moved into these new deep traps explained. One would assume that the UV irradiation would fill the deep traps, as well as transfer electrons to shallower traps, if as the authors note the peak is UV stimulated. If this is the case, then why does a logarithmic decay in TL intensity occur if the deep traps are filling with electrons as a result of UV exposure? One would expect the traps to be refilled as the specimen was exposed to more radiation. An alternative interpretation is that thermal pretreatment not only creates additional deep traps, but fills them as well. Plainly, the effects caused by thermal pretreatment and UV exposure are not well understood and require further investigation.

Driver (4) identified a low temperature TL peak (ca. 175 °C) for bone using thin slices. He used thin slices to avoid triboluminescence generated in crushed samples. Driver circumvents the chemiluminescence by soaking his slices in ethylene diamine and exposing them to excited oxygen. From different combinations of these two treatments, Driver obtains various results. Bone material exposed to β-irradiation exhibited a low temperature peak, as well as some luminescence at temperatures above 200 °C. On the other hand, bone not exposed to β-irradiation did not yield a low temperature peak, but did show drastic increase in NRI luminescence (when compared to the other specimen's glow curves) at temperatures above 200 °C. Bone given the ethylene diamine extraction, exposure to "rf" excited oxygen, seven days' annealing, and then β-irradiation showed a low temperature peak, as well as a considerable reduction in high temperature luminescence . Driver suggests treating this last result with some caution, because only one specimen was treated in this manner. Driver's analysis suggests luminescence above 200 °C is the result of either a chemical reaction or a phase change in the mineral.

Glimcher and Krane (9) note that bone mineral is not entirely hydroxyapatite. One hypothesis they review recognizes that octocalcium phosphate is a minor constituent of bone mineral, and when the bone is heated somewhere between 200 ° C and 600 °C, a pyrophosphate is formed (10 - 12).

Driver's explanation for the large amount of spurious luminescence in the glow curve above 200 °C (i.e., chemiluminescence) are similar to those given by Chapman et al. (3). In the latter study an increase in slope of the glow curve just before and up to 200° C is attributed to "the formation of a pyrophosphate-type phase" (3). In all of these studies, high-temperature, spurious luminescence was noted whether the bone was chemically deproteinized or incinerated and whether specimens for analysis were ground or thin sliced. Early studies (1) attributed high temperature spurious luminescence to "induced triboluminescence," but the subsequent studies identify the high temperature spurious luminescence as chemiluminescence. These studies thus suggest that, in spite of the general problems of bone TL, burned bone might prove a suitable TL dating subject.

Burned Bone. Studies of burned bone (9 - 11, 13 - 17) suggest that heating causes both chemical and physical alterations, which are related to specific temperature thresholds in bone. Because the changes are irreversible, the maximum temperature to which a bone has been heated can be determined.

Both Bonucci and Graziani (13) and Shipman et al. (17) characterize changes in bone material induced by heating using macroscopic, microscopic, and color analyses. In addition, these studies used X-ray diffraction (XRD) to characterize crystal morphology change. Bonucci and Graziani also used thermogravimetric (TG) and differential thermal (DT) analyses. Based on these studies, and the work of others (9 - 12, 15, 18), changes in bone as a consequence of heating can be summarized: (1) from 20 °C to 300 - 350 °C a loss of unbound water and some initial carbonization of the organic phase (19, 20); (2) from 350 °C to 500 - 600 °C complete combustion of the organic phase, loss of bound crystal water, and formation of pyrophosphates (9, 11, 18, 21); and (3) from 600 °C to 950 °C increase in crystal size of hydroxyapatite, based on XRD, and a fusing of the crystals themselves based on microscopic analysis (22, 23).

Elsewhere I report experiments which link this sequence of physical changes to the color of burned bones (16). The linkage of heating temperatures to bone color opens the possibility of identifying bones potentially suitable for TL analysis independently from TL characteristics. Bones burned in antiquity share the same thermally induced characteristics as bones burned experimentally. Primarily, the organic phase is removed, and the crystalline structure is altered. Although, organic matter can diffuse back into the bone it is no longer part of the original structure. More importantly, the crystalline bound water is combusted, resulting in the formation of pyrophosphates. The significance of pyrophosphate production occurring in antiquity is taken up below; if bone, when burned in antiquity, gives off all pyrophosphates possible, they cannot be a source of spurious TL during TL analysis.

Methods and Results

The present study was undertaken to examine the TL of bone heated to 600 °C or greater, specifically to determine if heated bone was more TL sensitive (as suggested by the annealing results) and whether burning eliminates the spurious TL signal and if so at what temperatures.

Experimental Bones. Thermoluminescent analysis is a destructive analytical technique. Since a TL analytical protocol was not yet constructed for dating bone material, the use of archaeological specimens was not warranted. Raw deer bone was thus chosen for experiments in order to construct an archaeological TL bone dating protocol. Bones chosen for testing were cut from the diaphysis of a tibia and burned at 650 °C and 950 °C. A total of three burned samples (Specimens 1, 2, and 3) were then used in the following TL analysis. An unheated bone sample (Specimen 4) was included to assess chemiluminescence. Each specimen (1, 2, 3, 4) underwent at least two glow-outs.

TL Analysis. The following equipment was used in the TL analysis: Daybreak 510 system controller, Daybreak 520 oven temperature control, Daybreak 540 ratemeter, Moseley Division Hewlett Packard plotter. A water cooled glow oven was made in the Civil Engineering machine shop, University of Washington. During analysis, the sample chamber is evacuated to 5 microns Hg pressure, oxygen free (5 ppm) argon flows through the oven at a fixed rate (10 cc/min), and the sample is heated at a rate of 5 °C/second. An EMI 9635QA photomultiplier tube plus Daybreak 530 amplifier and discriminator and Corning 7-59 and 4-69 filters were all located in the photomultiplier tube housing. A ^{90}Sr β-source with a dose rate 0.912 Gy/minute was used for irradiating samples.

Experimental Results. An initial experiment used a thin slice (1.0 mm) cut from a bone burned at 950 °C (Specimen 1), that resulted in a glow curve composed entirely of spurious TL (Figure 1a). The sample was then given a ten minute β-irradiation and a low temperature TL peak was observed (Figure 1b), after which a background reading was taken that showed high temperature spurious luminescence (Figure 1b). The sample was then held under vacuum for ca. 10 minutes in an effort to draw off any remaining water and oxygen;, however, the level of luminescence in the second background curve was higher than the first (Figure 1b). This result is unexplained.

All subsequent analyses used powdered specimens to reduce the chemical reaction associated with the combustion of water and oxygen during glow-out. In order to circumvent potential triboluminescence resulting from grinding, the grains were given a 10 minute bath in glacial acetic acid (5). The sample was settled onto aluminum discs for TL analyses. Each disc is designated as an unique specimen for analysis. Specimens 2, and 3 were prepared from the thin slice of bone burned experimentally at 950 °C. In the present study, β-irradiation was used after the natural dose was recorded in order to assess the response of TL peaks to dose.

UV Exposure and Annealing. In an effort to replicate results of Chapman et al. (3), Specimen 2 was given a 10 minute exposure to long wavelength ultraviolet light and a

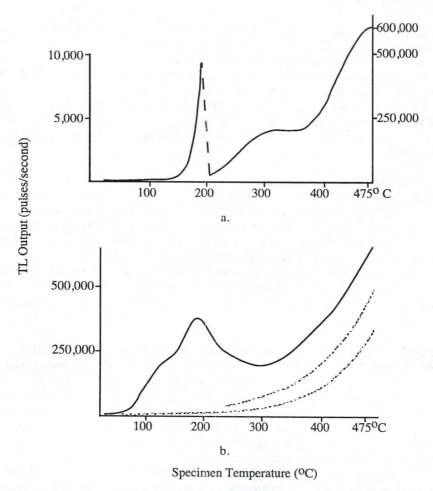

Figure 1. TL Glow Curves of Specimen 1. (a) Natural TL signal of 1mm thin slice;
(b) TL signal after 10 minute irradiation.   Stippled curves represent background
signal. Note that the second background signal in curve (b) is higher than the first.

short annealing to 300 °C, and then re-exposed to UV light. The use of UV irradiation in this manner is known as phototransfer (*24 - 26*), in which the exposure to β-irradiation gives the specimen some radioluminescence (simulating a natural or paleo-dose) and the following exposure to UV light transfers electrons from deep traps to shallow traps. Annealing to 300° C served to empty the shallow traps, as well as potentially create additional deep traps (e.g., *3*). The low temperature peak can be assessed further by phototransferring deep trap electrons to shallow traps, thereby avoiding the high temperature NRI luminescence. Aitken (*6*) also regards the phototransfer technique as advantageous to TL analysis of bone and shell.

In the first glow-out of Specimen 2, a small, undefined low temperature TL peak was noted, however only spurious TL was noted in the upper temperatures (Figure 2a). The background curve yielded low temperature spurious luminescence at a greater intensity than initial glow-out; this result is unexplained. Specimen 2 was then given a 10 minute UV exposure and a glow curve was generated again; a TL peak around 150 °C appears among much spurious luminescence (Figure 2b). After giving Specimen 2 an additional 1 minute exposure to UV light the peak became better defined; low temperature spurious luminescence was reduced (Figure 2c). With a low temperature peak isolated, Specimen 2 was given an additional 10 minute β-irradiation, heated to 300 °C, and exposed to 1 minute of UV light (Figure 2d); the low temperature TL peak intensity increased slightly.

Specimen 3 was used to investigate further the relationship between the UV exposure and the annealing process that results in a low temperature. The specimen was irradiated with a β-source for 5 minutes, heated to 500 °C for 5 minutes, and then exposed to UV light for 1 minute. The subsequent glow-out yielded a poorly defined, low temperature peak, i.e., it was largely masked by spurious TL (Figure 3a). An identical β-irradiation, UV exposure, and annealing was performed to replicate results illustrated in Figure 3a. The subsequent glow curve exhibited a reduction in NRI luminescence and a more well defined peak (Figure 3b).

In an effort to get the identified low temperature peak to respond to dose, the specimen was exposed to 15 minutes of UV light. With a longer UV exposure more electrons are expected to be phototransferred to the shallow traps, which should result in a more intense low temperature peak. The peak intensity did increase as expected and the spurious TL decreased after the peak, but as the higher temperatures were reached the spurious TL increased over the previous glow curves (Figure 4a). Finally, the specimen was exposed to 5 minutes β-irradiation, a 5 minute 500 °C annealing, and a 15 minute and 15 second UV exposure. This produced a slight increase in peak intensity and a reduction in the high temperature spurious TL (Figure 4b).

At this point, what the low temperature peak is related to other than UV exposure alone is not understood. In subsequent analyses, different combinations of UV exposure and annealing were tried in order to get some insight into the processes that are occurring during these two pretreatments. A low temperature peak is noted when the specimen is irradiated, annealed, and exposed to UV light initially. Upon repeating this procedure the peak becomes better defined and the spurious TL in the upper and lower temperatures of the glow curve is reduced.

Identical results were obtained when specimens of bone burned experimentally at 650 °C were treated in a similar manner. Specimen 4, unheated, undeproteinized

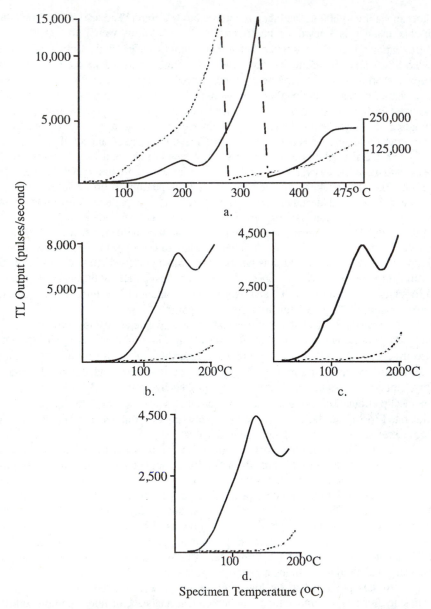

Figure 2. Glow Curves of Specimen 2. Solid line represents TL signal and stippled line represents background output. (a) Specimen 2 received 10 minutes UV, short annealing at 300 °C, and then was re-exposed to UV; (b) Specimen 2 received another 10 minute UV; (c) Specimen 2 received an additional 1 minute UV exposure, the spurious luminescence is reduced; (d) Specimen 2 received another 10 minute β-irradiation, heated to 300 °C, and exposed to 1 minute UV.

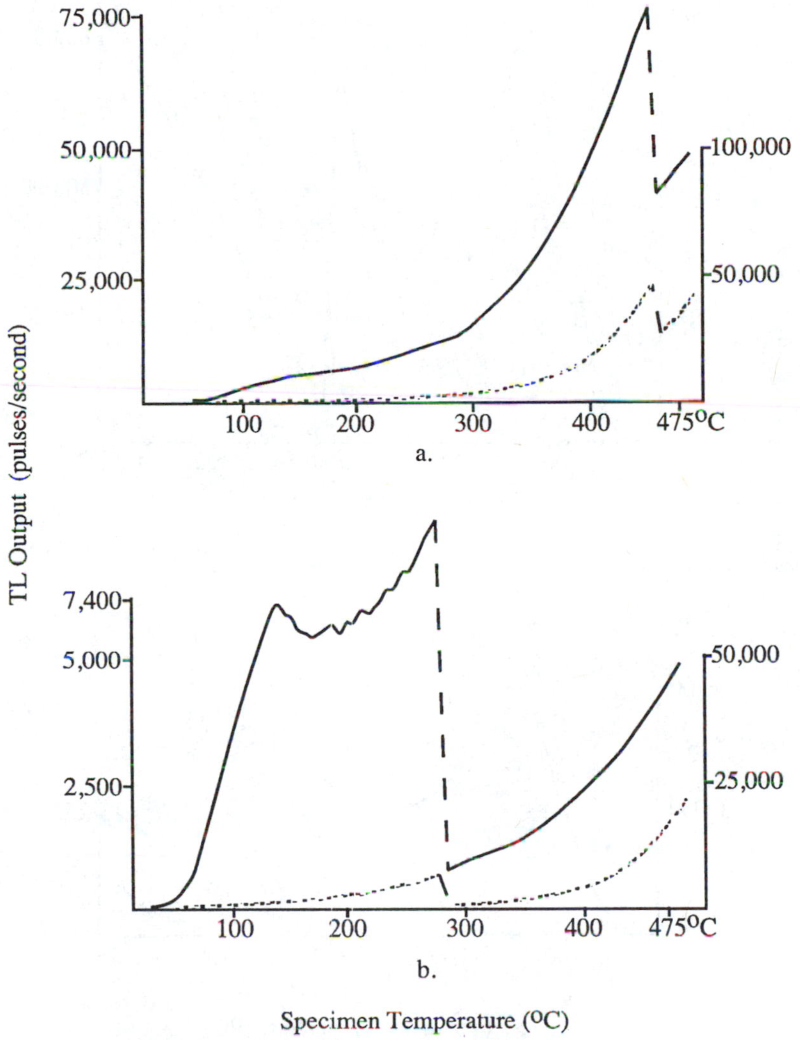

Figure 3.   Glow Curves of Specimen 3: Phototransfer and Annealing Techniques. Solid lines represent initial TL signal and stippled lines represent background signal. (a) 5 minute β-irradiation, heated to 500 °C for 5 minutes, and then 1 minute UV exposure; (b) 5 minute β-irradiation, annealed at 500 °C for 5 minutes, then a 1 minute UV exposure.

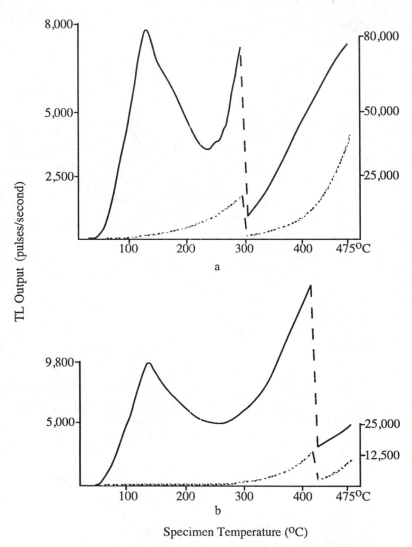

Figure 4. Glow Curves of Specimen 3: Phototransfer and Annealing Techniques.
Solid lines represent initial TL signal and stippled lines represent background signal.
(a) exposed to 15 minutes of UV in order to elicit an increase in peak intensity; (b) 5
minute β-irradiation, 500 °C annealing for 5 minutes, and a 15 minute and 15 second
exposure to UV.

bone, was given identical treatment and gave similar results, i.e., the initial glow curve is composed of mostly NRI luminescence, but upon identical treatment, a second glow curve shows a better defined TL signal, as well as a reduction in the NRI luminescence (Figure 5a and 5b).

Interpretation

These analyses combined with previous TL studies of bone material suggest some general interpretations. First, the large amount of spurious luminescence (Figure 1) is attributed to chemical reactions that take place when bone has been heated. Approximately 1/3 of bone is organic and when burned above 600 °C the organic phase is combusted completely, resulting in a very hygroscopic material. The surface or hydration layer of hydroxyapatite crystals contains phosphates bound with hydroxls (HPO_4). When bone mineral is heated to 600 °C or greater pyrophosphates are formed resulting in a loss of water (*9 - 11, 18, 27*). This leaves the surface layer of the crystals anhydrous. In this condition, burned bone readily absorbs atmospheric water and oxygen during TL analysis and these absorbed materials might produce NRI luminescence.

Second, once a low temperature TL peak is identified, additional exposures to β-irradiation result in only a small increase in signal intensity (compare figures 2c and 2d), however, there was a reduction in NRI luminescence and the peak was better defined (Figure 3b). To further investigate this result, Specimen 3 was exposed to 15 minutes of UV light. The expectation was an increase in intensity with the larger exposure to UV light, i.e., more electrons phototransferred. Peak intensity did increase as expected and the spurious TL emitted at temperatures just above the peak temperature decreased in intensity, but as higher temperatures were reached spurious TL increased over previous glow curves (Figure 4a). An increase in the high temperature TL output may be a response to increased dose. A possible interpretation is that the spurious TL in the upper temperatures is masking a high temperature TL peak.

One final interpretation is suggested from the results of TL analysis on unburned, undeproteinized bone. Spurious TL is not necessarily related only to the organic phase of bone. A more plausible explanation may be that the production of NRI luminescence is either a function of the presence of atmospherically absorbed oxygen and water in the bone itself, or the production of a pyrophosphate phase when bone is heated. If the source of spurious luminescence is pyrophosphate production, this problem should be circumvented by using specimens of bone burned at temperatures greater than 600 °C. However, as the results above exhibit, using bone burned at temperatures greater than 600 °C does not eliminate NRI TL. This suggests two problems: (1) each time a bone is heated repeatedly, some amount of pyrophosphates are produced; and (2) burned bone is hygroscopic and when exposed to air it absorbs water and oxygen. In an effort to identify which of these problems prevails, differential thermal and thermogravimetric analyses were carried out on fresh bone.

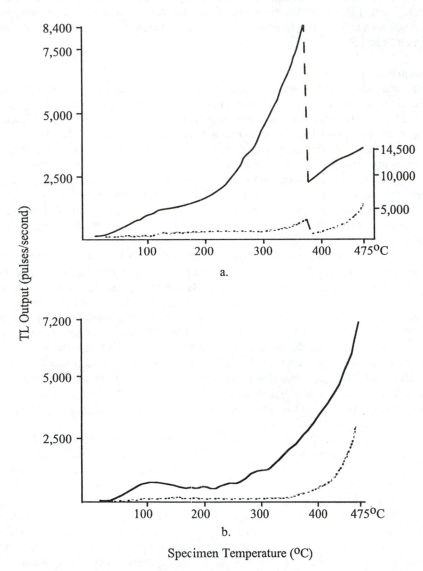

Figure 5. Glow Curves of Specimen 4: Phototransfer/Annealing Techniques. Solid lines represent initial TL signal and stippled lines represent background signal. (a) 5 minute β-irradiation, heated to 500 °C for 5 minutes, and then 15 minute UV exposure; (b) 5 minute β-irradiation, annealed at 500 °C for 5 minutes, and then a 15 minute UV exposure. Note the decrease in spurious TL from 5a to 5b.

Differential Thermal and Thermogravimetric Analyses of Fresh Bone

Problem. Identifying the source of spurious TL observed in previous experiments is the key to judging the feasibility of TL dating of bone. In order to eliminate the spurious TL the source must be identified. From my previous work, there appears to be two possible sources. Nothing in the bone chemistry literature suggests a repeated production of pyrophosphates with each burning event. On the other hand, when preparing specimens for analysis the sample is continually exposed to atmospheric oxygen and water. In the TL oven, the chamber is pumped down to 5 microns Hg pressure and held there momentarily under vacuum. Then the oven is filled with argon gas, after which a glow curve is generated. Holding the specimen under vacuum does not reduce the spurious TL, but when a specimen was annealed for 10 minutes some of it was reduced.

These results, as well as those of other researchers, suggest the source of spurious TL is the production of pyrophosphates with each heating event. Differential Thermal (DT) and Thermogravimetric (TG) analyses should register this chemical reaction each time a bone is heated. To preclude conflation with the other probable source of spurious TL, atmospheric water and oxygen, the bone specimen was kept in an argon gas environment between heating events.

Materials and Techniques. Fresh deer bone was used in the experiment. The bone was crushed in a porcelain mortar and pestle to a fine powder. The DT and TG instruments had the following parameters: Scan rate = 10 °C/minute; atmosphere = argon at flow rate of 50 cc/minute. Bone samples for each of the instruments weighed approximately 50.0 milligrams for the DT analysis and 66.0 milligrams for TG analysis. For both of the analyses three independent measurements were made on each of six samples of fresh deer bone. The sample was loaded into the instrument and then sealed in the heating chamber. The chamber was then filled with argon gas for the duration of the experiment. Each instrument was programmed for two identical heating events. A heating event consisted of starting from room temperature and heating to 1000 °C. The specimen and chamber were then allowed to cool to room temperature before the second heating event took place.

Results. The TG and DT results were practically identical for each of the three samples. Figure 6 shows curves typical of TG and DT analyses. For the DT analysis, the results are interpreted as follows (see Figure 6b): (1) from 20 °C to 350 °C an exothermic peak response corresponds to driving off unbound water from both the organic and mineral phases, in addition to the initial combustion of the organic phase; (2) from 350 °C to 600 °C the complete combustion of the organic phase, the loss of crystal bound water, and the production of pyrophosphates are exhibited; (3) from 600 °C to 865 °C a slight increase in slope occurs resulting from either oxidization of trace elements and/or a phase change; and (4) the endothermic peak after 865 °C may be decomposition of $CaCO_3$ constituent (*20*).

Thermogravimetric analysis results can be interpreted as follows (Figure 6a): (1) initial weight loss occurs from room temperature to 275 °C, resulting from the loss of unbound water; (2) a large weight loss from 276 °C to 650 °C, due to the

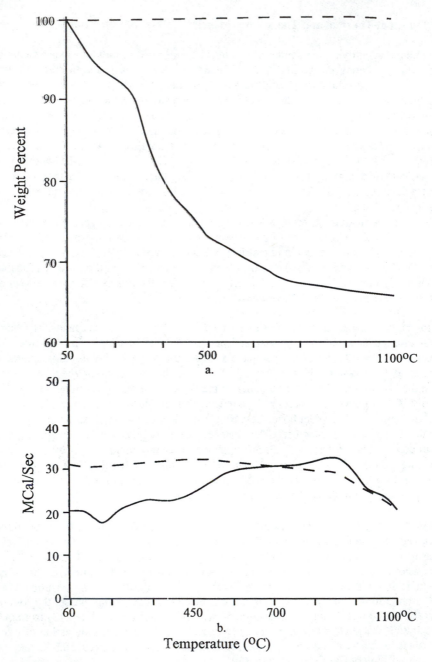

Figure 6. Differential Thermal and Thermogravimetric Analysis. Solid lines represent initial DT or TG analysis, dashed lines represent second heating event. Note lack of weight loss (a) and endo-/exothermic peak (b) response during second heating event (dashed lines).

combustion of organic phase, the loss of bound crystalline mineral water, and the production of pyrophosphates; and (3) at 651 °C a slight weight loss can be detected through 875 °C, most likely due to the phase change in inorganic constituent of the bone and/or oxidation of trace elements. These results are similar to other published DT and TG analyses performed on bone materials (*19, 20*).

The second heating event to which all specimens were subjected resulted in no appreciable weight loss during TG analysis and there was no endo- or exothermic peak registered from the specimens during DT analysis (Figure 6).

Summary.

Based on my TL analysis, there appears to be a direct relationship between the occurrence of a low-temperature peak and the thermal pretreatment and UV exposures. A better defined peak and reduced spurious luminescence results from repeated annealing and UV exposure. In the Chapman and et al. (*3*) study, the samples were exposed to UV irradiation only. The logarithmic decrease in intensity they noted was not identified in the present study, probably as a result of the β-irradiation given to the burned bone specimens. Given that the spurious luminescence decreases with additional annealing and UV exposures, a plausible explanation is that the pretreatments (UV and annealing) appear to reduce atmospheric oxygen and water in the bone mineral, concomitantly reducing the spurious TL. The better defined peak may be a result of reduced spurious TL, thereby allowing an independent assessment of the low temperature peak. As noted above, another source of NRI luminescence might be the production of pyrophosphates. As bone is continually exposed to heat and UV light the pyrophosphate phase is gradually combusted, resulting in reduced NRI luminescence and a better defined low temperature peak. However, this peak exhibited only a slight increase in intensity with β-irradiation. There appears to be no relationship between radiation dose and TL intensity in the low-temperature peak.

The results of the DT and TG analyses are not consistent with the continual production of pyrophosphates as the source of spurious luminescence. The results are compatible with the alternative hypothesis, i.e., the hygroscopic nature of burned bone attracts atmospheric oxygen and water, which are responsible for the spurious luminescence that prevents dating.

If the source of spurious luminescence is atmospheric, it can be eliminated by an appropriate sample preparation protocol. It would thus seem that the problem with TL dating bone is a procedural problem, not a physical limitation of the bone mineral itself. The issue of TL dating bone, burned or otherwise, is by no means a closed case.

Acknowledgments

I would like to thank Anthropology, Material Science and Engineering, and Geology academic Departments, University of Washington, for permitting access to the necessary equipment to undertake this research. Specifically, I wish to thank Drs. R. C. Dunnell, M. Readhead, and T. G. Stoebe for guidance during research and write-up of results. Finally, use of the TL facilities was supported by NSF grant BNS 8504394 to Dunnell and Stoebe.

Literature Cited

(1) Jasinska, M. and Niewiadomski, T. *Nature*, **1970**, *227*, 1159-1160.
(2) Christodoulides, C. and Fremlin, J. H. *Nature*, **1971**, *232*, 257-258.
(3) Chapman, M. R., Miller, A. G., and Stoebe, T. G. *Medical Physics*, **1979**, *6*, 494-499.
(4) Driver, H. S. T. *PACT*, **1979**, *3*, 290-297.
(5) Huxtable, J. *PACT*, **1982**, *6*, 346-352.
(6) Aitken, M. J. *Thermoluminescent Dating*; Academic Press: New York, NY, 1985.
(7) Chapman, M. R., Miller, A. G., Burnell, J. M., and Stoebe, T. G. Thermoluminescence in the bone mineral system. 1977, Proceedings of the 5th International Conference Luminescence Dosimetry, San Paulo, Brazil.
(8) Termine, J. D., Eanes, E. D., Greenfield, D. J., and Nylen, M. U. *Calcified Tissue Residue* **1973**, *12*, 73-90.
(9) Glimcher, M. J. and Krane, S. M. In *Treatise on Collagen*; Gould, B. S., Ed.; Academic Press: London, 1968; pp. 68-252.
(10) Armstrong, W. D. and Singer, L. *Clinical Orthopedics* **1965**, *38*, 179-194.
(11) Knubovyets, R. G., Zubkova, T. V., Cherenkova, G. I., and Gnedenkova, V. T. *Zhurnal Neorganicheskoi Khimii* **1979**, *24*, 2072-2076.
(12) Legros, R., Bonel, G., Balmain, N., and Juster, M. *Journal of Chim. Phys. Phys.-Chim. Biol.*, **1978**, *75*, 761-766.
(13) Bonucci, E. and Graziani, G. *Atte della Accademia Nazionale dei Lincei, Rendiconti Sci. Ris. Mat. Nat. Series* 8, **1975**, *59*, 517-532.
(14) Buikstra, J. E., and Swegle, M. In *Bone Modification*, Sorb, M. and Bonnichsen, R. Ed.; University of Maine Press: Orono, 1989; pp. 247-248.
(15) McConnell, D. *Apatite: Its Crystal Chemistry, Mineralogy, Utilization, and Geologic and Biologic Occurrences;* Springer Verlag; New York, 1973.
(16) McCutcheon, P. T. In, *Deciphering Shell Middens*, Stein, J. K. Ed.; . Academic Press: Orlando, 1992; pp.192-210.
(17) Shipman, P., Foster, G., and Schoeninger, M. *Journal of Archaeological Science* **1984**, *11*, 307-325.
(18) Posner, A. S. *Physiological Review* **1969**, *494*, 760-793.
(19) Civjan, W. J., Selting, W. J., De Simon, L. B., Battistone, G. C., and Grower, M. F. *Journal of Dental Research*, **1971**, *512*, 539-542.
(20) Holager, J. *Journal of Dental Research*, **1970**, *49*, 546-548.
(21) Dean, W. E., Jr.. *Journal of Sedimentary Petrology*, **1974**, *44*, 242-248.
(22) Franchet, L. *Bulletin Societe Prehistorique Francaise*, **1934**, *4*, 177-194.
(23) Herrmann, B. *Journal of Human Evolution*, **1977**, *6*, 101-103.
(24) Bailiff, I. K. *Nature* **1976**, *264*, 531-534.
(25) Mobbs, S. F. Low temperature optical re-excitation in thermoluminescence dating. Manuscript on file, Faculty of Physical Sciences, Oxford University, 1978.
(26) Mobbs, S. F. *PACT*, **1979**, *3*, 407-413.
(27) Betts, F., Blumenthal, N. C., and Posner, A. S. *Journal of Crystal Growth*, **1981**, *53*, 63-73.

RECEIVED October 9, 1995

Chapter 23

Trace Elements in Bone as Paleodietary Indicators

James H. Burton

Department of Anthropology, University of Wisconsin,
1180 Observatory Drive, Madison, WI 53706

Because bone levels of barium and strontium decrease with an organism's increasing position in the food-chain, they are used by archaeologists to estimate past consumption of plants versus meat. While the Ba/Ca and Sr/Ca ratios in bone reflect their dietary ratios, factors other than trophic position affect dietary and, hence, bone ratios. Dietary Sr/Ca and Ba/Ca are affected not only by the Ba and Sr content of any specific food but also by the relative contribution of that food to the pool of bone forming cations: Ca+Sr+Ba@Ca. This implies that foods high in calcium, such as leafy vegetables, will contribute most of the bone-forming minerals, and dominate bone composition, even when present in minor amounts. Thus bone composition should not be interpreted as a proportional meat/plant measure but as a characterization of whatever foods contribute the most calcium.

Although bone composition accurately reflects dietary Sr/Ca and Ba/Ca ratios, these are not a simplistic index of the dietary plant/meat ratio. Because herbivores usually have more Sr and Ba in their bones than do carnivores (1), archaeologists and others have tried, although problematically, to interpret bone strontium levels as a measure of the dietary plant/meat ratio, or "trophic level" (2 – 15). Calculations of the compositions of diets with varying plant/meat ratios, however, show that the dietary Sr/Ca ratio does not vary in a simple, proportional relationship to the plant/meat ratio (16). Since dietary strontium and barium levels themselves do not reflect plant/meat ratios, bone strontium levels can not be expected to reveal these ratios. However, bone composition can place useful constraints on possible prehistoric diets once one recognizes what aspects of diet it reveals.

Reinterpretation of Bone Ba and Sr

In the conventional interpretation, foods have been interpreted as contributing to the chemistry of bone in proportion to their fraction of the diet. However, as Runia (17) and later Burton and Wright (16, 18) pointed out, foods high in bone-forming cations - essentially high-calcium foods - will contribute significantly to the composition of bone while foods low in calcium will not. Addition of any given dietary item causes the Sr/Ca ratio of the diet to shift towards its own Sr/Ca ratio, but the effectiveness of that food in causing this shift is directly proportional to its contribution to the total dietary pool of calcium. Thus individual foods do not affect

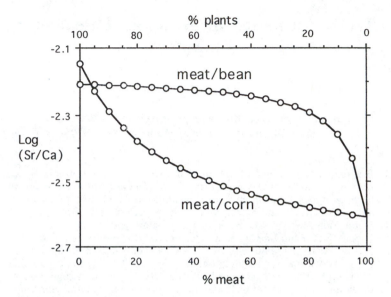

Figure 1: Variation of the dietary Sr/Ca ratio as a function of the plant/meat ratio for two plants with different calcium contents. Data sources: Elias et al. (1), Watt and Merril (19), Shacklette (20).

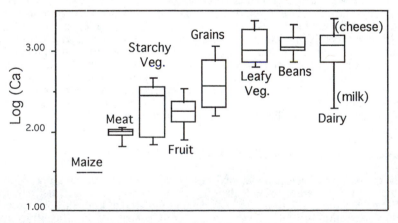

Figure 2: Typical calcium contents of various foods. Note the scale is logarithmic. Data source: Watt and Merril (19).

the composition of diet, or bone, in proportion to their weight percent in the diet, but in proportion to their contribution of bone-forming cations. This fraction, itself, is dependent upon both the calcium content of the specific item and the total amount of calcium in the diet. To be precise, the relevant parameter for "bone-forming cations" is the sum of calcium plus barium plus strontium plus other cations that substitute for calcium, but calcium is greatly in excess so that this sum may be approximated by the calcium content (e.g., 380,000 ppm Ca + 300 ppm Sr + 80 ppm Ba @380,000 ppm).

While high-calcium foods dominate diet composition and hence that of bone, low-calcium foods may be invisible. Nonetheless, because the calcium contribution of any given item is relative to the total amount of dietary calcium, low-calcium foods could still affect bone in the absence of high-calcium foods. Thus to interpret bone strontium levels appropriately, we must have some understanding of what foods most likely contribute the majority of the calcium to the diet as well as their Sr/Ca and Ba/Ca ratios.

Figure 1 shows this effect of calcium content upon the composition of diet. Dietary Sr/Ca ratios are shown as a function of the dietary plant/meat ratios for two plants: beans and corn, which have different mineral (calcium) contents. Beans have much more calcium than meat and are more effective at dominating the dietary composition. With beans as the plant source, a significant change in the amount of meat results in virtually no change in the dietary Sr/Ca ratio untill meat becomes by far the major constituent - and thus the major mineral source. Corn, in contrast to beans, is low in calcium, even lower than meat. Thus, against corn as the plant source, even a small amount of meat significantly lowers the dietary Sr/Ca and Ba/Ca ratios, but not in proportion to its dietary abundance. Conversely, at high meat levels there is little change in the Sr/Ca ratio since meat is the major source of calcium regardless of the amount of corn.

Figure 1 also reveals the problem of equivocality in trying to assess plant/meat ratios. Because corn and beans vary greatly in their effect upon diet composition, a given Sr/Ca ratio could be produced by a wide range of plant/meat ratios. For example, a dietary Sr/Ca ratio of 0.005 (log = -2.3), corresponding to a typical bone concentration of approximately 400 ppm., could be produced by a diet with 10% meat or 80% meat, depending upon whether beans or corn were the dominant plant counterpart. For more realistic, multi-component diets, any intermediate plant/meat ratio could produce the same 400 ppm in bone simply by varying the ratio of beans to corn. Because a wide range of possible diets can potentially have the same Sr/Ca and Ba/Ca ratios, one must recognize that bone composition can not reveal specific amounts of any food but can only indicate its significance as a source of bone-forming mineral.

Calcium content, in general, varies inversely with moisture, starch, and fat content. Foods high in these, such as fresh fruits, root vegetables, and meat, respectively, tend to be low in mineral content while leafy vegetables tend to have high calcium (*19, 20*) (Figure 2). Some foods such as cheese, fishmeal, and mineral-enriched foods such as cornmeal to which lime or ash has been added as a culinary practice, can be extraordinarily higher in calcium by an order of magnitude or more. Such high-calcium foods are likely to dominate the chemistry of the diet regardless of what other foods may be consumed. Because dairy products reflect biopurification through both digestive discrimination (*1, 21 – 27*) and fractionation in the mammary gland (*28 – 30*), these should produce low barium (tens of ppm) and strontium in bone (~100-200 ppm or less), comparable to or lower than levels found in carnivores. Anchovies and fishmeal from marine species can generate extremely low barium concentrations in bone (<10 ppm) even at relatively low dietary percentages (*31, 32*), although strontium could still be moderately high.

Non-marine meat products that include bonemeal produce low Sr/Ca and Ba/Ca ratios characteristic of carnivores, but the barium will not be as low as that indicative of marine resources.

In the absence of these extraordinarily high calcium foods, plants other than starchy cultigens will likely be the major contributors to the mineral content of diet. Such diets will have Sr/Ca and Ba/Ca ratios like those of purely herbivorous diets even when they include significant quantities of meat (33). Thus changes in the amount of meat in such diets are not likely to show corresponding changes in bone composition. This is probably the most common pattern for archaeological samples and is likely to be the least diagnostic. This pattern precludes dairy products, fishmeal, and anchovies as major contributors of calcium, but does not rule out seafood or meat without bones as a significant, but mineral-poor, dietary component. Variability in plant Sr/Ca and Ba/Ca ratios might affect diet composition, which could then appear as differences in bone composition, but such differences should not be interpreted as differences in trophic position.

Low calcium foods, including most meats and starchy plant foods, will not greatly affect the chemistry of diet or bone unless they are present in quite high proportions (>70%) or unless high-calcium foods are virtually absent. In this case, the "low"-calcium foods will in fact be the major contributors of bone-forming mineral.

Examples

A particularly revealing case is the effect of introducing maize into a diet in substitution for higher-mineral plants. If meat were a component of the original diet with leafy vegetables, this would not be visible as lower Sr and Ba against the contribution from high-calcium plants. Because meat would contribute relatively more calcium in the maize diet, its presence would appear as lower Ba and Sr even though the actual amount of meat in the diet might not have changed. The introduction of a high-starch cultigen such as maize could appear in the bones as a decrease in Ba and Sr, which would paradoxical suggest in the conventional interpretation that meat were increasing. The important detail is that, even though no more meat might be consumed, meat would be contributing a greater fraction of the total dietary calcium in the maize diet.

This effect has been noted for seafood. In Larsen's study of prehistoric populations of the Georgia Bight (34), introduction of maize at the expense of other plants decreased the mineral contribution from plants so that the low-barium signature for seafood became more apparent even though it did not increase in the diet (35, 36) It would be in error to interpret the decreased barium as an increase in the amount of seafood, but nonetheless appropriate to infer that more calcium was coming from seafood.

Note, however, that if maize is mineral-enriched through culinary practices, its effect upon diet composition might not be predictable. The culinary practices of treating maize with alkaline materials such as lime and plant ash can elevate its mineral content such that it can strongly affect bone levels (18, 37) but the Sr/Ca ratio will reflect that of the alkaline source, not that intrinsic to maize. The use of plant ashes will probably produce the high Sr/Ca and Ba/Ca ratios characteristic of plants, but lime derived from limestone or shell will vary in its composition depending upon its source.

As other examples of how paleodietary inferences should be reconsidered, we can compare several archaeological contexts where seafood was probably a component of diet. Human bones from archaeological sites along the coast of Kodiak Island, Alaska average approximately 20 ppm Ba. Likewise, human bones from archaeological sites along the arid coasts of Peru and Chile characteristically contain less than 10 ppm Ba (*31*). In contrast, human bones from Rio Viejo and Cerro de La Cruz, on the Oaxacan coast, contain an average of 300 ppm Ba, well within the terrestrial norm (*31, 38*).

Although the coastal Oaxacan communities were probably consuming some marine resources, they were agricultural and deriving their calcium primarily from plants. The trace-element data show that seafood was not a substantial calcium source but are otherwise equivocal. We can not rule out seafood but can rule out high-mineral marine resources such as anchovies & fishmeal. The extremely low Ba levels in bones from both the desert coast of South America and Kodiak Island, however, clearly show that virtually all of their dietary calcium was derived from marine resources. At the South American sites herring and anchovies, including bones, were a prominent dietary component (*39*) and thus would be the major calcium source regardless of what other foods were consumed in any quantity. Thus the South American data provide little constraint upon the dietary possibilities beyond an unspecified quantity of small fish. Kodiak Island data also imply that mineral-rich terrestrial resources were not a major part of the diet. High-starch, low-mineral plants such as root vegetables don't contain enough bone-forming minerals to overshadow the low Ba characteristic of marine resources and thus might have been a significant part of the diet, but high-calcium leafy vegetables can be excluded.

Although interpretation of the barium data as a direct percentage of seafood in these diets would be inappropriate, the data can be used, along with other dietary information, to place qualitative constraints upon possible diets. Proper interpretation requires some assessment of what were the possible dietary items, especially the high calcium components. Thus trace-element data are best used as ancillary data in conjunction with other archaeological evidence and, if available, stable isotopic data from the bones.

Summary

Because the Sr/Ca and Ba/Ca composition of diets themselves do not change in proportion to plant/meat ratios, bone composition can not be used as a simple estimator of trophic position. Nonetheless Ba and Sr levels in bone appear to faithfully reflect the dietary Ba/Ca and Sr/Ca ratios. With some understanding of what foods might have been, or might not have been, part of the diet, bone composition can be compared to that expected from theoretical diets. The key to using these elements is the realization that they reflect not the weight percent of various foods, but their relative contribution of calcium to the diet.

Literature Cited

1. Elias, R. W.; Hirao, Y.; Patterson, C. C. *Geochimica Cosmochimica Acta* **1982**, *46*, 2561-2580.
2. Rheingold, A. L.; Hues, S.; Cohen., M. N. *Journal of Chemical Education* **1983**, *60*, 233-234.
3. Brown, A. B. Ph.D. thesis, University of Michigan, Ann Arbor, 1973.
4. Ezzo, J. A. Ph.D. dissertation, University of Wisconsin-Madison, 1991.
5. Gilbert, R. I. Ph.D. dissertation, University of Massachusetts, Amherst, 1975.
6. Katzenberg, M. A. *Chemical Analysis of Prehistoric Human Bone from Five Temporally Distinct Populations in Southern Ontario*; Archaeological Survey of Canada / National Museum of Man, 1984.
7. Kavanagh, M. M.A. thesis, University of Wisconsin, Madison, 1979.

8. Price, T. D., Connor, M., & Parsen, J. D. *Journal of Archaeological Science* **1985,** *12*, 419-442.
9. Radosevich, S. In *Old Problems and New Perspective in the Archæology of South Asia*; Kenoyer, J. M., Ed.; Wisconsin Archæological Reports: Madison, 1989; Vol. 2, pp 93-102.
10. Schoeninger, M. J. *American Journal of Physical Anthropology* **1979,** *51*, 295-310.
11. Schoeninger, M. J. In *The Chemistry of Prehistoric Human Bone*; Price, T. D., Ed.; Cambridge: New York, 1989, pp 38-67.
12. Sillen, A. Ph.D. dissertation, University of Pennsylvania., 1981.
13. Sillen, A. *American Journal of Physical Anthropology* **1981,** *56*, 131-137.
14. Sillen, A. *Journal of Human Evolution* **1992,** *23*, 495-516.
15. Toots, H.; Voorhies, M. R. *Science* **1965,** *149*, 854-855.
16. Burton, J. H.; Wright, L. E. *American Journal of Physical Anthropology* **1995,** *96*, 273-282.
17. Runia, J. T. *Journal of Archaeological Science* **1987,** *14*, 599-608.
18. Wright, L. E. Ph.D. dissertation, University of Chicago, 1994.
19. Watt, B. K.; Merril, A. L. *Composition of Foods*; U.S. Government Printing Office: Washington D.C., 1963.
20. Shacklette, H. T. *Elements in Fruits and Vegetables from Areas of Commercial Production in the Conterminous United States*; U.S. Government Printing Office: Washington, D.C., 1980.
21. Ichikawa, R. a. Y. E. *Health Physics* **1963,** *9*, 717-720.
22. Kostial, K.; Gruden, N.; Durakovic, A. *Calcified Tissue Research* **1969,** *4*, 13-19.
23. Lengemann, F. W. In *Transfer of Calcium and Strontium Across Biological Membranes*; Wasserman, R. H., Ed.; Academic Press: New York, 1963, pp 85-96.
24. Rosenthal, H. L.; Chochran, O. A.; Eves M. M. *Environmental Research* **1971,** *5*, 182-191.
25. Wasserman, R., and Comar, C. L. *Proceedings of the Society for Experimental Biological Medicine* **1956,** *101*, 314-317.
26. Wasserman, R., Comar, C. L., and Papadopoulou, D. *Science* **1957,** *126*, 1180-1182.
27. Taylor, D. M.; Bligh, P. M.; Duggan, M. H. *Biochemistry Journal* **1962,** *83*, 25-29.
28. Ruhmann, A. G.; Stover, B. J.; Brizzer, K. R.; Atherton, D. R. *Radiation Research* **1963,** *20*, 484-492.
29. Twardock, A. R. In *The Transfer of Calcium and Strontium Across Biological Membranes*; Wasserman, R., Ed.; Academic Press: New York, 1963, pp 327-340.
30. Kollmer, W. E.; Kriegel, H. *Nature* **1963,** *200*, 187-188.
31. Burton, J. H.; Price, T. D. *Journal of Archaeological Science* **1990,** *17*, 547-557.
32. Burton, J. H.; Price, T. D. In *Proceedings of the 27th International Symposium on Archaeometry, Heidelberg, 1990*; Pernicka, E., Wagner, G. A., Eds.; Berkhauser Verlag: Basel, 1991, pp pp.787-795.
33. Elias, M. *American Journal of Physical Anthropology* **1980,** *53*, 1-4.
34. Ezzo, J. A.; Larsen, C. S.; Burton, J. H. *American Journal of Physical Anthropology* **in press.**
35. Larsen, C. S.; Schoeninger, M. J.; Merwe, N. J. v. d.; Moore, K. M.; Lee-Thorp, J. *American Journal of Physical Anthropology* **1992,** *89*, 197-214.

36. Schoeninger, M. J.; Merwe, N. J. v. d.; Moore, K. M.; Lee-Thorp, J.; Larsen, C. S. In *The Archaeology of Mission Santa Catalina de Guale. 2: Biocultural Interpretations of a Population in Transition*; Larsen C.S, Ed.; American Museum of Natural History: New York, 1990; Vol. 68, pp 78-93.
37. Kuhnlein, H. V. *Journal of Ethnobiology* **1981**, *1*, 84-94.
38. Joyce, A. Ph.D. dissertation, Rutgers University, 1991.
39. Reitz, E. *American Anthropologist* **1988**, *90*, 310-322.

RECEIVED August 15, 1995

Chapter 24

An Electron Microprobe Evaluation of Diagenetic Alteration in Archaeological Bone

Diana M. Greenlee

Department of Anthropology, University of Washington, Box 353100, Seattle, WA 98195–3100

Backscattered electron imaging and wavelength dispersive spectrometry documented variability in the structure and composition of archaeological human bone samples from the central Ohio and Mississippi River valleys. Structural alterations to bone histomorphology are consistent with those known to reflect microorganismic activities, with hydroxyapatite dissolution followed by remineralization involving ions from both dissolved bone mineral and the soil solution. Patterns in the distribution of structurally intact bone were observed to vary with the local post-depositional environment. Differences in the mean concentration and variability of elements (Ca, P, Sr, Ba, Mn, Fe, Zn, Cu and V) between structurally intact and structurally altered areas were examined; structurally altered areas frequently displayed higher elemental concentrations and greater variability than structurally intact areas.

The structure and chemistry of bone virtually guarantee that it will interact with most geochemical environments. Bone's porous histological structure (Figure 1) allows easy invasion by soil microorganisms and weathering-induced cracks aid in penetration by ground water. Additionally, its microstructure, with structurally imperfect nonstoichiometric hydroxyapatite [$Ca_{10}(PO_4)_6(OH)_2$] microcrystals embedded in an organic fibrous matrix, gives bone a huge surface area and high reactivity (*1*), and thus great potential for chemical interaction with the burial environment. Indeed, bone is recognized as being quite susceptible to diagenesis, the physical and chemical alteration of materials resulting from interaction with the post-depositional environment. The exact nature of diagenetic alterations to bone will depend on local factors such as pH, temperature, moisture conditions, geochemistry, and microbiology.

Despite these potential modifications, researchers who use archaeological bone chemistry to make inferences about the age, diet or environment of prehistoric organisms must assume that the elemental and isotopic levels they measure correspond

0097–6156/96/0625–0334$12.25/0

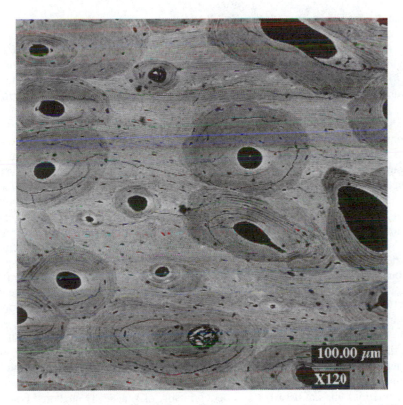

Figure 1.   Backscattered electron (BSE) micrograph of an undecalcified transverse section of a modern human femur, illustrating histomorphological structure and variable mineralization surrounding Haversian canals. The lighter areas have a higher mean atomic density than darker areas. The cracks are artifacts of sample preparation.

to those at death. Given the near certainty and complexity of chemical interaction between bone and the post-depositional environment, this assumption is problematic, especially for trace element studies of the mineral hydroxyapatite fraction of bone. The challenges, then, are to determine what manner of post-depositional alterations has occurred, how such alterations might affect our conclusions and how the effects might be minimized.

Strategies for Evaluating Diagenetic Alteration

Histomorphological analysis using microscopic, radiographic, and electronic imaging has shown that many kinds of diagenetic alteration can be identified visually (2-10). Areas of structurally intact bone are often present, even in bones showing extensive damage, with the distribution of diagenetic alterations apparently determined by both bone structure and post-depositional context. Because the diagenetic processes that alter the structure of bone probably also alter its chemical composition, histomorphological analysis would seem a useful way to ascertain the likelihood of some kinds of chemical alteration.   However, bone chemists seldom use histomorphological examination to identify diagenetic alteration; they choose instead to focus on patterns in elemental concentrations in the bone. If patterning in elemental concentrations meets certain expectations, the elemental levels are assumed to reflect dietary consumption; if concentrations do not conform to expected patterns, those elements are considered to be diagenetically altered (11-16).

Most compositional analyses rely on analytical techniques which employ powdered, homogenized bulk samples of bone. While researchers have shown that various pretreatments designed to remove contaminants, e.g., cleaning with ultrasonic baths, physical removal of 1-3 mm of the periosteal and endosteal surfaces of the bone (17-19) and chemical washes (19-26), often change elemental concentrations in anticipated directions, they have yet to show conclusively that the diagenetically altered, and only the diagenetically altered, bone has been removed.   Indeed, experimental data suggest that, with the chemical wash approach, at least some biogenic bone may be removed (19, 27) and some diagenetically altered bone may remain (28).   The results of bulk sample chemical analyses, then, are "average" concentrations that are *assumed* to reflect primarily diagenetically unaltered bone, but may potentially include diagenetically altered bone, as well.   Yet, successful documentation of prehistoric diets requires certain knowledge that diagenesis has not altered significantly the elemental concentrations measured.

The Electron Microprobe Approach

The potential contribution of electron microprobe technology to studies of archaeological bone chemistry lies in its ability to identify the structural, if not chemical, integrity of the bone from which compositional information is obtained. The electron microprobe (a.k.a. electron probe microanalyzer) uses a focused beam of electrons to irradiate a sample. As the beam strikes the specimen, elements in the sample gain energy and produce several types of secondary signals. Of particular importance to this study are the backscattered electron (BSE) and characteristic X-ray signals. The BSE signal permits the sample to be imaged, while X-rays permit characterization of its elemental composition. Thus, the electron microprobe allows documentation of the relationship between the microstructure of archaeological bone and its elemental composition.

Backscattered Electrons. Backscattered electrons are former primary beam electrons that have collided with electrons in the sample and bounced back through the sample surface. BSE generation is dependent on the mean atomic number of the sample; as the atomic number of the sample increases, the number of electrons in the sample with which the primary beam electrons can potentially collide increases, and thus, the number of emitted BSEs increases. In BSE images, areas of higher mean atomic number will appear brighter than areas of lower mean atomic number. Images generated by using the BSE signal provide nonspecific compositional information to a depth of 1-2 μm in the sample, with differences in mean atomic number as low as 0.1% being distinguishable in topographically flat specimens (29).

BSE imaging has been used successfully by researchers studying mineralization and histological structure of both modern and prehistoric bone (2, 30-34). BSE images are particularly useful in studies of diagenetic change in archaeological bones because they allow the difference in atomic density resulting from microorganismic destruction and mineral replacement to be evaluated with respect to sample histomorphology. BSE images do not, however, allow particular elements to be identified, nor their concentrations to be specified.

X-rays. Characteristic X-rays are produced when the primary electron beam dislodges an electron from an inner shell of an atom and an electron from an outer shell of that atom moves in to fill the vacancy. To fill the vacancy, the outer shell electron must lose energy in an amount equal to the difference in Coulomb force between the two orbitals, the emitted energy being within the X-ray portion of the electromagnetic spectrum. Because each element has a unique number of electrons and protons, the Coulombic field is unique to each element and each orbital of each element. Therefore, all X-rays emitted have energies (and wavelengths) characteristic of that element and the particular orbital-vacancy transition. At the same time, there are primary beam electrons which fail to collide with other electrons and instead randomly lose variable amounts of energy proportional to the distance from atomic nuclei that they pass by. This lost energy forms a continuous range of X-ray wavelengths, known as bremsstrahlung or background radiation, which must be subtracted from the intensity of characteristic X-rays before quantitative analysis can be obtained (29).

X-rays are generated within a small interaction volume (usually ~5 μm^3), the exact size and shape of which is determined by characteristics of the target sample. Even though the number of characteristic X-rays generated is proportional to the concentration of the element in the sample, the number of X-rays actually emitted will differ because of sample matrix effects. Sample composition and density influence both X-ray generation and emission and hence corrections to the X-ray intensities due to the effects of Z (mean atomic number), A (X-ray absorption) and F (secondary X-ray fluorescence) must be included in each quantitative determination (29). Thus, quantitative analyses depend on our ability to account for matrix differences and ZAF iterative correction procedures have been developed to do that. One aspect of the sample that ZAF corrections cannot account for, though, is sample topography. Topographic variation in a sample significantly influences both X-ray generation and absorption; therefore, flat samples are imperative for quantitative analyses (29).

The composition of biological specimens also introduces additional considerations during microprobe analyses as compared to analyses of most inorganic samples. As the electron beam deposits a very large amount of energy into a relatively small analytical volume, a significant increase in temperature occurs. This can lead to vaporization of components such as water, carbon-containing compounds (e.g., proteins and lipids), chlorine and fluorine, as well as migration of elements such as sodium, potassium and phosphorus (29). Vaporization and migration of some sample components will affect the relative concentration of other materials in the sample. Thus, microprobe studies of biological tissues must be sensitive to, and minimize the effects of, beam damage. Fortunately, in this regard, archaeological bone is more similar to geological specimens than fresh bone; archaeological bone is more stable under the beam due to the diagenetic loss of water, lipids and other organic components. My own beam/sample interaction studies indicate that, unlike fresh specimens, archaeological bones require no unusual analytical conditions to ensure their stability under the beam.

Wavelength Dispersive and Energy Dispersive Spectrometry. By measuring the net intensity (counts/second) of characteristic X-ray emissions (known as X-ray lines), the relative abundance of each element can be determined. This determination can be made in two ways, by wavelength dispersive spectrometry (WDS) or by energy dispersive spectrometry (EDS). Both compare the intensity of characteristic X-rays in the unknown sample with that of a standard of "known" composition; their differences lie in which aspect of the X-rays they measure, their wavelength or their energy.

In the case of WDS, diffracting crystals are arranged so as to selectively diffract X-rays of a particular wavelength to a detector, where they are subsequently counted. Consequently, a WDS spectrometer counts only one specified element at a time; however, instruments may have one or more detectors that can be used simultaneously. In contrast, EDS, which uses a solid state semiconductor detector to measure X-ray energy, receives and counts X-rays of all energies simultaneously. WDS is superior to EDS for trace element analysis because it has greater resolution and lower detection limits (under favorable conditions, on the order of 10 ppm vs. 1000 ppm for EDS), even if counting times are considerably longer (29).

The advantage of using quantitative electron microprobe analysis in the study of materials is that precise measurements of elemental concentrations can be obtained from small volumes at known locations. Surprisingly, the use of the microprobe to examine variability in elemental concentrations within bones by medical researchers (35-38) has been somewhat limited. No studies of trace element chemistry in archaeological bone have used the microprobe in a systematic quantitative manner; microprobe analyses are typically presented as secondary investigations (7, 39-41) and generally lack sufficient descriptive detail to evaluate the results.

X-ray Maps. X-ray maps display the distribution and concentration of a particular element across an area of the sample. Images are produced when the primary electron beam, rastering over an area of the sample, generates characteristic X-rays which are detected through either WDS or EDS spectrometry. The location of each X-ray is displayed on a monitor as a dot; after multiple raster passes, an image of dots appears that records distributional information about the element of interest. Such maps have been used to document the distribution of elements incorporated into reprecipitated hydroxyapatite matrix and appearing in voids and along fractures in archaeological and paleontological bones (*17, 32, 42-45*). Digital X-ray images, like those collected here, provide quantitative information about the concentration, as well as the spatial distribution, of particular elements. Thus, elemental composition can be directly linked with the histomorphological structure of the bone.

Implications. Structurally intact areas of bone have greater potential for retaining biological elemental concentrations than areas altered by post-depositional processes. From this, it follows that we should concentrate on determining the compositional integrity of structurally intact bone. If structurally intact bone is also chemically intact, analyses of these specific areas would provide secure dietary information.   If structurally intact areas are determined to be chemically altered, the alterations may be less severe than in visibly altered areas; thus, chemical pretreatments might be more effective on structurally intact areas.   Electron microprobe technology, because it permits both visual inspection of bone histomorphology and precise elemental analysis of very small volumes of material, may be useful in evaluating the chemical integrity of structurally intact archaeological bone.   The remainder of this paper describes preliminary work toward identifying and analyzing both structurally intact and altered areas of archaeological bone, with considerations of the diagenetic processes that have influenced them and the kinds of data necessary to evaluate fully the potential of this approach.

Archaeological Application

The archaeological research described here is part of a larger project aimed at documenting dietary variability through time and across different environments in the central Ohio and Mississippi River valleys.   For this preliminary study, samples of archaeological human bone representing eight individuals were selected randomly from a range of archaeological contexts and post-depositional environments (Table I). The goal is to explore the structural and compositional variability in archaeological bones, establishing that the electron microprobe is an appropriate instrument for this kind of study and determining what kinds of additional information are needed to explain patterns in trace element chemistry.

Sample Preparation. Transverse sections approximately 1 cm thick were removed from archaeological bones with a steel hacksaw. All sections originate from the mid-shaft region of long bones; several individuals were represented by multiple samples, in some cases from different bones and in others from different locations on the shaft of the same bone. Beyond a brief sanding of the cut surfaces with 240-grit silicon carbide

Table I. Archaeological Contexts and Post-Depositional Environments Represented

Archaeological Deposit	Approximate Age[a]	Archaeological Context	Sample Size	Soil pH	Permeability Index (in./hr)	References
Chilton (CHLT)	LW	Limestone mound	1	6.6 - 8.4[b]	0.06 - 0.6	46-47
C. L. Lewis (CLL)	MW	Limestone mound	2	5.6 - 7.3[b]	0.6 - 6.3	48-49
Cleek-McCabe (CMC)	ELP	Earthen mound	1	6.1 - 7.3	0.6 - 2.0	50-51
Poverty Point Object (PPO)	LA	Midden	1	5.0 - 7.0[c]	0.06 - 0.2	32, 52-54
Hankins (HAN)	LA	Midden	1	5.0 - 7.0[c]	0.06 - 0.2	32, 52, 54
Childers (CHLD)	LW	Pit burial	1	5.4 - 6.2	0.2 - 10.0	55-56
Slone (SLO)	MLP	Pit burial	1	4.0 - 5.0	0.8 - 5.0	57-58

[a]Deposits have been partitioned into the following age groups: LA = Late Archaic (4000 B.C. - 500 B.C.); MW = Middle Woodland (0 - A.D. 400); LW = Late Woodland (A.D. 400 - A.D. 1000); ELP = Early Late Prehistoric (A.D. 1000 - A.D. 1200); MLP = Middle Late Prehistoric (A.D. 1200 - A.D. 1400); [b]the pH of these archaeological deposits is probably neutral to alkaline due to dissolution of overlying limestone slabs; [c]through time, the dissolution of bone and shell in these deposits has changed the soil chemistry to allow preservation of these materials by altering the pH of the local microenvironment from acidic to neutral to slightly alkaline.

grinding paper to remove any contamination from the saw, no efforts (e.g., ultrasonic cleaning, surficial abrasion) were made to clean the bone prior to impregnation with Epo-thin resin.   After curing, the resin blocks were sanded flat, mounted to a petrographic slide and cut, leaving a thick section of bone approximately 250 μm thick. Each sample was polished with a series of silicon carbide and levigated alumina grits to 5 μm on a lap wheel and to 0.3 μm with a Buehler Petropol automated polishing system. Between grits, samples were carefully rinsed in distilled water and examined microscopically to determine if they required ultrasonic cleaning before proceeding to the next grit. Prior to analysis, samples were wiped with ethanol and a 25 μm-thick conductive coating of carbon was sputtered onto the polished sections. To ensure proper conduction, lines of carbon paint were extended across each slide from the sample holder to the bone.

Analysis. Each transverse section was examined with BSE imaging to firstly, identify qualitatively any patterning in the nature and location of diagenetic alterations to the histomorphology and secondly, to locate structurally intact and altered areas of bone. Compositional information was obtained for elements (Ca, P, Sr, Ba, Cu, V, Mn, Fe and Zn) often assumed to reflect bone preservation and dietary consumption (*11-16*). Both structurally intact and diagenetically altered areas along radially oriented (periosteal to endosteal) transects were analyzed. X-ray maps were acquired to examine the spatial distribution of elements of high concentration.

Analytic parameters (beam current, beam diameter and counting time) for the quantitative WDS analyses were established through a series of beam/sample interaction experiments. X-ray counts and absorbed current were monitored at different beam currents and beam diameters to determine under which conditions there were no significant X-ray intensity variations through time resulting from beam damage to the archaeological specimens.

BSE, X-ray imaging and WDS analyses were conducted on a JEOL 733 SuperProbe in the Department of Geological Sciences at the University of Washington using the following settings: 15 kV accelerating voltage; 40° take-off angle; <1 μm beam diameter for images and 10 μm beam diameter for elemental analyses; and 4 nA beam current for BSE images, 7-100 nA for X-ray maps, 10 nA for Ca and P WDS analyses, and 100 nA for trace element WDS analyses. The microprobe has four WDS spectrometers, enabling data from a maximum of four elements to be obtained simultaneously. Table II contains other pertinent information regarding the WDS elemental analyses. Elemental concentrations were determined by calibration against homogeneous standards with high concentrations of the elements of interest. Minimum detection limits under these analytic conditions were found to be adequate for detecting these elements in modern bones. Matrix effects were corrected with a ZAF correction routine. X-ray images were captured over periods of six to thirty hours.   Both the X-ray and BSE images were collected using a GATAN Digitalmicrograph imaging system.

Error and Variability. The randomness inherent in the generation of X-rays during elemental analysis allows error associated with each measurement to be calculated using Poisson statistics. Using this method of error estimation, the analytic precision

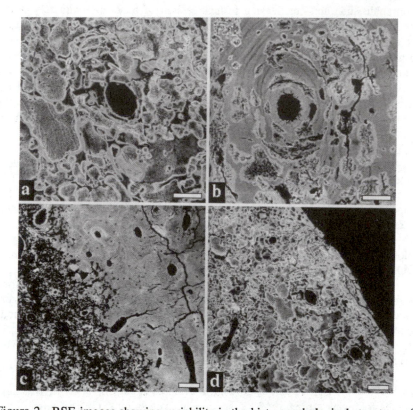

Figure 2. BSE images showing variability in the histomorphological structure of archaeological specimens. (a) an osteon in a specimen from SLO (X300 magnification; bar = 60 μm); note the hypermineralized borders surrounding diagenetically altered foci that are probably the result of microorganismic activities. (b) an osteon in a specimen from PPO (X300 magnification; bar = 60 μm); diagenetically altered foci surround the canal, reflecting the structure of the osteonal bone. (c) near the exterior surface of a specimen from CLL (X120 magnification; bar = 100 μm); note that the bone to the exterior (right) is structurally intact, while that to the interior (left) is severely damaged by diagenesis. (d) the exterior surface of the specimen from SLO in 2(a) (X120 magnification; bar = 100 μm); structurally intact bone is present as a thin strip along the extreme outside edge and as very small discrete areas in the interior.

(2 SD) of the trace elements of over 670 points analyzed in this study ranged from approximately 30 to 80 ppm. Duplicate analyses suggest that this statistical error is an underestimate of the actual error involved in the analysis of these complex materials; doubling the statistical error provides a better estimate of the analytic precision.

Table II. Parameters for WDS Elemental Analyses

Element	Diffracting Crystal[a]	Standard	X-ray Line	Counting Time (s)	Minimum Detection Limits (ppm)[b]
Ca	PET	Apatite	K_α	25-30	232
P	TAP	Apatite	K_α	25-30	145
Sr	TAP	Strontianite	L_α	200	27
Ba	PET	Benitoite	L_α	200	62
Mn	PET	Garnet	K_α	200	31
V	PET	Vanadium metal	K_α	200	23
Cu	LiF	Chalcopyrite	K_α	200	58
Zn	LiF	Willemite	K_α	200	80
Fe	LiF	Hematite	K_α	200	41

[a]Crystal composition: TAP, thallium acid phthalate; PET, pentaerythritol; LiF, lithium fluoride; [b]Minimum detection limits (MDLs) were calculated for each element as follows:

$$\text{MDL} = \frac{4 \times C(b)}{\sqrt{N}}, \text{ with } C(b) = C(std)\frac{I(bkg)}{I(std)} \text{ and } N = I(bkg) \times 2,$$

where C(std) is the elemental concentration in the standard, I(std) is the characteristic X-ray intensity (counts/second/nA) in the standard, and I(bkg) is the background intensity (counts/second/nA) in the sample.

Results and Discussion

BSE images (Figure 2) illustrate the highly variable nature of diagenetic change, with areas of relatively intact histomorphological structure often found adjacent to areas showing considerable damage. As shown most clearly in Figures 2a and 2b, altered areas typically appear as rounded, irregularly-shaped blobs, often with sharp, hypermineralized borders and varying interior density. Known as destructive foci, these kinds of alterations are characteristic of mineral dissolution associated with microorganismic activities, followed by remineralization with incorporation of ions from both dissolved bone mineral and the soil solution (*2-3, 5-6, 59-60*). The BSE images also provide evidence of other diagenetic processes in the form of precipitates and mineral inclusions in osteonal canals.

Diagenetic alterations are not distributed randomly within these bones, but rather are influenced by structural features inherent in the bones, proximity to exposed surfaces and the post-depositional environment. In samples from neutral to alkaline contexts (Figure 2c), structurally intact bone was frequently found at the periosteal and endosteal surfaces, with extensive diagenesis located mid-cortically (cf. *6*). In

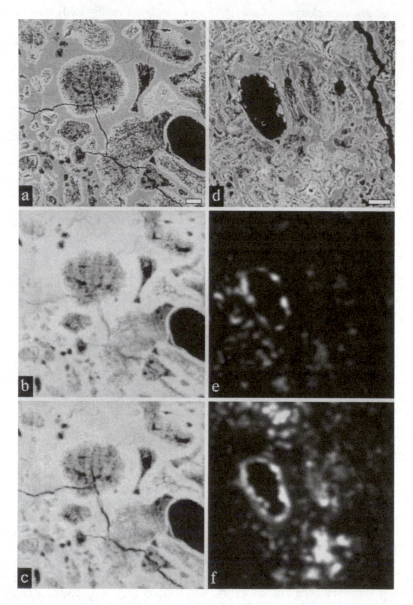

Figure 3. Two BSE images at (a) X780 (bar = 20 μm) and (d) X600 (bar = 30 μ m) magnification and their associated X-ray maps, showing the distribution of (b) Ca, (c) P, (e) Mn, and (f) Fe across different structural features in archaeological bones. In the X-ray maps, white areas represent areas of high concentration of the element of interest.

contrast, specimens from acidic environments occasionally had tiny areas of structurally-intact bone along either surface (Figure 2d), with larger areas of recognizable bone often remaining around Haversian canals in the mid-cortical region (Figure 3b; cf *2, 5*). Similarities in BSE gray-scale levels (which reflect mean atomic number), give reason to suspect that other areas too small to be morphologically recognizable (e.g., Figure 2a) may be structurally intact bone, as well.

BSE images suggest that the mean composition of structurally intact areas is closer to that of modern unaltered bone than the obviously diagenetically altered portions. Clearly, before arguments about the chemical integrity of structurally recognizable areas can be made, such areas of bone must be shown to be chemically different from obviously altered areas and, further, the elemental composition of the structurally intact bone must lie within the range of modern, unaltered bone (cf *61*). Sufficient data to compare the microstructural distribution of trace elements in archaeological bone with modern bone are currently lacking. The remainder of this paper explores the ways in which structurally intact areas are chemically different from structurally altered areas.

X-ray Maps. Comparisons of BSE images and associated X-ray maps demonstrate the relationship between histomorphological structure and elemental composition in archaeological bone samples. When Ca and P X-ray maps (Figures 3b and 3c, respectively) are compared to their corresponding BSE image (Figure 3a), they show that both the structurally intact and altered areas of bone are primarily calcium phosphate, with slightly increased Ca and P concentrations in the hypermineralized borders. Indeed, EDS spectra rarely register the presence of other elements, suggesting that the bulk of the remaining material is probably largely water or organic material, undetectable with the microprobe.

Comparisons of trace element X-ray maps with their BSE images show considerable variability in the composition of archaeological bone. The distributions of post-depositionally added elements, such as Mn and Fe (Figures 3e and 3f, respectively), are not simply restricted to the periosteal and endosteal surfaces of bone. Elements are localized along cracks and canals in the interior (cf *17, 42, 45, 62*) and are concentrated in areas throughout the bone visible in BSE images (Figure 3d) as structurally altered. Here, the structurally intact areas are relatively deficient in these two elements; the altered areas are quite variable, some showing significant concentrations of Mn and Fe, while others do not. This suggests that these post-depositional alterations occurred at different times, under different geochemical or microbiological influences.

Compositional Analysis. The results of compositional analyses comparing structurally intact and structurally altered areas of archaeological bone frequently show significant differences between the two. In general, diagenetically altered areas have greater compositional variability and higher mean elemental concentrations than areas that appear structurally intact. The supposition that the recrystallized material of the diagenetically altered areas is primarily composed of redeposited, formerly-dissolved hydroxyapatite with varying additions from the soil solution and microbial metabolites (*2, 63-65*) is borne out by compositional studies.

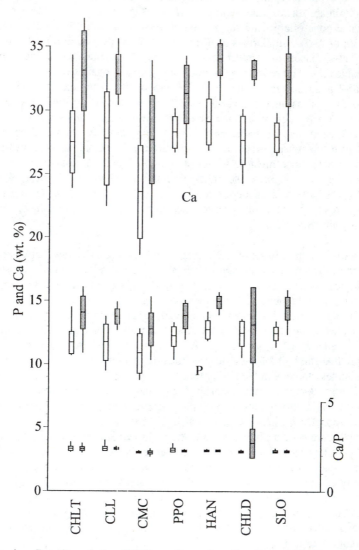

Figure 4.   Results of the WDS analyses for Ca, P and Ca/P, comparing structurally intact and altered areas in bone from each of the archaeological deposits (identified across the bottom). The first of each pair (white) represents structurally intact bone, the second (gray) represents structurally altered bone. For each symbol, the length of the block shows one standard deviation around the mean (horizontal line), while the length of the vertical line reflects the range of values obtained beyond the standard deviation.

A comparison of calcium and phosphorus concentrations for structurally intact and diagenetically altered areas (Figure 4) shows significant differences. The diagenetically altered areas have greater mean weight % Ca and P and greater variance, while at the same time, Ca/P ratios do not differ between the two. The relatively low concentration of P in most soils, coupled with an overall Ca/P mean for the structurally altered bone (± 1 SD) of 2.311 ± 0.169, only slightly higher than the theoretical ratio of 2.157, indicates that the bulk of Ca and P in the diagenetically altered areas is recrystallized hydroxyapatite.

Patterns in the distribution of trace elements (Sr, Ba, Cu, V, Mn, Fe and Zn) across structurally altered and intact areas of bone are less obvious than for the major elements. Certainly, the concentrations of these elements within and between bones are far more variable than Ca and P, with some elements frequently occurring in levels below the minimum detection limits for this study. This must correspond, in part, to the geology, acidity and redox potential of the post-depositional environment; these factors are particularly influential in determining the solubility of soil minerals (*66-68*) and their subsequent interaction with bone (*69*).

Strontium and barium, which are known to substitute for Ca in calcium phosphates, tend to occur in higher mean concentrations and with greater variability in altered than in unaltered areas (Figures 5a and 5b, respectively). The concentrations in the altered areas are higher than would be expected relative to the increased density of Ca and P there. Presumably, then, some of the Sr and Ba in the altered areas must be exogenous. Worth noting here is the apparent difference in Sr levels between two sets of these deposits (CHLT, CLL, CMC vs. PPO, HAN, CHLD, SLO). This probably reflects differences in the geochemistry of the post-depositional environments involved; however, there are insufficient data to allow evaluation of this hypothesis.

Like Sr and Ba, the mean elemental concentrations of Cu and V were consistently lower in intact areas, although not statistically significantly so (Figure 5c and 5d, respectively). The variances, however, were significantly different in five cases, with the intact always having less variability and being overlapped by the variance of the altered. Cu and V were concentrated above minimum detection limits in less than half of the specimens, though.

On the other hand, Mn, Fe and Zn are far more variable both within and between bones (Figures 6a, 6b and 6c). Areas near the surfaces and around large canals in the interior of bones had elevated concentrations that raised the mean and variance of both structurally intact and structurally altered areas. Given that diagenetic alterations occur at different times and are subject to changing geochemical conditions, differing contributions from the soil solution are probably incorporated into remineralizing bone.

Overall, the structurally intact areas appear to be more homogeneous chemically than the structurally altered areas. Depending upon the nature of the original bone feature (e.g., lamellar vs. osteonal bone), bone within a small area may be expected to be more homogeneous than post-depositionally modified bone that has been added at different times and under different geochemical conditions. Of course, systematic studies of microstructural variability in the composition of modern, unaltered bones are lacking; this research is currently under way.

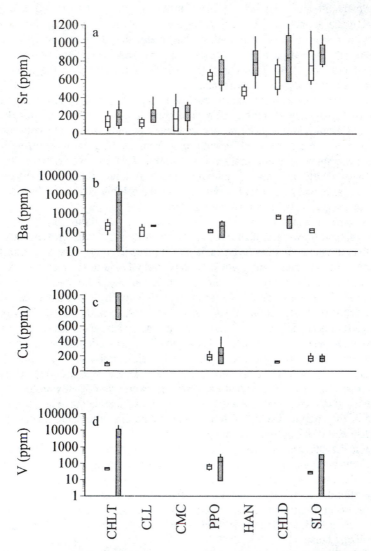

Figure 5. Results of the WDS analyses for (a) Sr, (b) Ba, (c) Cu and (d) V. See Figure 4 caption for explanation of symbols. The absence of symbols in any graph indicates that elemental levels were below minimum detection limits in bones from that deposit.

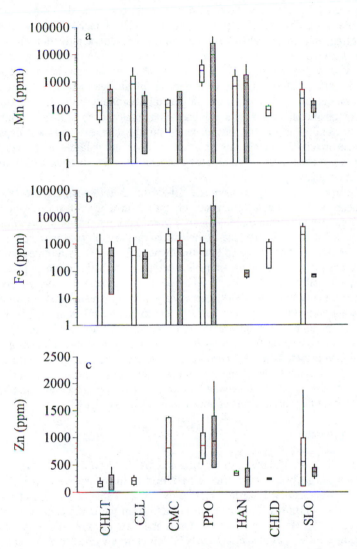

Figure 6. Results of the WDS analyses for (a) Mn, (b) Fe and (c) Zn. See Figure 4 caption for explanation of symbols. The absence of symbols in any graph indicates that elemental levels were below minimum detection limits in bones from that deposit.

Conclusion

In this paper, I have argued that electron microprobe analysis holds promise for studying the trace element chemistry of archaeological bones. The combination of BSE imaging and compositional analysis allows the characterization of individual structural features, both diagenetically altered and intact, thus allowing us to know precisely what is being measured. The results of compositional analyses comparing structurally intact and structurally altered areas of archaeological bone suggest the approach merits further consideration. In general, morphologically altered areas show greater compositional variability and higher elemental concentrations than areas that appear structurally intact, but this depends on the element and post-depositional environment involved. While this observation does not guarantee that structurally intact areas are chemically unaltered, they appear, at least, to have been exposed to less severe diagenetic processes.

Clearly, to actually explain the effect of these specific post-depositional environments and diagenetic processes on the distribution of trace elements will require larger samples from these contexts. More information about the composition of the soil solution, reflecting the composition of the soil parent material and the pH/redox potential of the local soil environment (27, 70-71), is required in order to determine the products resulting from its interaction with the calcium phosphate of bone. Additionally, future research is needed to characterize the distribution of elements in modern unaltered bone, particularly from organisms raised on controlled diets. Microstructural variability in the composition of bone is likely to be documented, given the temporal variability in bloodstream levels of elements, varying distance of mineralization fronts from the blood supply and the abilities of different hydroxyapatite crystal sizes to incorporate or adsorb ions. Once this information is obtained, the variability in elemental composition of archaeological bone can be evaluated relative to documented biological levels of variability and the utility of analyzing structurally intact areas of bone can be better assessed.

Acknowledgments

R. C. Dunnell, F. E. Hamilton, N. Justice and M. L. Powell made bone specimens available for this analysis. S. M. Kuehner advised on many aspects of the microprobe analysis. A. King and B. J. Carter kindly provided a sample of modern human femur for comparative purposes. This paper benefitted from the comments of B. J. Carter, R. C. Dunnell, S. M. Kuehner, M. V. Orna and two anonymous reviewers. Thin section equipment was available through a GSRF grant awarded to R. C. Dunnell; the Buehler Petropol polishing system was possible through NSF grant, SBR-9319278, awarded to L. Newell and R. C. Dunnell. Portions of this research were supported by NSF Dissertation Improvement Grant, SBR-9310137.

Literature Cited

(1)   Posner, A. S. In *Bone and Mineral Research/5*; Peck, W. A., Ed.; Elsevier: New York, NY, 1987; pp 65-116.
(2)   Bell, L. S. *J. Archaeol. Sci.* **1990**, *17*, 85-102.

(3) Garland, A. N. *App. Geochem.* **1989**, *4*, 215-229.
(4) Graf, W. *Acta Anatomica* **1949**, *8*, 236-250.
(5) Hackett, C. J. *Med., Sci. Law.* **1981**, *21*, 243-265.
(6) Hanson, D. B.; Buikstra, J. E. *J. Archaeol. Sci..* **1987**, *14*, 549-563.
(7) Hassan, A. A.; Ortner, D. J. *Archaeometry* **1977**, *2*, 131-135.
(8) Hedges, R. E. M.; Millard, A. R.; Pike, A. W. G. *J. Archaeol. Sci.* **1995**, *22*, 201-209.
(9) Salomon, C. D.; Haas, N.; *Israel J. Med. Sci.* **1967**, *3*, 747-754.
(10) Schoeninger, M. J.; Moore, K. M.; Murray, M. L.; Kingston, J. D. *App. Geochem.* **1989**, *4*, 281-292.
(11) Buikstra, J. E.; Frankenberg, S.; Lambert, J. B.; Xue, L. In *The Chemistry of Prehistoric Human Bone;* Price, T. D., Ed.; School of American Research Advanced Seminar Series; Cambridge University Press: Cambridge, UK, 1989; pp 155-210.
(12) Edward, J. B.; Benfer, R. A. In *Investigations of Ancient Human Tissue;* Sandford, M. K., Ed.; Gordon and Breach: Langhorne, PA, 1993; pp 183-268.
(13) Price, T. D. In *The Chemistry of Prehistoric Human Bone;* Price, T. D., Ed.; School of American Research Advanced Seminar Series; Cambridge University Press: Cambridge, UK, 1989; pp 126-154.
(14) Sandford, M. K. In *Skeletal Biology of Past Peoples: Research Methods*; Saunders, S. R.; Katzenberg, M. A., Eds.; Wiley-Liss: New York, NY, 1992; pp 79-103.
(15) Sandford, M. K. In *Investigations of Ancient Human Tissue;* Sandford, M. K., Ed.; Gordon and Breach: Langhorne, PA, 1993; pp 3-57.
(16) Whitmer, A. M.; Ramenofsky, A. F.; Thomas, J.; Thibodeaux, L. J.; Field, S. D.; Miller, B. J. In *Archaeological Method and Theory, Vol. 1;* Schiffer, M. B., Ed.; University of Arizona Press: Tucson, AZ, 1989; pp 205-273.
(17) Lambert, J. B.; Xue, L.; Buikstra, J. E. *J. Archaeol. Sci.* **1991**, *18*, 363-383.
(18) Lambert, J. B.; Xue, L.; Buikstra, J. E. *J. Archaeol. Sci.* **1989**, *16*, 427-436.
(19) Lambert, J. B.; Weydert, J. M.; Williams, S. R.; Buikstra, J. E. *J. Archaeol. Sci.* **1990**, *17*, 453-468.
(20) Krueger, H. W.; Sullivan, C. H. In *Stable Isotopes in Nutrition;* Turnlund, J. R.; Johnson, P. E., Eds.; American Chemical Society Symposium Series 258; American Chemical Society: Washington, D. C., 1984; pp 205-220.
(21) Price, T. D.; Blitz, J.; Burton, J.; Ezzo, J. A. *J. Archaeol. Sci.* **1992**, *19*, 513-529.
(22) Sillen, A. *Paleobiology* **1986**, *12*, 311-323.
(23) Sillen, A. In *The Chemistry of Prehistoric Human Bone*; Price, T. D., Ed.; School of American Research Advanced Seminar Series; Cambridge University Press: Cambridge, UK, 1989; pp 211-229.
(24) Sillen, A.; LeGeros, R. *J. Archaeol. Sci.* **1991**, *18*, 385-397.
(25) Sillen, A.; Sealy, J. C. *J. Archaeol. Sci.* **1995**, *22*, 313-320.
(26) Sullivan, C. H.; Krueger, H. W. *Nature* **1981**, *292*, 333-335.
(27) Pate, F. D.; Hutton, J. T.; Norrish, K. *App. Geochem.* **1989**, *4*, 303-316.
(28) Tuross, N.; Behrensmeyer, A. K.; Eanes, E. D. *J. Archaeol. Sci.* **1989**, *16*, 661-672.

(29) Goldstein, J. I.; Newbury, D. E.; Echlin, P.; Joy, D. C.; Romig, A. D., Jr.;
     Lyman, C. E.; Fiori, C.; Lifshin, E. *Scanning Electron Microscopy and X-Ray
     Microanalysis*; Plenum Press: New York, NY, 1992.
(30) Boyde, A; Jones, S. J. *Metab. Bone Dis. & Rel. Res.* **1983**, *5*, 145-150.
(31) Boyde, A; Maconnachie, E.; Reid, S. A.; Delling, G.; Mundy, G. R. *Scanning
     Electron Microscopy* **1986**, 1537-1554.
(32) Greenlee, D. M.; Dunnell, R. C. In *Materials Issues in Art and Archaeology
     III*; Vandiver, P. B.; Druzik, J. R.; Wheeler, G. S.; Freestone, I. C., Eds.;
     Symposium Proceedings 267; Materials Research Society: Pittsburgh, PA,
     1993; pp 883-888.
(33) Jackes, M. In *Skeletal Biology of Past Peoples: Research Methods*; Saunders,
     S. R.; Katzenberg, M. A., Eds.; Wiley-Liss: New York, NY, 1992; pp 189-
     224.
(34) Reid, S. A.; Boyde, A. *J. Bone Min. Res.* **1987**, *2*, 13-22.
(35) Baud, C. A.; Bang, S.; Lee, H. S.; Baud, J. P. *Calcif. Tiss. Res., Suppl.* **1968**,
     2, 6.
(36) Mellors, R. C. *Lab Invest.* **1964**, *13*, 183-195.
(37) Wergedal, J. E.; Baylink, D. J. *Am. J. Physiology* **1974**, *226*, 345-352.
(38) Wollast, R.; Burny, F. *Calcif. Tiss. Res.* **1971**, *8*, 73-82.
(39) Gilbert, R. I. *Trace Element Analyses of Three Skeletal Amerindian
     Populations at Dickson Mounds;* Ph.D. Dissertation, University of
     Massachusetts; UMI Dissertation Services: Ann Arbor, MI, 1975.
(40) Radosevich, S. C. *Diet or Diagenesis? An Evaluation of the Trace Element
     Analysis of Bone;* Ph.D. Dissertation, University of Oregon; UMI
     Dissertation Services: Ann Arbor, MI, 1989.
(41) Schoeninger, M. J. *Dietary Reconstruction at Chalcatzingo, a Formative
     Period Site in Morelos, Mexico;* Contributions in Human Biology, 2;
     Museum of Anthropology, University of Michigan: Ann Arbor, MI, 1979.
(42) Lambert, J. B.; Simpson, S. V.; Buikstra, J. E.; Hanson, D. *Am. J. Phys.
     Anthrop.* **1983**, *62*, 409-423.
(43) Parker, R. B. *Geol. Soc. Am. Prog. Abstr. 1966.* **1968**, *101*, 415-416.
(44) Parker, R. B.; Toots, H. *Geo. Sci. Am. Bull.* **1970**, *81*, 925-932.
(45) Waldron, H. A. *Am. J. Phys. Anthrop.* **1981**, *55*, 395-398.
(46) Funkhouser, W. D.; Webb, W. S. *The Chilton Site in Henry County,
     Kentucky;* Reports in Archaeology and Anthropology; University of
     Kentucky: Lexington, KY, 1937; *3*, 168-206.
(47) Whitaker, O. J.; Eigel, R. A. *Soil Survey of Henry and Trimble Counties,
     Kentucky;* United States Department of Agriculture, Soil Conservation
     Service, in cooperation with Kentucky Natural Resources and Environmental
     Protection Cabinet and Kentucky Agricultural Experiment Station:
     Washington, D. C., 1992.
(48) Brownfield, A. H. *Soil Survey of Shelby County, Indiana;* United States
     Department of Agriculture, Soil Conservation Service, in cooperation with
     Purdue University Agricultural Experiment Station: Washington, D. C., 1974.
(49) Kellar, J. H. *The C. L. Lewis Stone Mound and the Stone Mound Problem;*
     Prehistory Research Series; Indiana Historical Society: Indianapolis, IN, 1960;
     3, 367-474.

(50) Rafferty, J. E. *The Development of the Ft. Ancient Tradition in Northern Kentucky;* Ph.D. Dissertation, University of Washington; UMI Dissertation Services: Ann Arbor, MI, 1974.

(51) Weisenberger, B. C.; Dowell, C. W.; Leathers, T. R.; Odor, H. B.; Richardson, A. J. *Soil Survey of Boone, Campbell, and Kenton Counties, Kentucky;* United States Department of Agriculture, Soil Conservation Service, in cooperation with Kentucky Agricultural Experiment Station: Washington, D. C., 1973.

(52) Brown, B. L. *Soil Survey of Pemiscot County, Missouri;* United States Department of Agriculture, Soil Conservation Service, in cooperation with Missouri Agricultural Experiment Station: Washington, D. C., 1971.

(53) Dunnell, R. C.; Whittaker, F. H. *Louisiana Archaeology* **1990,** *17,* 13-36.

(54) Hamilton, F. E. *Relative Dating of Bone in Surface Assemblages Using Fluorine Content: An Archaeological Study of the Little River Lowland, Southeast Missouri;* Ph.D. Dissertation, University of Washington; In preparation.

(55) Gorman, J. L.; Rayburn, J. B. *Soil Survey of Jackson and Mason Counties, West Virginia*; Series 1957, No. 11; United States Department of Agriculture, Soil Conservation Service, in cooperation with West Virginia Agricultural Experiment Station: Washington, D. C., 1961.

(56) Shott, M. J., Ed. *Childers and Woods: Two Late Woodland Sites in the Upper Ohio Valley, Mason County, West Virginia*; Program for Cultural Resource Assessment Archaeological Report 200; University of Kentucky: Lexington, KY, 1990.

(57) Dunnell, R. C.; Hanson, L. H., Jr.; Hardesty, D. L. *Southeastern Archaeological Conference Bulletin* **1971,** *14,* 1-102.

(58) McDonald, H. P.; Blevins, R. L. *Reconnaissance Soil Survey of Fourteen Counties in Eastern Kentucky*; Series 1962, No. 1; United States Department of Agriculture, Soil Conservation Service, in cooperation with Kentucky Agricultural Experiment Station: Washington, D. C., 1965.

(59) Marchiafava, V.; Bonucci, E.; Ascenzi, A. *Calcif. Tiss. Res.* **1974,** *14,* 195-210.

(60) Garland, A. N. In *Death, Decay and Reconstruction*; Boddington, A; Garland, A. N.; Janaway, R. C., Eds.; Manchester University Press: Manchester, UK, 1987; pp 109-126.

(61) Radosevich, S. C. In *Investigations of Ancient Human Tissue;* Sandford, M. K., Ed.; Gordon and Breach: Langhorne, PA, 1993; pp 183-268.

(62) Lambert, J. B.; Simpson, S. V.; Szpunar, C. B.; Buikstra, J. E. *Archaeometry* **1984,** *26,* 131-138.

(63) Grupe, G.; Piepenbrink, H. *App. Geochem.* **1989,** *4,* 293-298.

(64) Grupe, G.; Piepenbrink, H. In *Trace Elements in Environmental History*; Grupe, G.; Herrmann, B., Eds.; Proceedings in Life Sciences Series; Springer-Verlag: New York, NY, 1988; pp 104-112.

(65) Piepenbrink, H. *App. Geochem.* **1989,** *4,* 273-280.

(66) Cresser, M.; Killham, K.; Edwards, T. *Soil Chemistry and its Applications;* Cambridge Environmental Chemistry Series 5; Cambridge University Press: Cambridge, UK, 1993.

(67) Kabata-Pendias, A.; Pendias, H. *Trace Elements in Soils and Plants*, 2d ed;
        CRC Press: Boca Raton, FL, 1992.
(68) Lindsay, W. L. *Chemical Equilibria in Soils;* John Wiley and Sons: New
        York, NY, 1979.
(69) Williams, C. T. *App. Geochem.* **1989,** *4,* 247-248.
(70) Pate, F. D.; Hutton, J. T. *J. Archaeol. Sci.* **1988,** *15,* 729-739.
(71) Pate, F. D.; Hutton, J. T.; Gould, R. A.; Pretty, G. L. *Archaeol. Oceania*
        1991, *26,* 58-69.

RECEIVED December 6, 1995

Chapter 25

Stable Isotope Analysis of Bone Collagen, Bone Apatite, and Tooth Enamel in the Reconstruction of Human Diet

A Case Study from Cuello, Belize

R. H. Tykot[1], N. J. van der Merwe[1], and N. Hammond[2]

[1]Archaeometry Laboratories, Department of Anthropology,
Harvard University, Cambridge, MA 02138
[2]Department of Archaeology, Boston University, Boston, MA 02215

Stable isotope analysis of bone collagen is now a well-established method of studying ancient human diet. Carbon isotope values distinguish between C_3 and C_4 plants in the terrestrial food web; nitrogen values can indicate marine resource exploitation, terrestrial climate, and trophic level. Unfortunately, the relative contributions of the protein, carbohydrate, and fat portions of the diet to bone collagen and bone apatite are still not fully understood. Stable isotope data for human burials from the Preclassic Maya site of Cuello, Belize demonstrate that isotopic analysis of both tissues is necessary for proper dietary reconstruction of all but the simplest ancient food webs. Equally important are isotopic analyses of the fauna and flora available for human exploitation, and the integration of these data with archaeological evidence. At Cuello, it appears that maize-eating dogs may have been a significant dietary component, but there is no evidence that deer were tamed or loose-herded as ethnohistoric accounts suggest.

Archaeological excavations conducted at Cuello, Belize by Norman Hammond and co-workers have uncovered the earliest known Preclassic Maya site, with its Swasey ceramic phase dating to 1200 BC (*1*). The 180 human burials discovered so far have provided a wealth of information on Preclassic Maya ritual and ideology (*2, 3*), population characteristics and health (*4*), and diet (*5, 6*).

At Cuello, the presence of maize cobs of varying size (*7, 8*), along with manioc and perhaps other root crops (*9*), and large numbers of deer, dog and turtle remains (*10, 11*), raises questions about the nature of the Preclassic Maya diet. Cupule sizes suggest that early, small-cobbed maize was replaced by progressively larger cobs, inferring increased production and consumption. Stable carbon isotope ratio ($^{13}C/^{12}C$, relative to a standard) determinations for several unidentified animal bones from Cuello, for the purpose of radiocarbon date correction (*12*), also indicated a significant C_4 component (one of two dominant photosynthetic pathways, in which C_3 and C_4 refer to the number of carbon atoms in a molecule formed during the first stage of photosynthesis) to their diet and thus the possibility of dog-raising and/or deer-taming, both observed in historic times by Bishop Landa, Francisco Hernande, and others (*13-15*). Finally, the large number of human burials at Cuello permits the study of gender- or status-based differences in diet, including a comparison between sacrificial victims from two mass burials and those from regular, individual graves.

Stable Isotopes and Diet

Stable isotope analysis of bone collagen is now a well-established method of studying ancient human diet (*16-18*). Carbon isotope values distinguish between C_3 and C_4 plants in the terrestrial food web; nitrogen values can indicate marine resource exploitation, terrestrial climate, and trophic level. Stable isotope ratios of carbon in bone apatite and tooth enamel are also reflective of diet, provided that appropriate pretreatment procedures are employed to remove adsorbed and diagenetic carbonate (*19-23*).

Unfortunately, allocation of the protein, carbohydrate, and lipid portions of the diet to bone collagen and bone apatite are still not fully understood (*24-26*). It now appears that $\delta^{13}C$ values for collagen represent mostly dietary protein while those for apatite reflect the whole diet (*27*), rather than just the energy components (*28*). For herbivores, isotopically similar plant proteins and carbohydrates supply carbon to collagen and apatite, respectively. For omnivores and carnivores, lipids from animal foods are also important contributors to apatite. Furthermore, under conditions of nutritional stress carbohydrates and lipids may be routed to collagen growth and maintenance.

We argue that analysis of both collagen and apatite fractions should be routinely done to avoid over-estimating the importance of the protein portion of prehistoric diets.

Finally, interpretation of analytical data must take into consideration the ontogeny and turnover rates of collagen and apatite tissues in bone and teeth. Bone collagen certainly reflects long-term diet for adults, although high-protein diets may stimulate higher rates of bone turnover. Elevated metabolism and growth rates of juveniles may also result in more rapid collagen turnover, while turnover rates are much higher for trabecular than for cortical bone (*29*). Bone apatite is likely to turn over faster than bone collagen, but still probably represents diet over several years, while tooth enamel apatite is laid down rapidly and is not replaced.

Analytical Methods

Collagen was extracted from the poorly preserved Cuello bones by slow dissolution of the mineral component in dilute hydrochloric acid, neutralization of humic acids with sodium hydroxide, and separation of fatty residues with a 2:1:0.8 chloroform, methanol and water mixture. Collagen pseudomorphs were freeze-dried and percent yields calculated prior to combustion in closed quartz tubes with granular Cu and CuO wire. The resulting carbon dioxide and nitrogen gases were cryogenically purified and isolated on a vacuum line in the Archaeometry Laboratories at Harvard University. Samples consistently produced carbon and nitrogen gases with an elemental C:N ratio between 2.9 and 3.6, carbon weight percents around 40, and nitrogen weight percents around 15, all indicative of the integrity of the original collagen sample (*30*).

Apatite was produced from cleaned bone or tooth enamel by dissolving the organic phase in Clorox and removing diagenic and adsorbed carbonate contaminants with dilute 1 M acetic acid. Carbon dioxide gas was liberated in 100% phosphoric acid at 90 °C in a sealed and evacuated reaction vessel. Weight percent yields of carbon dioxide were consistent with our in-house enamel and apatite standards.

The carbon and nitrogen isotope ratios were measured on a VG Prism 2 ratio mass spectrometer. Cryogenically distilled gas samples were introduced through the manifold, while some collagen samples were introduced directly into the Carlo Erba CHN analyzer, which combusts the organic material on-line and sends the sample gas directly to the mass spectrometer source. The latter method has the advantage of rapidly producing reliable results with sample sizes on the order of 0.3 milligrams.

Isotope ratios are reported using the delta notation in parts per thousand or per mil (‰), relative to the PDB (Peedee Formation, South Carolina, *Belemnitella americana* marine fossil limestone) standard in the case of carbon, and to AIR (atmospheric N_2) in the case of nitrogen. C_3 plants average about -26‰ for carbon, and C_4 plants about -12‰. Herbivores typically show a positive fractionation in their bone collagen of about 5‰, to about -21‰ and -7‰ respectively for pure feeders. Nitrogen isotope ratios are a function of climatic effects, trophic levels, and marine vs. terrestrial food consumption. For accurate interpretation of both carbon and nitrogen isotope ratios it is necessary to establish baseline values for the region and time period under investigation.

Isotopic Ecology

Both archaeological fauna from Cuello and modern forest-dwelling fauna from the Orange Walk, Belize area were extensively tested to establish a baseline for the interpretation of the human remains. Archaeological deer (*Mazama americana, Odocoileus virginianus*), peccary (*Tayassuidae sp.*), and mud turtle (*Kinosternon sp.*) have an average $\delta^{13}C$ of -20.8 ± 1.4‰, while modern deer, peccary, tapir (*Tapirus bairdii*), kinkajou (*Potos flavus*), paca (*Agouti paca*) and nine-banded armadillo (*Dasypus novemcinctus*) average -22.4 ± 1.1‰. The Industrial Effect on atmospheric carbon dioxide accounts for the 1.6‰ negative shift for modern organisms relative to Preclassic fauna. Additional care must be taken with the selection of modern organisms, since the modern landscape includes less forest than the Preclassic, not to mention extensive stands of sugar cane, a C_4 crop. The only wild archaeological fauna with a C_4 component to the diet is the insect-eating armadillo, with an average $\delta^{13}C$ of -16.4 + 2.8‰. Armadillos frequently make their burrows in cornfields, and would have been easily caught there.

The isotopic values for a number of marine organisms were also determined, despite the minimal evidence at Cuello for the consumption even of riverine fish. Bone collagen values for several fish species (n = 9) from Caye Caulker, about 20km offshore, have average $\delta^{13}C$ values of -7.3 ± 2.0‰ and average $\delta^{15}N$ values of 6.8 ± 1.4‰, while the flesh of conch, whelks, winkles, and flat tree oysters (n = 11) averages -13.3 ± 1.8 and 3.5 ± 1.3 respectively. These values are characteristic of the Caribbean marine foodweb, which is based in part on extensive stands of flowering marine grasses like *Thalassia testudinum* (turtle grass) which have enriched carbon and depleted nitrogen isotope ratios (*31-33*). The differences between Caribbean values and those for other parts of the New World (*34*) emphasize the importance of establishing baseline values for each food source in all studies of human diet.

Deer and Dogs

The collagen $\delta^{13}C$ values for all archaeological deer specimens analyzed are consistent with those of C_3 plant eaters. Five brocket deer (*Mazama americana*) average -22.0 ± 0.8‰, while five white-tailed deer (*Odocoileus virginianus*) average -20.5 ± 0.9‰. These values are consistent with those for 49 white-tailed deer from several Maya sites published by Gerry (*35*). There is thus no isotopic evidence of human maintenance of deer populations, which presumably would have involved supplying them with maize feed or at least tolerating some access to maize fields. At Dos Pilas, a cache of subadult deer limbs was excavated (*36*), with one individual exhibiting a completely healed femoral fracture. It has been considered unlikely that this individual would have survived without human intervention (*37*), while isotopic analysis did not show any enrichment in $\delta^{13}C$ (*38*). The animal apparently survived for several months, but perhaps not long enough for its bone isotopic composition to change significantly if it were fed maize. As Wright (*38*) correctly notes, evidence of maize consumption would support the semi-domestication hypothesis while the converse is not necessarily true.

In contrast, small dogs (*Canis familiaris*) averaging less than 20 pounds (*11*) usually had a substantial C_4 component to their diet, although usually not as much as their human masters. For 12 individuals, collagen $\delta^{13}C$ values average -15.6 ± 3.9‰ and $\delta^{15}N$ values 7.5 ± 2.0‰. In most cases, the variation in canine diet probably reflects a combination of household scraps and scavenging for food, rather than a stable household diet. The unimodal distribution of their age at death (about 1 year), along with the chopped-up and broken nature of their limb bones (presumably for marrow extraction) (*11*), suggests that dogs were utilized as a low-cost meat resource, as observed in the 1500's by Francisco Hernande (*14*). Two dogs have very positive $\delta^{13}C$ (-10‰) and $\delta^{15}N$ (9‰) values, indicating significantly more maize consumption than was typical of the average human diet ($\delta^{13}C$ = -13‰), and suggestive perhaps of fattening for sacrificial purposes (*13, 39*). Three others have no significant C_4 component ($\delta^{13}C$ = -20 to -23‰).

Gerry's (*35*) isotopic data (ave. $\delta^{13}C$ = -9‰, n=19) for dogs from three Classic Maya sites, on the other hand, show greater reliance on C_4 food sources than the human population there (ave. $\delta^{13}C$ = -10‰), perhaps as a result of scavenging on human fecal material in addition to eating maize.

Human Diet at Cuello

Based on the quantity of faunal remains, white-tailed deer were the most important meat source at Preclassic Cuello, followed by freshwater turtle and dogs (*10*). Peccary, armadillo, and an assortment of mammalian, reptilian and avian species constitute only a small percentage of the faunal remains; fish and shellfish are negligible in number, but a wide variety of marine and estuarine species are represented. Maize was domesticated long before the Preclassic period, and would probably be the major plant staple consumed.

Fifty-four human individuals have been isotopically tested, most for more than one tissue type (i.e., bone collagen, bone apatite, tooth enamel). There are slight indications that the percentage of C_4-based food items in the Preclassic adult male diet at Cuello actually decreased over time, from a high in the Early Middle Preclassic of -11.2‰ (n = 2), to -12.1‰ (n = 3) in the Late Middle Preclassic, to -13.4‰ (n = 7) in the Late Preclassic (Table I). This would appear contrary to the commonly-held belief that maize consumption intensified during this period; in fact this may reflect changes in dietary protein sources rather than the maize contribution to the diet. There does not seem to be any change in female diet throughout the Preclassic (n = 11). Our small sample hints that in the Early Middle Preclassic (n = 5), males may have had greater access to C_4-based food sources, averaging about 2‰ more positive in their bone collagen than females; in the Late Preclassic a larger sample (n = 13) shows no gender-based difference.

Three sacrificed individuals from the Early Cocos Period mass burial have significantly enriched $\delta^{13}C$ collagen values (-10.7 ± 0.2‰, n=3), strongly inferring that they were not native to the Cuello area, or had substantially different diets for a significant part of their later life (Table II).

Overall, the Preclassic inhabitants of Cuello have average $\delta^{13}C$ collagen values of -12.9 ± 0.9‰ (n = 28); $\delta^{15}N$ values of 8.9 ± 1.0‰ (n = 23); $\delta^{13}C$ bone apatite values of -9.8 ± 1.0‰ (n = 16); and $\delta^{13}C$ tooth enamel values of -8.7 ± 2.3‰ (n = 33). The C_4 contribution to human tissues may be calculated by simple interpolation. Using collagen $\delta^{13}C$ end-members of -21‰ for pure C_3 diets and -7‰ for pure C_4 diets, we can estimate the C_4 contribution to human collagen carbon was approximately 58% at Preclassic Cuello. Using apatite end-members of -14‰ and 0‰, we can similarly estimate that the C_4 contribution to human bone apatite carbon was approximately 30%. This latter figure may be taken as the average C_4 content of the diet, while the consumption of $\delta^{13}C$-enriched dog meat would have enhanced the collagen C_4 carbon content, which overemphasizes the protein portion of the diet. The even more positive tooth enamel values probably reflect both a high-maize juvenile diet, as well as the pre-weaning diet where the child is a trophic level above the lactating mother.

Alternatively, one may use the experimentally-derived model produced by Ambrose and Norr (*27*) in their controlled-diet study using rats:

$$\delta^{13}C_{carbonate} = 8.3733 + 0.93957 \times \delta^{13}C_{diet}$$

$$\delta^{13}C_{diet-collagen} = 4.6061 + 0.52263 \times \delta^{13}C_{diet-protein}$$

The average $\delta^{13}C$ value of the whole diet, based on the Cuello human bone apatite value of -9.8‰, would then be -19.3‰; using the Cuello human bone collagen value of -12.9‰, the average $\delta^{13}C$ value of dietary protein would be -15.9‰. With C_3 plant foods averaging -26‰, maize about -12‰, wild animal (deer, turtle) flesh about -23‰ (flesh, with a high lipid component, is about 2‰ more negative than collagen) and

Table I. Stable Isotope Data for Preclassic Maya from Cuello, Belize

Phase	Group		$\delta^{13}C_{collagen}$	$\delta^{15}N_{collagen}$	$\delta^{13}C_{apatite}$	$\delta^{13}C_{enamel}$
Swasey?	Female		-11.8		-9.1	-7.3
Bladen	Males	ave.	-11.2	9.8	-8.7	-8.8
		s.d.	0.2	0.1	0.2	1.4
		n	2	2	2	2
Bladen	Females	ave.	-13.6	8.2	-10.0	-12.3
		s.d.	1.1	0.9	1.3	3.8
		n	3	3	2	2
Bladen	All adults	ave.	-12.6	8.8	-9.3	-10.6
		s.d.	1.4	1.0	1.1	3.4
		n	5	5	4	4
Lopez	Males	ave.	-12.1	8.4	-9.9	-6.9
		s.d.	0.4	0.4	0.6	0.0
		n	3	3	3	1
Lopez	All adults	ave.	-12.3	8.5	-9.8	-9.3
		s.d.	0.5	0.4	0.5	2.3
		n	4	4	4	2
Middle	All adults	ave.	-13.2	8.0	-11.2	-9.3
Preclassic		s.d.	0.3	0.3	1.5	1.8
		n	4	3	2	4
Cocos	Males	ave.	-13.4	8.9	-9.4	-8.5
		s.d.	0.6	1.1	0.1	2.9
		n	7	6	2	6
Cocos	Females	ave.	-13.3	9.4	-10.4	-9.5
		s.d.	0.8	0.7	0.0	1.2
		n	6	4	1	5
Cocos	Juveniles	ave.	-12.5	10.9	-10.2	-7.7
		s.d.	0.0	0.0	0.0	0.8
		n	1	1	1	9
Cocos	All	ave.	-13.3	9.3	-9.8	-8.4
		s.d.	0.7	1.1	0.5	1.9
		n	14	11	4	20

ave. = average of n individuals; s.d. = standard deviation of the mean

Table II. Stable Isotope Data for Individual and Mass Burials at Cuello, Belize

Group	Subgroup		$\delta^{13}C_{collagen}$	$\delta^{15}N_{collagen}$	$\delta^{13}C_{apatite}$	$\delta^{13}C_{enamel}$
Single Burials	Males	ave.	-12.8	8.8	-9.8	-8.5
		s.d.	0.9	0.9	1.3	2.5
		n	15	13	8	11
Single Burials	Females	ave.	-13.2	8.9	-9.8	-9.7
		s.d.	0.9	0.9	0.9	2.6
		n	11	8	5	9
Single Burials	Juveniles	ave.	-12.5	10.9	-9.7	-8.4
		s.d.	0.0	0.0	0.0	1.5
		n	1	1	1	11
Single Burials	All	ave.	-12.9	8.9	-9.8	-8.7
		s.d.	0.9	1.0	1.0	2.3
		n	28	23	16	33
Mass Burials	All	ave.	-10.7	9.7	-9.4	-9.6
		s.d.	0.2	0.8	0.6	2.7
		n	3	3	6	6

ave. = average of n individuals; s.d. = standard deviation of the mean

dog meat as positive as -12‰ to -16‰, it is clear that maize did not constitute an overwhelming percentage of the whole diet. Although legumes and some leafy vegetables and nuts contain over 20% protein, maize is only about 10% protein, and there is some evidence that plant protein is not assimilated in bone as well as animal protein (*27*). The average value for Cuello human dietary protein, enriched by 3.4‰ relative to the whole diet average, can thus best be explained by the regular consumption of C_4-enriched dog meat and the occasional armadillo. Small but measurable differences between adult males and females in the Preclassic may potentially be explained by greater male consumption of C_4-enriched meat sources or considerably more maize and/or maize beer.

Maya Diet

These results complement the isotopic data available from other Maya sites in Belize (*35, 40, 41*) and demonstrate overall an historic trend towards increasing reliance on maize and other $\delta^{13}C$-enriched food sources, followed by a reduction in the modern era (Figure 1). Data from Maya sites elsewhere in Mesoamerica (*35, 38, 42*) show that residents of the Peten area of Guatemala and in Honduras were considerably more dependent on maize than the Belize Maya, with those at Copan truly subsisting on a corn and beans diet (Figure 2). In contrast to Belize, the wildlife of the Copan valley were rapidly eradicated by growing populations. In the Peten, more meat was

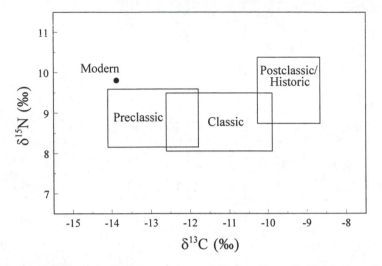

Figure 1. Chronological Comparison of Human Bone Collagen Data (5, 6, 35, 40) from Belize

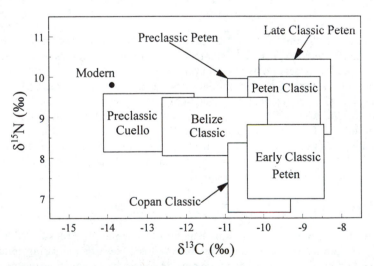

Figure 2. Comparison of Human Bone Collagen Data from Different Maya Regions (5, 6, 35, 38, 40, 42)

consumed, but the residents there were still more dependent on maize than their contemporaries in Belize who had access to a wider range of ecozones and who presumably had a lower population density. Scholars have conflicting opinions over whether dogs were an important food resource for the Maya (*11, 35, 43*), but as the only domesticated animal available in areas with diminished wild game, it is likely that they were eaten particularly when other resources were scarce or out-of-season. If their bones were crushed and boiled to extract marrow, they may be under-represented in the faunal record, or lumped together with other unidentified mammals (*10, 11*). At Cuello, the quantity of dog bones found in midden and domestic contexts with other food remains, the evidence of cut marks and bone crushing, and the average mortality age of one year, all enhance the isotopic interpretation of their regular consumption by the Preclassic Maya.

Conclusion

It would appear that geography and local ecology played the greatest role in determining the diet of the ancient Maya, with relatively minor local differences in terms of gender or status. The historic practice of raising dogs for food is now documented at Preclassic Cuello, and illustrates long-term continuity in this cultural practice. Ethnohistoric accounts of taming or loose-herding deer are not supported by the isotopic data, and reminds us to exercise caution in the use of historic documents to reconstruct ancient Maya lifeways. This study emphasizes the importance of establishing an isotopic baseline from local fauna and flora, of analyzing both bone collagen and bone apatite in humans, and integrating these data with other archaeological information to reconstruct prehistoric diet.

Literature Cited

1   Hammond, N., Ed. *Cuello: An Early Maya Community in Belize*; Cambridge University Press: Cambridge, 1991.

2   Robin, C. *Preclassic Maya Burials at Cuello, Belize;* British Archaeological Reports International Series 480: Oxford, 1989.

3   Robin, C.; Hammond, N. In *Cuello: An Early Maya Community in Belize*; Hammond, N., Ed.; Cambridge University Press: Cambridge, 1991; pp 204-225.

4   Saul, F. P.; Saul, J. M. In *Cuello: An Early Maya Community in Belize*; Hammond, N., Ed.; Cambridge University Press: Cambridge, 1991; pp 134-158.

5   van der Merwe, N. J.; Tykot, R. H.; Hammond, N. In *Proceedings of The Fourth Advanced Seminar on Paleodiet, Banff, Alberta, Canada, September 4-9, 1994*, Ambrose, S.; Katzenberg, A., Eds.; Plenum: New York, in press.

6   Tykot, R. H.; van der Merwe, N. J.; Hammond, N. From Deer to Dogs: A Dynamic View of Subsistence, Status and Sociopolitical Development at Cuello, Belize. Paper presented at the 93rd Annual Meeting of the American Anthropological Association, Atlanta, Georgia, November 30, 1994.

7   Miksicek, C. H.; Bird, R. M.; Pickersgill, B.; Donaghey, S.; Cartwright, J.; Hammond, N. *Nature* 1981, *289*, 56-59.

8   Miksicek, C. H. In *Cuello: An Early Maya Community in Belize*; Hammond, N., Ed.; Cambridge University Press: Cambridge, 1991; pp 70-84.
9   Hather, J. G.; Hammond, N. Belize. *Antiquity* **1994**, *68*, 330-335.
10   Wing, E. S.; Scudder, S. J. In *Cuello: An Early Maya Community in Belize*; Hammond, N., Ed.; Cambridge University Press: Cambridge, 1991; pp 84-97.
11   Clutton-Brock, J.; Hammond, N. *Journal of Archaeological Science* **1994**, *21*, 819-826.
12   Hedges, R. E. M.; Housley, R. A.; Bronk, C. R.; van Klinken, G. J. *Archaeometry* **1991**, *33*, 121-134.
13   Tozzer, A. M., Ed. *Landa's Relacion de las Cosas de Yucatan*. Papers of the Peabody Museum of Archaeology and Ethnology, Harvard University 18: Cambridge, 1941.
14   Allen, G. M. *Bulletin of the Museum of Comparative Zoology, Harvard University* **1920**, *63*, 431-517.
15   Pohl, M. E. D.; Feldman, L. H. In *Maya Subsistence*; Flannery, K. V. Ed.; Academic Press: New York, 1982; pp 295-311.
16   van der Merwe, N. J. *American Scientist* **1982**, *70*, 596-606.
17   van der Merwe, N. J. *Proceedings of the British Academy* **1992**, *77*, 247-264.
18   Ambrose, S. H. In *Investigations of Ancient Human Tissue: Chemical Analyses in Anthropology*; Sandford, M. K. Ed.; Gordon and Breach Science Publishers: Langhorne, PA, 1993; pp 59-130.
19   Sullivan, C. H.; Krueger, H. W. *Nature* **1983**, *301*, 177.
20   Lee-Thorp, J. A. *Paleoecology of Africa* **1986**, *17*, 133-135.
21   Lee-Thorp, J.A.; van der Merwe, N. J. *South African Journal of Science* **1987**, *83*, 712-715.
22   Lee-Thorp, J. A.; van der Merwe, N. J. *Journal of Archaeological Science* **1991**, *18*, 343-354.
23   Krueger, H. W. *Journal of Archaeological Science* **1991**, *18*, 355-361.
24   Lee-Thorp, J. A.; Sealy, J. C.; van der Merwe, N. J. *Journal of Archaeological Science* **1989**, *16*, 585-599.
25   Sillen, A.; Sealy, J. C.; van der Merwe, N. J. *American Antiquity* **1989**, *54*, 504-512.
26   Hare, P. E.; Fogel, M. L.; Stafford, T. W., Jr.; Mitchell, A. D.; Hoering, T. C. *Journal of Archaeological Science* **1991**, *18*, 277-292.
27   Ambrose, S. H.; Norr, L. In *Prehistoric Human Bone. Archaeology at the Molecular Level*; Lambert, J. B.; Grupe, G., Eds.; Springer Verlag: Berlin, 1993; pp 1-37.
28   Krueger, H. W.; Sullivan, C. H. In *Stable Isotopes in Nutrition*; Turnlund, J. E.; Johnson, P. E., Eds.; American Chemical Society Symposium Series 258, 1984; pp 205-222.
29   Klepinger, L. L.; Mintel, R. In *Proceedings of the 24th Archaeometry Symposium*; Olin, J. S.; Blackman, M. J., Eds.; Smithsonian Institution Press: Washington, D.C., 1986; pp 43-48.
30   Ambrose, S. H. *Journal of Archaeological Science* **1990**, *17*, 431-451.
31   Fry, B.; Lutes, R.; Northam, M.; Parker, P. L.; Ogden, J. *Aquatic Botany* **1982**, *14*, 389-398.

32   Fry, B.; Anderson, R. K.; Entzeroth, L.; Bird, J. L.; Parker, P. L. *Contributions in Marine Science* **1984**, *27*, 49-63.
33   Keegan, W. F.; DeNiro, M. J. *American Antiquity* **1988**, *53*, 320-336.
34   van der Merwe, N. J.; Lee-Thorp, J. A.; Raymond, J. S. In *Prehistoric Human Bone. Archaeology at the Molecular Level*; Lambert, J. B.; Grupe, G., Eds.; Springer Verlag: Berlin, 1993; pp 63-98.
35   Gerry, J. P. *Diet and Status among the Classic Maya: An Isotopic Perspective.* PhD Dissertation, Department of Anthropology, Harvard University. University Microfilms International: Ann Arbor, 1993.
36   Chinchilla Mazariegos, O. In *Proyecto Arqueologico Regional Petexbatún*; Demarest, A. A.; Inomata, T.; Escobedo, H.; Palka, J., Eds.; Informe Preliminar #3, Tomo II; Tercera Temporada, 1991; pp 69-122.
37   Emery, K. In *Proyecto Arqueologico Regional Petexbatún*; Demarest, A. A.; Inomata, T.; Escobedo, H.; Palka, J., Eds.; Informe Preliminar #3, Tomo II; Tercera Temporada, 1991; pp 813-828.
38   Wright, L. E. *The Sacrifice of the Earth? Diet, Health, and Inequality in the Pasión Maya Lowlands.* PhD Dissertation, Department of Anthropology, University of Chicago. University Microfilms International: Ann Arbor, 1994.
39   Hamblin, N. L. *Animal Use by the Cozumel Maya.* University of Arizona Press: Tucson, AZ, 1984.
40   White, C. D.; Schwarcz, H. P. *Journal of Archaeological Science* **1989**, *16*, 451-74.
41   White, C. D.; Wright, L. E.; Pendergast, D. M. In *In the Wake of Contact. Biological Responses to Conquest*; Larsen, C. S.; Milner, G. R., Eds.; Wiley-Liss: New York, 1994; pp 135-145.
42   Reed, D. M. In *Paleonutrition: The Diet and Health of Prehistoric Americans*; Sobolik, K. D., Ed.; Southern Illinois University, Center for Archaeological Investigations, Occasional Paper 22: Carbondale, 1994; pp 210-221.
43   Crane, C. J.; Carr, H. S. In *Paleonutrition: The Diet and Health of Prehistoric Americans*; Sobolik, K. D., Ed.; Southern Illinois University, Center for Archaeological Investigations, Occasional Paper 22: Carbondale, 1994; pp 66-79.

RECEIVED August 15, 1995

Chapter 26

Amino Acid Racemization and the Effects of Microbial Diagenesis

A. M. Child

Department of Chemistry, University of Wales, Cardiff, P.O. Box 912, Cardiff CF1 3TB, United Kingdom

Proteins extracted from both bones and teeth found in archaeological contexts are used for radiocarbon dating, amino acid racemization age at death determinations and genetic and dietary studies. This archaeometric information is only reliable when it is certain that the diagenetic changes inherent in the extracted proteins are minimal, or at least, defined. The changes induced in these proteins as a result of decomposition by known microorganisms have been little studied, however, and definitions of the limits of possible diagenetic changes cannot therefore be certain. This paper attempts to address the possibility that amino acid residues in bone collagen are susceptible to an increased rate of aspartic acid racemization due to microbial degradation. Using microorganisms previously isolated from bones and teeth taken from archaeological sources, decomposition studies were done and the results are outlined below.

Bones and teeth are highly specialized composite materials combining both an organic phase, composed mostly of collagen, with a mineral phase, composed chiefly of calcium hydroxyapatite (HAP). They are both composed of Type I collagen, which comprises two different collagen chains that make up the triple-helix; there are two 1(I) chains and one 2(I) chain (*1*). Collagen is characterized by being high in glycine (33%), with proline and hydroxyproline together making 20% of the total residues. HAP $(Ca_{10}(PO_4)_6.(OH)_2)$ is the basis of the inorganic component of mineralized collagen, but it is not stoichiometric with respect to hydroxyapatite, so that the apatite requires substantial amounts of carbonate ions, lesser amounts of pyrophosphate, magnesium, sodium and potassium (*2*). For a comprehensive description of mineralized collagen structure and chemistry, the reader is referred to Hukins (*3*), Miller (*4*) and Triffitt (*5*).

Amino Acid Racemization

The amino acid racemization reaction converts one optically active form of an amino acid into its enantiomer. The reaction is both pH and temperature dependent and

0097–6156/96/0625–0366$12.00/0

typically is measured in the direction of conversion of L- to D- enantiomer, since all higher animals and plants synthesize proteins containing L-amino acids. With time, metabolically isolated proteins will slowly racemize *in vivo* and D-forms will accumulate. Proteins that are not isolated will also form D-amino acid residues, but these are removed and broken down by the liver. Sites where proteins are metabolically isolated include the eye lens, tooth dentine and tooth enamel.

Amino acids present in any of these proteins may be extracted and analyzed for enantiomeric purity, which might yield valuable chronological information (*6*). Aspartic acid is easily extracted from a hydrolyzed protein mixture of amino acids (by anion-exchange chromatography). It has a relatively fast racemization rate, and has been used to estimate the age of fossil bone (*6-8*) for shell proteins (*9, 10*), and for both forensic and archaeological tooth specimens (*11-13*). It has been used to estimate the age at death in human populations within a London crypt burial (*14*). In this latter work, an improved method for the detection of % D-aspartic acid was applied to check the validity of the assumption that the relationship seen between %D-aspartic acid and age in modern teeth can be held to be true for archaeological teeth (*15*). The results of this blind trial application showed that the ages of the archaeological teeth were not close to their expected values, the results falling into two groups with the majority of samples having a higher than expected % D-aspartic acid content. The authors could not present a chemical explanation for this apparent increase in racemization rate, and suggested that the cause may be microbiological.

Wide variation in D/L aspartic acid values in a series of bone samples has been reported (*16*). The samples were assumed to be of a similar age as they were all associated with a suite of ^{14}C determinations from a stratified context in a late Holocene site. The D/L ratios better fitted the amino acid nitrogen content and glycine/glutamic acid ratios and thus reflected the condition of the samples rather than their age. Similar variation has been seen in samples taken from one fossil bone (*17*).

Possible Microbiological Mechanisms of Amino Acid Racemization in Archaeological Material

Microbial Racemases. Bacteria and actinomycetes are the only groups of organisms which contain both D- and L- amino acids as constituents of their proteins during life. To enable them to fulfill this requirement, some bacteria have stereoselective dietary abilities, that is, they will select only one form of an amino acid for ingestion from a mixture of both enantiomers. A few bacteria and actinomycetes have enzymes which enable them to convert one enantiomeric form of an amino acid to the other (*18*). These enzymes are called amino acid racemases and can occur both inside the bacterium on internal membranes or on the external cell wall. Racemases have been the subject of several reviews (*19-21*). Their mode of action is poorly understood, but it is thought to mimic that of the non-enzymic process catalyzed by pyridoxal-5'-phosphate (PLP) because most racemases have PLP as their co-enzyme (*19*). The most commonly reported racemase is alanine racemase, present in a variety of bacteria. Racemases do not always require PLP, but those that do present a high substrate specificity. Generally, non-alanine racemases appear to be less selective in their substrate binding capacity (*22-24*), and catalyze racemization of various types of amino acids except the acidic and aromatic ones.

Microbial Collagenases. Microbial collagenases are able to hydrolyze efficiently the triple helical region of collagen under physiological conditions. The group includes enzymes with a wide range of molecular weights [60 - 130 kilodaltons, (*25*)] and calcium is an essential co-enzyme. Microbial collagenases have active sites along the length of the collagen molecule. These sites are dictated by amino acid sequence rather than by whole molecule configuration. They are metalloproteinases containing a zinc ion at the active site, and can be reversibly inhibited by the removal of this zinc ion using 1,10-phenanthroline at pH 7.5 (26). Collagenases are complexes of several different collagenolytic enzymes, hence defining the sites of action of individual collagenases is laborious, for it is difficult to separate them. The action of one of the six collagenases produced by *Clostridium histolyticum* has been characterized however (27). The sites of action of this collagenase have been shown to be multiple, chain scission occurring between the X- and the Gly- residues in any of the following sequences:

Gly-Y-X-Gly-Pro-Hyp

Gly-Y-X-Gly-Ala-Arg

Gly-Y-X-Gly- Z -Ala

where Y-, X- and Z- are any of the other amino acids present in collagen except glycine. Microorganisms possessing these enzymes are more likely than others initially to decompose collagen. If the collagen structure has been opened up by the action of collagenases, then it is not inconceivable that the rate of racemization could be increased, since at neutral pH values, N-terminal amino acids racemize faster than any other residues (6).

Survival of Mineralized Materials in Archaeological Environments

Bone and tooth taphonomy (that is all degradative changes occurring in a substrate after death) is usually considered in the broadest of terms. Much work is concerned with displacement by ancient humans, animals and environments and also with the gross modification of both bone and tooth by humans for tool production among other needs. The role of microbes in the taphonomy of mineralized collagen is more rarely discussed. It is surprising, perhaps, that bone survives at all since it is a source of both organic (carbon) and inorganic (nitrogen and phosphorus) nutrients for microbes. Laboratory studies using bone as the substrate (Tuross, N. Smithsonian, personal communication, 1993) are typified by rapid demineralization and subsequent digestion of the collagen.

It is the close association of collagen with mineral that is thought to promote the survival of the bone. It is an association so intimate that only the water molecule can enter the spaces between the organic and the mineral, ethanol is too large (28). If these micropore spaces are too small to allow ethanol, it is not likely that collagenases (or proteases) would gain access to the fully mineralized bone, a fact supported by the mesopore protection hypothesis (29). It is well established that demineralization of the matrix must occur before collagen decomposition is initiated (30). To test the premise that microorganisms increase the rate of racemization during decomposition of the organic matrix of bone, decomposition studies were initiated. The microorganisms selected produced a "collagenase", microbial isolation procedures to find an aspartic acid racemase had met with failure.

Experimental

Enzyme Studies. The microorganisms used (see Table I) were isolated from the archaeological bone and soil samples (31), selection was made on the basis of collagenolytic activity. For enzyme characterization, each organism was grown in 10 dm^3 of simple collagen medium (0.2 M tris(hydroxymethyl aminomethane) (TRIS)/HCl buffer, 0.5% glycerol, 0.85% NaCl, 1% collagen, pH 7.5) using a scaling up method. Each stage was incubated at 10 °C for 3 days.

The 10 dm^3 of culture fluid was filtered through a 0.2 µm pore membrane to remove all microorganisms. The filtrate was concentrated by dialysis against 20% polyethylene glycol, (M$_r$ 20000), and passed down a Sephadex G-75 column. The eluent was passed down a Sephadex G-200 column developed with 1.0 M NaCl in 0.2 M TRIS/HCl buffer pH 7.5. The fractions which contained collagenase (determined by azocoll assay (*32*) in the presence of 5 mM N-ethylmaleimide) were pooled, dialyzed against 0.1 M TRIS/HCl 0.1 M NaCl buffer pH 7.5, then lyophilized. These preparations were used to perform the enzyme inhibition tests using azocoll as the substrate in the presence of a variety of protease inhibitors (see Table I). Azocoll is a protease substrate of short peptide chains which have an amino acid sequence similar to collagen. Additionally, there is an azo dye attached to the glycine residues which is liberated when the glycine-X bone is cleaved. The degree of color formation in the supernatant fluid reflects the activity of the enzyme. Whole cell and cell fragment extracts of all the bacteria used in the decomposition studies were also screened for the presence of aspartic acid racemases (*33*).

Bone Cultures. Compact bone was extracted from the metacarpals of freshly slaughtered pigs. Right and left metacarpi from the same animal were used as test and control to reduce the individual variation. The bone was defleshed, cut into pieces, labeled, sampled and gamma-sterilized. The bone pieces were placed into the reaction vessels with sterile collagen medium. Random samples of bone were placed into brain heart infusion broth and incubated at 10°C and 25°C as a check of sterility. All samples checked proved to be sterile under the conditions of this test.

The microorganisms used were grown in the collagen medium at 10°C and added to the bone vessels. Decomposition studies used organisms inoculated onto the bone samples singly and in combination; all combinations comprised at least two different organisms. All cultures were set up in duplicate and were incubated at 10°C. Samples were taken at intervals for testing. Preliminary studies had shown that for ideal decomposition, the culture medium had to be changed regularly. For this reason, variables such as the microbial counts and the concentration of calcium in the culture fluid could not be estimated. The change in the rate of racemization of aspartic acid (*14*) was measured.

The number of aspartic acid residues that were N-terminal were estimated by dansylation, but only for those samples which showed an increased rate of racemization . The powered bone samples were soaked in 10% (v/v) NaOCl, demineralized in 1.0 M HCl and thoroughly washed prior to dansylation by the method of Gray (*34*). This pre-treatment was also used prior to the measurement of the enantiomers of aspartic acid.

Results

Enzyme Inhibition Studies   The results can be seen in Table I. A strong correlation was found in the enzyme inhibition results among microorganisms of the same genus. The fungi gave the most diverse results, which is perhaps to be expected as they were the most disparate group represented. The "collagenases" produced by the unidentified hyphomycetes, *Penicillium chrysogenum* and *Trichinella cerebriformis* were sensitive to the protease inhibitor phenylmethylsulfonyl fluoride, but not to N-ethylmaleimide. Further, 1-10 phenanthroline and ß-mercaptoethanol did not

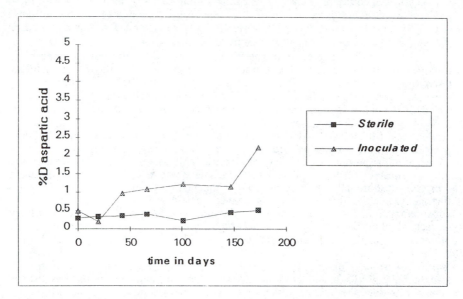

Figure 1. Rate of %D-aspartic acid induction with unidentified hyphomycete III

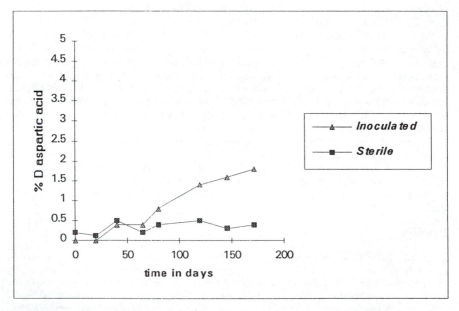

Figure 2. Rate of %D-aspartic acid induction with *Pseudomonas fluorescens* I

completely restrict the action of these enzymes. None of these isolates produced an aspartic acid racemase.

Table I : Enzyme Inhibition Studies

Source of Enzyme	1 mM EDTA	1 mM 1-10-PHE	10 mM β-MCE	5 mM NEM	1 mM PMSF
Clostridium histolyticum	-	-	-	+	+
Pseudomonas fluorescens I	-	-	-	+	+
Pseudomonas aureofaciens	-	-	-	+	+
Pseudomonas fluorescens II	-	-	-	+	+
Pseudomonas fluorescens III	-	-	-	+	+
Pseudomonas fluorescens IV	-	-	-	+	+
Xanthomonas maltophilia	-	-	-	+	+
Unidentified hyphomycete I	-	+	(+)	+	+
Unidentified hyphomycete II	-	+	(+)	+	+
Unidentified hyphomycete III	-	+	(+)	+	+
Trichinella cerebriformis	-	+	(+)	+	+
Penicillium chrysogenum	-	+	(+)	+	+
Penicillium expansum	-	-	-	+	+
Cladosporium species	-	-	-	+	+
Cladosporium cladosporoides	-	-	-	+	+
Fusarium culmorum	-	-	-	+	+

Key: EDTA - ethylenediaminetetraacetic acid; NEM - N-ethylmaleimide; PMSF - phenylmethylsulfonyl fluoride; MCE - mercaptoethanol; PHE - phenanthroline; "-" - no azocoll digestion; "(+)" - limited azocoll digestion; "+" - azocoll digestion

Decomposition Studies. Most of the combinations of microbes resulted in no increase in the rate of aspartic acid racemization. Only those organisms and combinations which did are presented in Table II; the figure given in the table is the %D-aspartic acid at the end of the incubation period. Labeling with dansyl chloride showed aspartic acid residues were N-terminal in the same order as the degree of their racemization, with two exceptions. The rate of production of %D-aspartic acid by various organisms can be seen in Figures 1-4.

Discussion

Enzyme Studies. The number of microorganisms known to produce collagenase is growing (31,35-36); this becomes significant when considering the proposed role of these enzymes in the degradation of archaeological bones and teeth. It must be emphasized that collagenase is not a single enzyme, but a complex of several. To be recognized as a true collagenase, demonstration of resistance to a variety of protease inhibitors is required.

If the enzyme is not resistant to protease inhibitors, it is termed "collagenolytic". It is interesting that an organism proven not to produce a true collagenase still had the ability to induce an increase in the rate of production of %D-aspartic acid. There is little in the literature regarding the collagenolytic action of enzymes other than collagenases, but one enzyme reported is a particular -chymotrypsin (37), which has been shown to

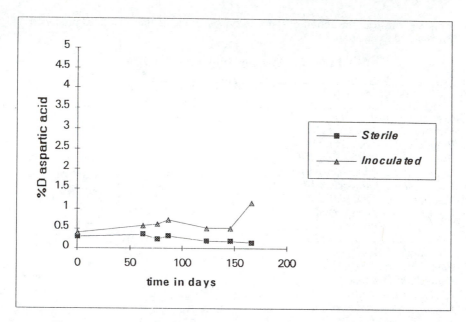

Figure 3. Rate of %D-aspartic acid induction with a *Cladosporium* sp.

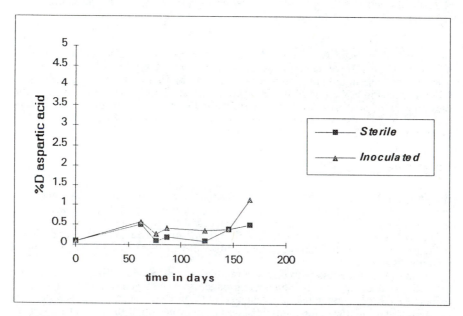

Figure 4. Rate of %D-aspartic acid induction with a *Cladosporium* sp. and *Fusarium culmorum*.

Table II: Increase in the rate of racemization of aspartic acid in bone collagen. Values given are means of at least two estimations.

Organism/Organism Combination	Aspartic Acid % D	Aspartic Acid %N-terminal	Collagenase
Fusarium culmorum + *Cladosporium* sp.	0.81	1.49	+/+
Cladosporium sp.	0.90	1.70	+
Cladosporium sp. + *Pseudomonas fluorescens II*	1.76	<0.04	+/+
Pseudomonas fluorescens I	1.20	<0.04	+
Pseudomonas fluorescens II	0.89	1.53	+
Unidentified hyphomycete III	1.72	1.35	-
Fusarium culmorum + *Penicillium expansum* + *Trichoderma cerebriformis*	0.75	1.10	+/+/-
Escherichia coli + *Penicillium expansum*	1.37	1.08	-/+
Sterile Bone Control	0.30	0.12	

be similar in amino acid sequence at the N-terminal to crab tissue collagenase (*38*). The action of tissue collagenases is different from that of the microbial ones, in that only a single cleavage results from the action of the former, whilst multiple chain scissions result from microbial enzyme action. For the increase in amino acid racemization seen in the decomposition studies with the unidentified hyphomycete III, microbial collagenase-like activity must be assumed.

Decomposition Studies. Taphonomic changes in archaeological bone are the result of the sum of all the reactions present in the burial environment, and these reactions may not all be deleterious. Microbial taphonomy of archaeological bone presents many technical problems; simply recording the microbes present in archaeological materials is not sufficient proof of their implication in the microbial taphonomy of the substrate from which they were isolated. Further laboratory studies are always necessary to confirm or deny this correlation. The organisms used here were selected from a range of microbes isolated from archaeological bones and teeth. They were selected for these decomposition studies because of their ability to produce collagenolytic effects (*31*).

Initial studies showed that axenic microbial decomposition of bone could be followed by monitoring the %D-aspartic acid (*39*). This is not surprising, since the measurement of D amino acids is used as a monitor of food spoilage (*40, 41*) but not for food poisoning. Once the method has been established, the consortium approach used here (i.e., multiple inoculation) is the best way of substantiating the effects of groups of microorganisms. Any change in the established rate of racemization induced by an organism due to the presence of other microorganisms would help to establish the importance of the various microbial species in the decomposition of bone. The effects of mixed microorganisms on racemization are more relevant to archaeological science and the interpretation of racemization results than the effects of one bacterial strain.

Racemization and Dansylation. It is well known that in unmineralized globular proteins, specific aspartic acid residues are prone to racemization and isomerization (*42*). It has been postulated, however, that without chain scission the racemization of residues in intact bone collagen cannot proceed (*43*), due to steric hindrance. With two exceptions (see Table II), the concentrations of aspartic acid that were N-terminal were similar to the concentrations of D-aspartic acid. This supports the mechanism proposed here of racemization by microbial action. That is, the collagen chains are cleaved by enzyme action, and the terminal amino acids racemize either during decomposition or during extraction and measurement procedures. The mechanisms of racemization may vary. Di- and tripeptides are known to racemize via the formation of diketopiperazines (44), but short peptides would be removed during the sample preparation. Succinimide formation is another well-known mechanism for intra-chain aspartic acid isomerization and asparaginyl deamidation in unmineralized peptides and proteins (45). The carbanion intermediate is another proposed mechanism of racemization (46), but amino acids that are terminal racemize faster than those held in the chain (6).

The two exceptions noted above are difficult to explain. If the mechanism above is correct, then scission of the collagen chains must be assumed to achieve an increased racemization rate. If the N-termini were already blocked by some by-product of microbial metabolism, this might explain the low results. This explanation is difficult to realize, since the organisms involved apparently do not produce this effect in all cases. The *Ps. fluorescens* II and the *Cladosporium* sp. alone did not block the dansylation, yet in concert they did. The *Ps. fluorescens* I also blocked the dansylation. Further work is needed here to clarify these results.

Another explanation for the racemization results is possible. There could be contamination of the collagen extract with non-collagenous proteins (NCP) which are high in aspartic acid. Collagen comprises 90% of the bone protein (5), the remaining 10% being the non-collagenous proteins (47). Most of the NCP are associated with the mineral phase and are easily removed from the organic matrix following demineralization (48). In contrast, matrix Gla-protein (MGP), first reported in 1983 (49), is insoluble in water at neutral pH and is extractable from the collagen matrix only by denaturing agents such as urea or guanidine hydrochloride. In adults, it forms 0.1% of the total protein in bone (0.4mg/g dry bone); bovine MGP contains 79 residues, 10 of which (12.7%) are aspartic acid, 5 are γ-carboxyglutamic acid (Gla) and 9 are glutamic acid (50). This small globular protein would be more amenable to microbially-mediated chain scission, racemization and isomerization than the collagen triple helix. The presence of MGP in collagen extracts for racemization studies might present a contamination problem. The method used isolates and concentrates the total aspartic acid fraction. There is insufficient MGP present in the bone to affect the collagen signal, however, until more than 87% of the original collagen has gone. The likelihood of these results being affected by the presence of contaminating NCP is negligible.

Of archaeological significance is the nullification of this effect by the combined activities of a variety of organisms. Further, it is possible that prolonged incubation of these samples would result in the microbial digestion of the exposed (and racemized) parts of the collagen helix, setting the racemization clock back to zero at the end of the early diagenesis phase.

Microbial Interactions (Competition).   In nature (and in the laboratory), microbes are in constant competition with each other to succeed in a given environment. They achieve this by producing metabolic by-products which interfere with the metabolism of related (and non-related) organisms present in the environment. It is not unusual, therefore, for the effects of a single microbe to be nullified when it is present in a mixture of organisms.

This is borne out here by the fact that singly, the two *Pseudomonas fluorescens* species produced an increase in the rate of racemization of aspartic acid which was not sustained in concert.. Interestingly, the *Cladosporium* species acting alone induced an increase in the rate of racemization (see Figure 3) not nullified by *Fusarium culmorum* (see Figure 4) or one of the *Pseudomonas fluorescens* strains (not shown) though it was by the other and by several fungi also.

Conclusions

No simple relationship expressed the decomposition of collagen by a consortium of microorganisms. It is important that this effect was sometimes nullified by the effects of mixed cultures. This may help to explain the successful application of amino acid racemization for chronological assessment, given that one bacterium is able to induce amino acid racemization in the aspartic acid residues of insoluble collagen. Other microorganisms, some not producing collagenases, are also able to induce an increase in the rate; some of these organisms induce racemization whether in single culture or mixed. A few of these organisms, however, do not have "true" collagenases, therefore the link between induced racemization and collagenase action cannot be substantiated.

Acknowledgments

I would like to thank Prof. R. D. Gillard and Prof. A. M. Pollard for their helpful discussions throughout this project. I would also like to thank Dr. M. J. Collins for his helpful suggestions and comments, and Dr. J. Richter and Prof. D. Brothwell for their assistance with archaeological bone sample collection. Thanks are also due to Dr. J. Morgan and staff in Bacteriology at the Cardiff Royal Infirmary. Finally, I would like to thank NERC for the funding (Grant Nos. GR3/9571'A' and B/92/RF1537).

Literature Cited

1. Kucharz, E. J. *The Collagens: Biochemistry and Pathophysiology*; Springer-Verlag: Berlin, 1992.
2. Misra, D. N. In *Methods of Calcified Tissue Preparation*; Dickson, G. R., Ed.; Elsevier: Amsterdam 1984.
3. Hukins, D. W. *Calcified Tissue: Topics in Molecular and Structural Biology*; Macmillan: London, 1989.
4. Miller, E. J. In *Biology, Chemistry and Pathology of Collagen*; Fleischmajor, R.,; Olsen, B. R.; Kuhn, K., Eds.; *Ann. NY. Acad. Sci.* **1985**, *460*, 1-13.
5. Triffitt, J. T. In *Fundamental and Clinical Bone Physiology*; Urist, M. R., Ed.; Lippincott: Philadelphia, 1980.
6. Bada, J. L. *Ann. Rev. Earth Planetary Sci.* **1985**, *13*, 241-268.
7. Bada, J. L.; Hermann, B.; Payan, I. L.; Man, E. H. *Appl. Geochem.* **1989**, *4*, 325 - 327.
8. Csapo, J.; Csapo-Kiss, Zs.; Kolto, T.; Papp, I. In *Archaeometry '90*; Pernicka, E.; Wagner, G. A., Eds.; Birkhauser Verlag: Basel 1990; pp 627-635.
9. Masters, P. M.; Bada, J. L. In *Archaeological Chemistry II*; Carter, G. F., Ed.; Advances in Chemistry, American Chemical Society: Washington D.C.; 1978; pp 117-137.
10. Goodfriend, G. A. *Nature* **1992**, *357*, 399-401.
11. Ogino, T.; Ogino, H.; Nagy, B. *Forensic Sci. International* **1985**, *29*, 259-267.
12. Ohtani, S.; Yamamoto, K. *J. Forensic Sci.* **1991**, *36*, 792-800.
13. Masters, P. M. *Calc. Tissue International* **1983**, *35*, 43-47.
14. Gillard, R. D.; Pollard, A. M.; Sutton, P. A.;Whittaker, D.;K. *Archaeometry* **1990**, *32*(1), 61-70.

15. Gillard, R. D.; Hardman, S. M.; Pollard, A. M.; Sutton, P. A.;Whittaker, D. K. In *Archaeometry '90*; Pernicka, E.;Wagner, G. A. Eds.; Birkhauser Verlag: Basel, 1990 pp 637-644.
16. Taylor, R. E.; Ennis, P.J.; Slota, P.J.; Payen, L.A. *Radiocarbon* **1989**, *31*(3), 1048-1056.
17. Kimber, R.W.L.; Hare, P.E. *Geochim. Cosmochim. Acta* **1992**, *56*, 739-743.
18. Rose, I. A. In *The Enyzmes*; Boyer, P. D., Ed.; 3rd Ed., vol 2; Academic Press: London, 1970, pp 281-320.
19. Davis, L.; Metzler, D. E. In *The Enzymes 7*; Boyer, P. D., Ed; Academic Press: London, (1972), pp 33-74.
20. Soda, K.; Tanaka, H.; Tanizawa, K. In *Vitamin B₆ Pyridoxal phosphate; Chemical, Biochemical and Medical Aspects. Part B*; Dolphin, D.; Poulson, R.;Avramovic, O. Eds.; Wiley and Sons: New York, 1986; pp 223-251.
21. Soda, K.; Tanizawa, K.; Esaki, N.;Tanaka, H. In *Biochemistry of Vitamin B₆;* Birkhauser-Verlag: Basel, 1987; pp 453-444.
22. Asano, Y.;Endo, K. *Appl. Microbiol. Biotechnol.* **1988**, *29*, 523-527.
23. Asano, Y.; Endo, K.; Osanori, N.; Sei, K. *Jap. Kokai Tokk. Koho JP* **1988**, *63*, 126.
24. Inagaki, K.; Tanizawa, K.; Tanaka, H.; Soda, K. *Agric. Biol. Chem.***1987**, *51*, 173-180.
25. Bond, M. D.; Van Wart, H. E. *Biochemistry* **1984** *23*, 3085-3091.
26. Angelton, E. L.; Van Wart, H. E. *Biochemistry* **1988**, *27*, 7413-7418.
27. Steinbrink, D. R.; Bond, M. D.; Van Wart, H.;E. *J. Biol. Chem.* **1985,** *260*, 2771-2776.
28. Lees, S. In *Calcified Tissue: Topics in Molecular and Structural Biology;* Hukins, D. W. Ed.; Macmillan: London 1989; pp153-173.
29. Mayer, L.M. *Geochim. Cosmochim. Acta* **1994**, *58*(4), 1271-1284.
30. Krane, S. In *Dynamics of Connective Tissue Macromolecules*; Burleigh, P.M.C.; Poole, A.R. Eds.; North Holland Publishing Co: 1974 pp 309-326.
31. Child, A.M. *J. Arch. Sci.* **1995**, *22*, 165-174.
32. Peterkovsky, B. In *Methods in Enzymology; Extracellular Matrix 82*(A) Academic Press: London, 1982, pp 453-471.
33. Child, A. M.; Gillard, R.;D.; Pollard, A. M. *J. Arch. Sci.* **1993**, *20,* 159-168.
34. Gray, W. R. In *Methods in Enzymology 25,* Academic Press: London, 1972, pp 121-138.
35. Seifter, S.; Harper, E. In *Methods in Enzymology 19,* Academic Press: London, 1970 pp 613-634.
36. Vrany, B.; Hnatkova, K.; Letti, A. *Folia Microbiologicka* **1988**, *33*, 458-461.
37. Tsai, I. H.; Lu, P-J.; Chang, L-J. *Biochim. Biophys. Acta* **1991**, *1080*, 59-67.
38. Welgus, H. G.; Grant, G.bA.; Jeffery, J. J.; Eisen, A. Z. *Biochemistry* **1982**, *21,* 5183-5189.
39 Child, A. M. in preparation for *J. Arch. Sci.*
40. Friedman, M. In *Nutritional and Toxicological Consequences of Food Processing,* Friedman, M.; Ed.; Plenum Press: NewYork; 1991, 447-481.
41. Palla, G.; Marchelli, R.; Dossena, A.; Casnati, G. *J. Chromatogr.* **1989**, *475*, 45-53.
42. Fujii, N.; ishibashi, Y.; Satoh, K.; Fujino, M.; Harada, K. *Biochim. Biophys. Acta,* **1994,** *1204*, 157-163.
43. Julg, A.; Lafont, R.; Perinet, G. *Quartern. Sci. Rev.* **1987**, *6*, 25-28.
44. Steinberg, S. M.; Bada, J. L. *J. Org. Chem.* **1983,** *48*, 2295-2298

45. Capasso, S.; Kirby, A. J.; Salvadori, S.; Sica, F.; Zagari, A. *J. Chem. Soc. Perkin Trans.* **1995**, 2, 437-442.
46. Bada, J. L. *Interdisciplinary Sci. Rev.* **1982**, 7, 30-46.
47. Young, M. F.; Kerr, J. M.; Ibaraki, K. Heegaard, A-M.; Robey, P.G. *Clin. Orth. Rel. Res.* **1992**, 281, 275-294.
48. Linde, A. *Anat. Rec.* **1989**, 224, 154-166.
49. Price, P. A.; Urist, M. R.; Otawara, Y. *Biochem. Biophys. Res. Comm.* **1983**, 117, 765-771.
50. Price, P. A.;Williams, M. K. *J. Biol. Chem.* **1985**, 260, 14971-14975.

RECEIVED October 24, 1995

Chapter 27

Ancient DNA in Texas Rock Paintings

Ronnie L. Reese[1,4], Elmo J. Mawk[1], James N. Derr[2], Marian Hyman[1],
Marvin W. Rowe[1,5], and Scott K. Davis[3]

[1]Department of Chemistry, [2]Department of Veterinary Pathobiology,
and [3]Department of Animal Science, Texas A & M University,
College Station, TX 77843

The source of binder/vehicle(s) used by prehistoric North American artists to prepare their paints was previously unknown. Little DNA is expected to survive after several millennia; but even minute quantities of degraded DNA can be amplified by polymerase chain reaction and sequenced. An ancient fragment of pictograph *histone* DNA from two *ca.* 3,000 to 4200 year old Pecos River style pictograph samples was sequenced and compared to varied plant and animal sources to establish its origin. The sequences revealed that it was from an ungulate (even-toed, hoofed mammal). To our knowledge, this is the first biochemical determination of a binder/vehicle in a New World pictograph. We are sequencing the mitochondrial cytochrome c oxidase subunit II gene which should provide enough sequence variability to narrow the origin of the DNA in the pictograph to species.

All researchers involved in radiocarbon dating would agree that it is important to know the composition and source(s) of the organic matter being dated. We, in the Chemistry Department at Texas A&M University have developed a technique that, when combined with accelerator mass spectrometry, allows ancient pictographs to be ^{14}C dated (*1*). At present, we are dating whatever organic material has survived in the pictograph sample, without knowledge of its chemical nature. A review of the literature shows that although many organic substances have been suggested as possible binders and vehicles for prehistoric paints, e.g., blood, egg whites and yolks, seed oils, plant resins, plant juices, milk, urine and honey, no chemical or biochemical analyses have demonstrated their presence (*2*). We are now working to determine the exact

[4]Current address: Department of Chemistry, Southeastern Louisiana University, Hammond, LA 70402
[5]Corresponding author

0097–6156/96/0625–0378$12.00/0

chemical content of paint from the Pecos River style that we have dated. This may allow separation of contaminants that are detrimental to radiocarbon dating. Improvement of the dating technique is only one reason for studying the organic component of paints used by a prehistoric society. The technology involved in producing these enduring polychrome paints is also of interest.

In the past, rock art research had been relegated to a marginal position with regard to mainstream archaeology. That situation began to change due to two primary causes: (1) development of methods to directly radiocarbon date pictographs, at our laboratory and elsewhere (*1*) and (2) resurgence of interest in prehistoric symbolic, religious and artistic systems (e.g., *2*, *3*). Lewis-Williams and his colleagues have been successful in demonstrating that much rock art is ritualistic, depicting experiences while in an altered state of consciousness (*4-8*). Turpin (*9*) concluded from an empirical examination of imagery in Pecos River style pictographs that some of this art is also ceremonial. Thus, the choice of paint components themselves may have been dictated by ritualism and are of interest in that regard. But whether or not the rock art is deemed to be ritualistic, in any attempt to understand the behaviors and beliefs of ancient societies, it is essential to study their art as well as other artifacts.

The watersheds of the Pecos, Devils and Rio Grande rivers were inhabited from 10,500 years ago (*10, 11*) until after the arrival of the Spanish. Dissolution shelters line the walls of the canyons and contain the largest collection of rock art sites in the Hemisphere. Some were occupation shelters; others show no evidence of habitation. Polychromatic pigments were used in the oldest pictograph style, the Pecos River: dark to light reds, black, yellow and orange pigments from iron and manganese oxides and hydroxides. The rare white pigments have not been chemically characterized. Organic binder/vehicle(s) were added to the pigment, but there is no ethnographic evidence as to what these were. In general, the organic binder/vehicle(s) used in prehistoric native American paints were largely speculative. We now report a biochemical study of these components.

Six examples of Pecos River style pictographs have been dated by our laboratory and yielded ages of ~3,000 to >4,320 before present (BP) (see ref. *1*). Samples from two Pecos River style pictographs were collected for DNA analysis to aid in the identification of the organic binder/vehicle(s). Under favorable environmental conditions, DNA fragments have provided useful archaeological and geological genetic information (*14-33*). Although it was thought that DNA should have totally degraded after several millenia on a limestone wall, an outer accretionary mineral coating may have protected it from the effects of atmosphere and weather. The microstratigraphy of a pictograph on a rock surface is illustrated by the polished section shown in Figure 1. Three principal layers are: (1) limestone substrate, (2) pigment, and (3) naturally occurring accretion. We (*1*) identified whewellite ($CaC_2O_4 \cdot H_2O$) and calcite ($CaCO_3$) in the accretion layer of another pictograph from the same site.

We extracted ancient DNA from the two pictographs and used PCR to acquire multiple copies of a 106 base pair fragment from the highly conserved *histone 4* gene. The Sanger (*34*) dideoxy method was used to sequence this segment. These sequences were phylogenetically compared to known *histone 4* sequences (*35*) to determine the origin of the organic binder/vehicle. Because PCR can theoretically amplify even a single copy of the targeted

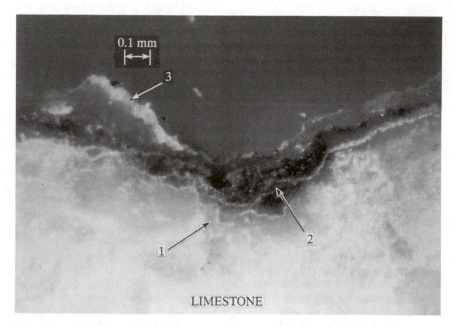

Figure 1. Polished section for the red-pigmented sample taken from site 41VV75. There are three principal layers visible: (1) limestone substrate, (2) red pigmented layer (hematite), and (3) naturally occuring mineral layer ($CaC_2O_4.H_2O$ and $CaCO_3$) which probably protected the DNA from total degradation.

DNA fragment, contamination is a severe problem. To guarantee that we were amplifying ancient DNA in the paint fragments, and not post-collection contamination DNA from other sources, we used negative controls (no DNA added) and positive controls (known DNA). We feel confident that our sequences are of ancient DNA fragments from the painted rock samples. Results indicate the presence of material from an ungulate (even-toed hoofed mammals).

We have now begun an investigation of a more polymorphic region of DNA that should produce a fragment that provides species identification. Mitochondrial DNA was chosen for this study. Aside from nuclei, mitochondria are the largest organelles in animal cells and are found in all cells of higher plants and animals. Mitochondrial DNA codes for some of the proteins used in energy production. It evolves over time within a given species at approximately ten times the rate of nuclear DNA and thus we anticipate that mitochondrial DNA will provide the necessary sequence variation for determination of the paint binder source. Mitochondrial DNA was also selected because each cell contains hundreds to thousands of mitochondria. Therefore, the probability of finding a mitochondrial gene fragment may be greater than for finding a single copy nuclear gene fragment in a pictograph after several millennia on an exposed wall.

The pictograph samples we selected for this study came from a Seminole Canyon State Historic Park rock shelter, 41VV75. That site was chosen since it is undergoing severe natural exfoliation. In one section of the shelter, less than half of the original painted surfaces remain and little further damage is introduced by judicious sampling. Deleterious visual effects to the pictures are minimized by choosing paint fragments that appear on the verge of exfoliating. The samples were taken from high on the wall, and thus were less likely to be contaminated by humans or animals. We wore rubber gloves and gently levered each fragment from the wall onto aluminum foil, wrapped the samples and placed them into a plastic bag. At the laboratory, we photographed the samples, re-wrapped them in fresh aluminum foil and placed them into an argon filled desiccator until ready for analysis.

Experimental Method - Histone 4 Gene.

To circumvent the lack of knowledge about the binder/vehicle origin, we selected eukaryotic *histone 4* gene regions as candidates for the first PCR/DNA analysis. The length of DNA in eucaryotic cells is so great that problems of entanglement and breakage become severe. Histones, proteins unique to eucaryotes, have evolved to bind to DNA and wrap it into a compact structure known as chromatin. All eucaryotes have histone bound to their DNA. The importance of the histones is reflected in the fact that they are remarkably similar from one species to the next. They also occur in multiple copies. Bacterial genomes contain no homologue to eukaryotic histone genes. Thus, the choice of *histone 4* genes provided an additional benefit of eliminating bacteria as a potential source of nucleotide contamination. The *histone 4* family was chosen over other histone genes because of the amount of DNA sequence data on *histone 4* of numerous species.

Primers, short segments of DNA, are needed as starting points for polymerase to begin synthesis of the targeted DNA template. Primers exactly complementary to the precursor region of targeted DNA fragment are desirable, but imperfect primer-template matches can work if the first six to nine bases are compatible. Typically, knowledge about the genetic target sequence to be amplified exists. In our case having neither ethnographic evidence nor previous knowledge as to the source(s) of ancient organic binder/vehicle(s) made primer selection more difficult. To circumvent this problem, we constructed universal eukaryotic primers that would detect and amplify ancient DNA of either plant or animal origin. We used GenBank data and MacVector 3.5 Sequence Analysis software (IBI/Kodak, New Haven, CT) to find the best fit alignment for the *histone 4* primer construction by comparing every published sequence of *histone 4*. The published sequences for the *histone 4* contained in the GenBank database are given in Table I. Following alignment, both forward and reverse primers were chosen based on oligonucleotide regions that were similar among all these sequences. Our *histone 4* forward and reverse primers bracket a 106 base pair region. Sequences for the *histone 4* gene primers are as follows:

forward   5'- CGC ATC TCC GGC CTC ATC TAC GAG GA - 3';
and reverse   5'- GTA GAG TGG GCG GCC CTG GCG CTT GA - 3'.

Both primers were synthesized on an Applied Biosystems, Inc., Oligonucleotide 392 RNA/DNA synthesizer. To verify their potential as broad based PCR primers, they were tested with PCR amplifications of *Bos taurus* (bovine), *Lycopersicon lycopersicum* (Better Boy tomato), *Homo sapiens* (human), and *Gallus gallus* (chicken) DNA. Each of these were found to amplify with the primers selected. Of these, only the bovine DNA fragment was later sequenced.

Ancient DNA present in the two 41VV75 samples (A and B) was extracted by conventional methods (*36-38*). The DNA solution was filtered through a Millipore 30,000 MW filter by centrifuging for 20 minutes at 10,000 rpm (Savant Microfuge HSC 10K). The DNA was recovered in a final volume of 50 µl and stored at 4°C.

Table I. *Histone 4* Genes from GenBank and from Our Work.

Common Name	Scientific Name	GenBank Accession Number
Human	*Homo sapiens*	X00038
House Mouse	*Mus muscalus*	J00422
Fish	*Tilapia niloticta*	X54078
Maize	*Zea maize*	M13370
Wheat	*Traiticum aestivum*	X00043
Neurospora	*Neurospora crassa*	X01611
Frog	*Xenopus laevis*	X00224
Chicken	*Gallus gallus*	J00866
American Bison	*Bison bison*	
Cattle	*Bos taurus*	
Goat	*Capra hircus*	
Rabbit	*Sylvilagus floridauus*	
White-tailed Deer	*Odocoileus virginanus*	
Better Boy Tomato	*Lycopersicon lycopersicum*	

SOURCE: Adapted with permission from ref. *39*. Copyright 1995.

PCR amplifications were performed using a Perkin Elmer-Cetus thermal cycler with the following protocol for one PCR cycle: denature DNA at 93°C for 0.5 minute, anneal primers at 50°C for 1 minute, extend template at 72°C for 1 minute. A total of 35-40 cycles was typical. Negative controls (no DNA added) and positive controls (known DNA added) were incorporated into every PCR experiment. Amplified PCR products were separated on 4% low melting Nusieve GTG agarose gels (FMC Bioproducts) by electrophoresis. Afterward, the gels were stained with ethidium bromide and the DNA bands visualized by UV fluorescence. The PCR products were extracted from the gels using sterile razor blades. The excised PCR bands were purified using the Magic PCR Preps DNA purification system kit (Promega) and PCR products were stored at -20°C. The PCR products were ligated, transformed and cloned using the TA cloning kit (Invitrogen) following the manufacturer's directions. Both strands of multiple clones were sequenced using the Sanger dideoxy method (*34*). Greater experimental detail can be found in a previous publication (*39*).

Results and Discussion - Histone 4 Gene.

The *histone 4* sequences determined for the two pictograph DNA samples are listed in Figure 2. These pictograph sequences were compared to aligned sequences found in GenBank (the first eight species listed in Table I, five different classes of vertebrates plus two plants and a fungus). Preliminary phylogenetic analysis using PAUP 3.1.1 (*35*) demonstrated that the 41VV75A and B sequences were most closely aligned with mammals (*39*).

To investigate the comparison with mammals more closely and further delineate the possible source of the organic binder/vehicle in the pictographs, we sequenced the same histone region of two likely Lower Pecos River area mammal fauna candidates: white-tailed deer, *Odocoileus virginanus*, and American bison, *Bison bison*. In addition, we sequenced cattle, *Bos taurus*, DNA to exclude it as a source of contamination, as bovine DNA is frequently handled in this laboratory. Each was subjected to similar PCR amplifications, cloning, and sequencing steps as described previously. Sequences found are listed in Figure 2. These sequences, plus selected (human, *Homo sapiens*; mouse, *Mus muscalus*; maize, *Zea maize*; wheat, *Traiticum aestivum*; chicken, *Gallus gallus*; sea urchin, *Paracentrotus lividus*; trout, *Salmo gairdneri*) GenBank sequences, were subjected to phylogenetic analysis, again using PAUP 3.1.1 (*35*). Results of a bootstrap majority-rule consensus tree for the 106 base pair PCR products as well as groups selected from GenBank are shown in Figure 3. We resolved the possibilities for the material added as binder/vehicle from kingdom down to the Order Artiodactyla. Order Artiodactyla contains the even-toed ungulates: e.g., cattle, sheep, goats, bison, deer, antelopes, camels, giraffe, llama and hippopotami. *Histone 4* gene fragments amplified from the Pecos River style rock art were most similar to bison.

Though the binder/vehicle used at 41VV75 contains material from a mammal phylogenetically related to bison and deer, the highly conserved 106 base pair fragment amplified from the *histone 4* did not provide the degree of resolution needed to make a decisive species identification. This limitation is not surprising given the conservation of this genetic region, and that the primers were chosen to amplify both animal and plant DNA.

41VV75A   5'- G   ACT   CGT   GGG   GTG   CTG   AAG   GTG   TTT
                   CTG   GAA   AAT   GTG   ATC   CGG   GAC   GCG
                   GTC   ACC   TAC   ACG   GAG   CAC   GCC   AAA
                   CGC   AAG   ACT   GTA   ACC   GCT   ATG   GAC
                   GTG   GTT   TAC - 3'

41VV75B   5'- G   ACC   CGT   GGG   GTG   CTG   AAG   GTG   TTT
                   CTG   GAA   AAT   GTG   ATC   CGG   GAC   GCG
                   GTC   ACC   TAC   ACG   GAG   CAC   GCC   AAA
                   CGC   AAG   ACT   GTA   ACC   GCT   ATG   GAC
                   GTG   GTT   TAC - 3'

Bison   5'- G   ACC   CGT   GGG   GTG   CTG   AAG   GTG   TTT
                   TTG   GAG   AAC   GTG   ATC   CGG   GAC   GCG
                   GTC   ACC   TAC   ACC   GAG   CAC   GCC   AAG
                   CGC   AAG   ACT   GTC   ACC   GCC   ATG   GAT
                   GTG   GTG   TAC - 3'

Cattle   5'- G   ACC   CGC   GGG   GTG   CTG   AAG   GTG   TTC
                   CTG   GAG   AAT   GTG   ATC   CGG   GAT   GCA
                   GTC   ACC   TAC   ACC   TAG   CAC   GCC   AAG
                   CGC   AAG   ACT   GTC   ACC   GCC   ATG   GAC
                   GTG   GTC   TAC - 3'

Deer   5'- G   ACG   CGC   GGC   GTC   CTG   AAA   GTG   TTT
                   CTG   GAG   AAT   GTG   ATC   CGG   GAT   GCA
                   GTC   ACC   TAC   ACC   GAG   CAT   GCC   AAG
                   CGG   AAG   ACT   GTC   ACC   GCT   ATG   GAT
                   GTG   GTG   TAC - 3'

Goat   5'- G ACC   CGC   GGG   GTG   CTG   AAG   GTG   TTC
                   TTG   GA?   AAT   GTG   ATC   CGG   GAT   GCA
                   GTT   ACC   TAC   ACA   GAG   CAC   GCC   AAG
                   CGC   AAG   ACT   GTC   ACC   GCC   ATG   GAC
                   GTG   GTC   TAC - 3'

Rabbit   5'- G   ACC   CGT   GGC   GTG   CTC   AAG   GTC   TTC
                   CTG   GAG   AAC   GTC   ATC   C?C   GAC   GCT
                   GTC   ACC   TAC   ACG   GAG   CAC   GCC   AAG
                   CGC   AAG   ACG   GTC   ATG   GC?   ATG   GAC
                   GTG   GTG   TAC - 3'

Figure 2. Aligned *histone 4* gene sequences for the ancient DNA fragments extracted from pictograph samples 41VV75A and 41VV75B, *Bison bison* (American bison), *Bos taurus* (cattle), *Odocoileus virginanus* (white-tailed deer), *Capra hircus* (goat) and *Sylvilagus floridauus* (rabbit). (Reproduced with permission from ref. *39*. Copyright 1995 Academic Press).

The lack of detection of plant DNA in the paint does not totally elimi-nate the possibility that plant matter was used; over time added plant matter may have degraded to the extent that it was not amplified. However, this study indicates that ancient Lower Pecos people most likely used non-human, mammalian products ~3,000 to 4,300 years ago to manufacture paints for this style of pictograph. Ungulate bone marrow, with its high concentration of DNA, would be a good candidate for a paint binder/vehicle that would retain traces of DNA after several thousand years.

Experimental Method - Cytochrome c Mitochondrial DNA.

We chose the mitochondrial gene, cytochrome c oxidase subunit II (COII), 684 base pairs long, to continue our search for the organic material used in the pictograph paints. COII is found in the mitochondria of all eukaryotic cells and should therefore be highly variable, variable enough to differentiate between species. Bacteria, an ever present possible source of DNA contam-ination, lack mitochondria and other organelles. Therefore, bacterial DNA present will not be amplified.

The paint sample to be used for this work was collected from the same site as the previous samples, shelter 41VV75, using the collection procedures described above. As before the sample was taken high on the shelter wall to minimize the chance of contamination from human and local fauna by casual contact. This sample was brought back to the laboratory, photographed, placed in a dessicator, and stored in a 4°C cold room. Having determined that the sequence obtained from the ancient samples, 41VV75A and B, likely came from an ungulate, Order Artiodactyla, the fauna we chose for further examina-tion were: white-tail deer (*Odocoileus virginianus*); mule deer (*Odocoileus hemionus*); elk (*Cervus elaphus*); American bison (*Bison bison*), pronghorn antelope (*Antilocapra americana*), javalina (*Tayasu tajacu*) and rabbit (*Oryctolagus cuniculis*). Several bison teeth (from *Bison antiquus*, and *Bison occidentalis Bison bison*), taken from Bonfire shelter (41VV218), the earliest known jump kill site, will be used as a direct source of ancient bison DNA. There are three time periods in which bison bones were present at Bonfire Shelter: the earliest occurred in ~12,500 years old strata; the second at ~10,500 years BP; and the last at ~2,500 years BP (*40*). Thus we can obtain COII sequences for bison that were present long before and nearer the time the pictograph was painted. Of interest is whether the larger, extinct *Bison antiquus* or *Bison occidentalis* was the ancestor stock for the smaller modern species, *Bison bison*. Pronghorn antelope was sequenced by Miyamoto *et al.* (*41*). Javalina and rabbit were included because they are now present in the area and rabbits were eaten by the ancient Lower Pecos River Indians (*42*). Unlike the situation with *histone 4*, it was necessary in this case to sequence the entire 684 base-pair (bp) COII gene to identify conserved regions of COII for primer development. After getting these sequences, we will then select a variable region of this gene <200 bp long surrounded on both ends by suit-able primer regions to sequence the pictograph DNA. Elk, mule deer, white tail deer and javalina sequences remain to be sequenced; the others are now available. These known sequences will then be used to design a primer set for use with the paint sample, after which the pictograph sample will be sequenced.

Primers and protocol for COII were developed previously for Order Artiodactyla (Honeycutt, R., Department of Wildlife and Fisheries Sciences, personal communication, 1993). The primers used were H-8086, L-7338, L-7251 and L-7252 which will be used to determine the entire 684 bp sequence of the COII gene for the species selected above. These primers will not be used on the paint sample because it is extremely unlikely that any DNA fragment >200 bp has survived. H-8086 is used in conjunction with either L-7338, L-7251 or L-7252 because some species will not anneal with the same primer, so a different primer in the same area is used. H-8086 is located in the tRNA^{LYS} gene and its sequence is 5'-CTC TTA ATC TTT AAC TTA AAA G-3'. L-7338 is located in the tRNA^{ASP} gene with the sequence 5'-AAC CAT TTC ATA ACT TTG TCA A-3'. L-7251 and L7252 are located in tRNA^{SER}. L-7251 and L-7252 have the following sequences, respectively:

5'-AAG AAA GGA   AGG AAT CGA ACC C-3'
5'-AGA AAG GA(G/A) GGA ATC GAA CCC C-3'.

L-7251 and L-7252 are shifted by one bp with respect to each other. This shift is sometimes necessary to get adequate amplification. In some cases, a primer will work better if it is, in reality, a mixture of two primers that have identical sequences except in one position. L-7252 is an example of this where the ninth base pair position (shown as (G/A) is a 50-50% mixture of the two bases, guanine and adenine.

We modified Honeycutt's thermocycling protocol yielding our three-step protocol as follows. (1) Solutions are held at 93°C for one minute to denature the DNA fragments. (2) They are then annealed at 45-50°C for one minute, and adjusted for maximum amplification. Occasionally, annealing temperatures between 50-60°C are used to minimize non-specific annealing that occurs when primer binds to sites that have similar bp sequences on the template DNA. During annealing, complementary bases on the DNA and the primer align and bind chemically by forming hydrogen bonds. When there is a perfect match of bases between primer and DNA, the maximum possible number of hydrogen bonds are formed, three for guanine and cytosine base pairing, two for adenine and thymine base pairing. These hydrogen bonds stabilise the complex between primer and template DNA. Mis annealing occurs when the primer attachs at the wrong site on the DNA. Thus when mis annealing occurs, not all the bases in the primer match perfectly; and the binding is not as strong as with a perfect match. After PCR amplification, mis annealing is detected by the presence of multiple bands of PCR product on the agarose gel. Mis annealing can be greatly reduced or even eliminated by raising the annealing temperature. Since a mis annealed primer does not have as many bases matching those of the DNA, it is more weakly bonded to the DNA. By raising the annealing temperature, the mis annealed primer/DNA complex becomes unstable and disassociates. The correct primer/DNA complex, which has the maximum possible hydrogen bonds, remains a stable product which is amplified. (3) Finally a linear temperature ramp takes the temperature from annealing temperature to extension temperature of 72°C, that is maintained for one minute. A total of 30-35 cycles of PCR is used. Negative controls (no DNA added) are run with all PCR amplifications.

Following PCR amplification, products are examined on a 1% agarose gel using a λ standard in blue tracking dye. The standard contains a mixture of DNA fragments of known bp length that are used to estimate bp lengths of PCR products. PCR products are visualized with ethidium bromide, which binds to DNA and fluoresces under UV irradiation. The PCR product of appropriate molecular weight is chosen for sequencing and is purified using QIAquick Spin PCR Purification Kit (QIAGEN Inc.) before cycle sequencing.

PCR products are directly sequenced, using four different dye-labeled nucleotides. These nucleotides are mixed with unlabeled nucleotides. When polymerase adds dye-labeled nucleotide to a growing DNA strand during extension, the label prevents further addition of nucleotides. Cycle sequence products are purified for sequencing using CentriSep Spin Column Kit (Princeton Separations). The sequence is read with an automated 373A DNA sequencer, with data storage on a MacIntosh Centris 650 (Applied Biosystems). The cycle sequencing kit used is the Prism Ready Reaction DyeDeoxy Terminator Cycle Sequencing (Applied Biosystems). The final product is two sequences, one for each primer used in the PCR reaction that amplified the template DNA. The two overlapping sequences, in this case for COII, are aligned using PAUP 3.1.1 (*35*) against a known sequence, the human COII gene.

Results and Discussion - Cytochrome c Mitochondrial DNA.

Sequence data for the 684 bp COII gene is being collected on human, domestic cow, mule deer, white-tail deer, pronghorn, rabbit and buffalo. Domestic cow, rabbit and human sequences were obtained from GeneBank, pronghorn antelope from (Honeycutt 1994). We are sequencing mule deer, white-tail deer and contemporary and ancient bison. From the data collected to date, the COII sequence appears to be sufficiently variable to give useful information about the source species of the paint binder. Figure 4 shows a portion of the COII sequence for all species sequenced, showing the variability between species.

The white-tail deer sequence we obtained so far still has 304 bps missing and an internal primer is being designed from the present sequence data. This primer will prevent mis reads of the sequence by the automated sequencer and will decrease the occurrence of missing sequence data. To guarantee that the primer will work among species and individuals in a species, a second sequence is being determined for each species except for human, pronghorn antelope and rabbit.

Conclusion

Inital work showed that the souce of organic paint binder came from a non human source, more specifially from the Order Artiodactyla. The use of *histone 4* did not provide enough sequence variation to get a good match to specific species. From initial sequence data, COII appears to be sufficiently variable to allow for the more precise determination of the binder source. To do this, a highly variable sequence portion of COII must be found flanked on each side by a highly conserved sequence. The whole length of the COII fragment must not exceed 200 bp, the typical size of DNA fragments extracted from ancient sources (*25*). Two primers will be designed to amplify the same sequence across all the species of interest. The new primers will be tested on the species under study before use on the ancient sample. These sequences will be used against the ancient sequence for analysis.

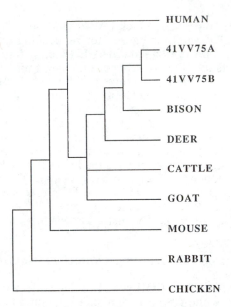

Figure 3. Consensus tree for a portion of the *histone 4* gene sequence comparing pictograph 41VV75 to selected Lower Pecos River region mammal fauna and selected GenBank *histone 4* sequences using the branch and bound method. *Homo sapiens* (human), pictograph samples (41VV75A, 41VV75B), *Bison bison* (American bison), *Odocoileus virginanus* (White-tailed deer), *Bos taurus* (cattle), *Capra hircus* (goat), *Mus muscalus* (house mouse), *Sylvilagus floridauus* (rabbit), and*Gallus gallus* (chicken). (Reproduced with permission from ref. *39*. Copyright 1995 Academic Press).

Cow	TAC ATT ATT TCA CTA ATA CTA ACG ACA
	AAG CTG ACC CAT ACA AGC ACG ATA
Mule Deer	TAT ATC ATT TCA TTA ATG CTA ACA ACA
	AAA CTA ACT CAC ACT AGT ACA ATA
White-tailed Deer	TGC ATT ATT TCA CTC GTG TTT ACA ACT
	AAG CTA ACA CAT ACA GGC ACA ATA
Human	TAT GCC CTT TTC CTA ACA CTC ACA ACA
	AAA CTA ACT AAT ACT AAC ATC TCA
Pronghorn	TAT ATT ATC TCC CTT ATA TTA ACA ACG
	AAA CTA ACC CAC ACT AGT ACA ATG
Rabbit	TAT ATT ATT TCT CTT ATA TTA ACT ACA
	AAG CTC ACT CAC ACA AGC ACA ATG
Bison	TAC ATT ATC TCA CTA ATA CTA ACA ACA
	AAA CTG ACT CAT ACA AGC ACA ATA

Figure 4. Portion of COII sequence showing variablity in bases between the species. Bold-face type indicates the positions of variable bases.

Acknowledgments

This work was supported in part by the Robert A. Welch Foundation, the Research Corporation, and the Petroleum Research Fund of the American Chemical Society. We are grateful to David Ing and Emmet Brotherton, Texas Parks and Wildlife Department, for permission to sample the pictographs at 41VV75.

Literature Cited

1. Ilger, W.; Hyman, M.; Southon, J.; Rowe, M. W. This Symposium; references 1, 5-20, 27 therein.
2. *Symbolic and Structural Archaeology*; Editor Hodder, I. Cambridge University Press: Cambridge, England, 1982.
3. Renfrew, C. *Towards an Archaeology of Mind*; Cambridge University Press: Cambridge, England, 1982.
4. Lewis-Williams, J. *Believing and Seeing: Symbolic Meaning in San Rock Paintings*; Academic Press: London, England, 1981.
5. Lewis-Williams, J. *Current Anthropology* **1982**, *23*, 429-450.
6. Lewis-Williams, J. In *New Approaches to Southern African Rock Art;* Editor Lewis-Williams, J.; Goodwin Series 4; South African Archaeological Society: Cape Town, South Africa, 1983; pp. 3-13.
7. Lewis-Williams, J. *Introductory Essay: Science and Rock Art*, 1983.
8. Lewis-Williams, J.; Dowson, T. *Current Anthropology* **1988**, *29*, 201-245.
9. Turpin, S. In *New Light on Old Art: Recent Advances in Hunter-Gather Rock Art Research*; Editors Whitley, D.; Loendorf, L. Monograph 36; UCLA Institute of Archaeology: Los Angeles, California, 1993, pp. 75-80.
10. Hester, T. *Bulletin of the Texas Archeoloical Society* **1988**, *335*, 774-.
11. Turpin, S. A. In *Papers on the Lower Pecos Prehistory, Studies in Archeology 8*, Editor Turpin, S. A., Texas Archeological Research Laboratory, University of Texas, Austin: Austin, Texas, 1991, pp.1-50.
12. Mullis, K.; Faloona, F. *Cold Spring Harbor Symposium on Quantum Biology* **1986**, *51*, 263-273.
13. Mullis, K.; Faloona, F.; Scharf, S.; Saiki, R.; Horn, G.; Erlich, E. *Methods in Enzymology* **1987**, *155*, 335-350.
14. Brown, T. A.; Brown, K. A. *Antiquity* **1992**, *66*, 10-23.
15. Cherfas, J. *Science* **1991**, *253*, 1354-1356.
16. DeSalle, R.; Barcia, M.; Wray, C. *Experimentia* **1993**, *49*, 906-909.
17. DeSalle, R.; Gatesy, J.; Wheeler, W.; Grimaldi, D. *Science* **1992**, *257*, 1933-1936.
18. Golenberg, E. M.; Giannasi, D. E.; Clegg, M. T.; Smiley, C. J.; Durbin, M.; Henderson, D.; Zurawski, G. *Nature* **1990**, *344*, 656-658.
19. Hagelberg, E.; Clegg, J. B. *Proceedings of the Royal Society of London Series B: Biological Sciences* **1991**, *244*, 45-50.
20. *Ancient DNA*; Herrmann, B.; Hummel, S. Eds.; Springer-Verlag; New York, New York, 1994.
21. Higuchi, R.; Bowman, B.; Freiberger M.; Ryder, O.; Wilson, A. C. *Nature* **1984**, *312*, 282-284.
22. Hummel, S.; Herrmann, B. *Naturwissenschaften* **1991**, *78*, 266-267.

23. Kocher, T. D.; Thomas, W. K.; Meyer, A.; Edwards, S. V.; Pääbo, S.; Villablanca, F. X.; Wilson, A. C. *Proceedings of the National Academy of Sciences, USA* **1989**, *86*, 6196-6200.
24. Lawlor, D. A.; Dickel, C. D.; Hauswirth, W. W.; Parham, P. *Nature* **1991**, *349*, 785-788.
25. Lindahl, T. *Nature* **1993**, *365*, 700.
26. Pääbo, S. *Nature* **1985**, *314*, 644-645.
27. Pääbo, S. *Proceedings of the National Academy of Science USA* **1989**, *86*, 1939-1943.
28. Pääbo, S. In *PCR Protocols: A Guide to Methods and Applications*, Editors, Innis, M. A.; Gelfand, D. H.; Sninsky, J. J.; White, T. J. Academic Press: San Diego, California, 1990, pp. 159-166.
29. Pääbo, S.; Gifford, J. A.; Wilson, A. C. *Nucleic Acids Research* **1988**, *16*, 9775-9787.
30. Pääbo, S.; Higuchi, R. G.; Wilson, A. C. *Journal of Biological Chemistry* **1989**, *264*, 9709-9712.
31. Poinar, H. N.; Cano, R. J.; Poinar, G. O. *Nature* **1993**, *363*, 677.
32. Richards, M.; Smalley, K.; Sykes, B.; Hedges, R. *World Archaeology* **1993**, *25*, 18-28.
33. Rollo, F.; Amici, A.; Salvi, A. *Nature* **1988**, *335*, 774.
34. Sanger, F.; Nicklen, S.; Coulson, R. *Proceedings of the National Academy of Science U.S.A.*. **1977**, *74*, 5463-5467.
35. Swofford, D. S. *PAUP: Phylogenetic Analysis Using Parsimony, version 3.1.1*, Computer Software; Smithsonian Institution: Washington, DC, 1993.
36. Bartlett, S.; Davidson, W. *Biotechniques* **1992**, *12*, 408-411.
37. Davis, L.; Dibner,M.; Battey, J. *Methods in Molecular Biology;* Elsevier: New York, NY, 1986; p. 131.
38. Maniatis, T.; Fristch, E.; Sambrook, J. *Molecular Cloning: A Laboratory Manual*; Cold Spring Harbor Publications: Cold Spring Harbor, Il, 1989.
39. Reese, R. Derr, J.; Hyman, M.; Rowe, M. Davis, S. *Journal of Archaeological Science* **1995**, in press.
40. Dibble, D. S.; *Bonfire Shelter: A Stratified Bison Kill Site, Val Verde County, Texas, Part 1, The Archeology*; Miscellaneous Papers No. 1; Texas Memorial Museum: Austin, TX, 1967; pp 7-76.
41. Miyamoto, M. M.; Allard, M. W.; Adkins, R. M.; Janeck, L. L.; Honeycutt, R. L. *Systematic Biology* **1994**, *43*, 236-249.
42. Lord, K. J. *The Zooarchaology of Hinds Cave (41VV456), Val Verde County, Texas*; Ph.D. Dissertation; The University of Texas: Austin, TX, June 1984.

RECEIVED August 28, 1995

Chapter 28

Ancient Nucleic Acids in Prehispanic Mexican Populations

R. Vargas-Sanders[1], Z. Salazar[1], and Ma. C. Enriquez[2]

[1]Laboratorio de Antropologia Molecular, Instituto de Investigaciones Antropológicas, Universidad Nacional Autónoma de México, C.P. 04510, México, D.F., México
[2]Departmento de Bioquimica Vegetal, Facultad de Quimica Universidad Nacional Autónoma de México, C.P. 04510, México, D.F., México

Ancient Mexican populations have been studied by archaeologists and physical anthropologists who have described cultural and some biological characteristics. Now, we report the isolation of high molecular weight DNA from bones of Mexican Tolteca culture. The results of the present study show a polymorphic pattern with the Elongation Factor 1-α (EF1-α) gene. These results open a new perspective on the knowledge of past Mexican populations and their relationships with present ethnic groups.

The physical anthropology of Mesoamerican prehispanic populations has been studied amply. Since the end of the last century interest in knowing who the ancestral inhabitants of the land of Mexico were, led to the early studies in osteology. Although these studies were initially descriptive, they were able to infer essential physical characteristics as well as the evaluation of other parameters, such as growth and nutrition, of ancient populations. More recent, anthropometric and paleodemographic investigations of Mesoamerican prehispanic bone remains have revealed data on sex, age, height and population features (1-4). Phenomena such as longevity graphs, mortality per age group and sex have been inferred, as well as some aspects of living conditions. Some examples are certain pathologies, nutritional characteristics and the appraisal of economic and social fluctuations (5-7). Furthermore, some studies have employed immunological techniques to attempt a classification of prehispanic bone remains based on blood groups (8, 9) and histological composition (10-12).
Presently, molecular archaeology is interested, in among other things, the search for nucleic acids present in human tissues from hundreds to thousands of years ago to attempt molecular genetics analysis, paleopathology, individual characterization, population studies , etc. (13-39). These investigations have demonstrated that genetic material may be preserved under certain environmental conditions. Therefore,

molecular studies of prehispanic Mexican populations presents the possibility of finding answers to old questions and proposing new ones.

In Mexico, this line of investigation started with the identification and characterization of human DNA recovered from Tomb 1, Peñón del Marquéz, Iztapalapa, Mexico wich is approximately 650-750 years-old (*11, 23*). The present study employed bone material from a prehispanic population of 130 humans found at the archaeological site of Tula, Hidalgo, Mexico aged approximately 750-1050 years old (*40*). The aim of this study was to isolate and characterize DNA and RNA for future systematic studies at the population level.

Materials and Methods

Human bone remains were obtained from the archaeological site of Tula, Hidalgo, Mexico. The areas studied were Malinche, Museum (Charnay square) from habitions, from civic centers and from palaces, Zapata II, Viaducto, Plazas and Juego de Pelota. The populations consisted of 130 individuals whose remains are about 750-1050 years old (*40*), belonging to Tolteca and Mexican cultures.

Histological Samples.   Bone remains were decalcified in formic acid in neutral formalin for 24 h, submerged in paraffin, and sectioned in 5 μm thicknesses with a manual microtome (American Optical). The paraffin was removed, and the sections were washed and stained with May-Greenwald-Giemsa stain.

DNA Isolation.   Bone remains consist of several anatomical pieces in diverse preservation states. The bones were always handled with gloves or forceps to avoid contamination by skin cells or perspiration. Excess soil was removed by scraping with a scalpel blade. Modern chicken bone and prehispanic osseous tissues were processed in parallel for comparative experiments. Bone samples of 1-3 g were crushed and genetic material extracted according to a modified version of Maniatis' method (*11, 12*).

Nucleic Acids Quantification.   The amount of nucleic acids was quantified following Spirin's method (*41*); DNA and RNA were measured by Scheinder's (*42*) and Burton's (*43*) methods, respectively.

Recovery of High Molecular Weight DNA.   DNA extracted from prehispanic bone was analyzed by agarose gel electrophoresis. The high molecular weight DNA was determined by a long wavelength (300-360 nm) from an UV lamp and was extracted following three methods: (1) Electroelution into dialysis bags, as described by Maniatis (*44*). (2) Separation by DEAE-81 (Whatman) or NA-45 (Schleicher & Schuell) membranes following a modification of the method described by Dretzen and collaborators (*45*). (3) Electroelution into troughs was prepared by a modification of the Maniatis method (*44*): the high molecular weight band was located on the 1% agarose gel. In front of the leading edge of the band a pit was made . This trough was filled with 50% glycerol in TBE 1X, and electrophoresis resumed. Every 5 or 8 minutes (min) the fluid was recovered. The trough was refilled with fresh solution and the electrophoresis was continued until all the DNA in the band had been removed from the gel and was recovered in the fluid taken from the trough. The high molecular weight band DNA was extracted once with phenol and once with chloroform before being precipitated by ethanol.

Southern Blot Analysis. Purified genomic high molecular weight DNA samples (approximately 10 µg/lane) were digested according to conditions recommended by manufacturers with EcoRI (Gibco BRL), at 37 °C overnight and subjected to electrophoresis in a 0.8% agarose gel in 0.089 M Tris-borate, 0.089 M boric acid and 0.002 M EDTA (TBE 1X). Phage lambda DNA preparations, digested with HindIII (Gibco BRL), were included as molecular size markers. Gels were run at 25V for 18-22 h and then stained with ethidium-bromide and photographed. Gels were denatured with 0.5 M NaOH/1.5 M NaCl for 60 min and neutralized with 1 M Tris-HCl pH 8.0, 1.5 M NaCl, and the DNA fragments were transferred to Nytran membranes (Schleicher & Schuell) using the method first described by Southern (46). After transfer, filters were rinsed in 2 X standard saline citrate (300 mM NaCl/30 mM sodium citrate) and then irradiated with a standard ultraviolet ligth for 5 min.

Nucleic Acid Hybridization. Filters were prehybridized for 16 h at 42 °C in 50% formamide, 5 X SSC, 10 X Denhardt's solution, 50 mM Na-phosphate (pH 7.2), sodium dodecyl sulfate (SDS) 0.1% and 100µg/ml denatured sonicated herring sperm DNA. Filters were hybridized overnight at 42 °C in the same solution with the addition of 1-2 x 10^6 cpm/ml ^{32}P probe labeled according to conditions recommended by manufactures with Random Primer Kit (BRL) to specific activity 1 x 10^9 dpm/µg. Filters were washed at room temperature once, with 2 X SSC, 0.1% SDS for 30 min at 50 °C, twice with 0.2 X SSC, 0.1% SDS for 30 min, and finally at 55 °C with 0.1 X SSC, 0.1% SDS for 15 min. Filters were then exposed to Cronex film (Dupont) with two intensifying screens at -70°C for 3-6 days. The probe corresponded to the gene HEF-1α (human elongation factor 1α) cloned at site PstI of PBr322 in *E. coli* C-600 supplied by Dr. Mario Castañeda from a donation by Dr.Win Möller.

Results and Discussion

Histological examination shows that the prehispanic bone tissue between 750-1050 years old preserves cellular debris, including osteocytes and remains of blood vessel epithelium (Figure 1). Nuclei were also observed, and their distinctness could be increased with a specific stain (May-Greenwald-Giemsa) for nucleic acids instead of hematoxylin-eosine. Similar results have been reported in mummified tissue and in ancient brain tissue, where various cell types and nuclei can be recognized (*17-20, 26*).

Romano and collaborators (*10*), using eosin-hematoxylin, have observed the presence of blood cells in different stages of differentiation in prehispanic bone remains dated to 500 years of age. Likewise, in a Mexican female skeleton osteocytes and blood vessels have been identified (*11*). These results indicated that some cells and their components can be preserved in prehispanic human remains under varying ecological conditions.

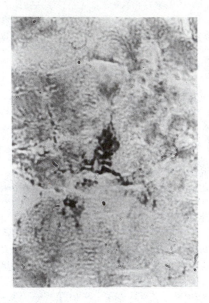

Figure 1. Tissue section from prehispanic bone. May-Greenwald-Giemsa staining. Magnification 200x.

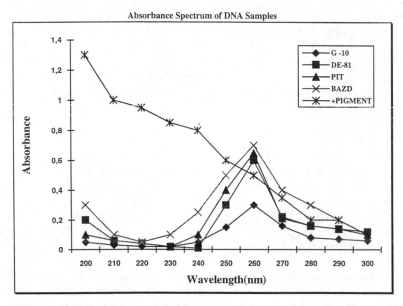

Figure 2. Absorbance spectrum of DNA from human fibroblast grown in culture (BAZD) and DNA from prehispanic bone remains.

Nucleic acids were estimated as described in the Materials and Methods section from two anatomical pieces of the same individual. DNA and RNA values obtained from rib, vertebra, phalanx, femur or tibia were similar and fairly reproducible. The recovery of nucleic acids obtained from 130 prehispanic bone remains collected in various sites of excavation have been summarized in Table I.

Table I. Nucleic Acids from Prehispanic Bones

Sample	Yield[a] (µg/g) dry tissue	Recovery (Percent) fresh tissue	Diphenylamine (µg/g) dry tissue	Orcinol (µg/g) dry tissue
rib/vertebra	159 ± 24	7-10		
rib/vertebra		1.0	13±4	
rib/vertebra		10.0		145±22
rib/vertebra	9.0[b]	2.5		

[a]Spirin's method quantification; [b]Treatment with Ribonuclease

The total amount of nucleic acids from different samples is 136-182 µg/g of dry tissue as determined by Spirin's method (*41*). However, only 7-10 percent is DNA and the rest is RNA. The amount of DNA in the presence of Ribonuclease during DNA extraction is 9 µ/g per dry tissue which corresponds to 6 percent. These results show, that on the order of 9-17 µg DNA are recovered typically from 1g from osseous tissue. Therefore the DNA yield obtained from prehispanic skeletons is similar to values reported for other ancient tissues (*18-20, 34, 37*). The recovery of DNA and RNA obtained from 1g from modern chicken bone is 1,400 µg/g and 1,200 µg/g respectively. Values obtained from prehispanic skeletons yield about 1 percent of DNA, and between 9-12 percent of RNA with respect to that recovered from fresh bone (Table I).

The presence of RNA in human bone remains had not been reported previously. The RNA recovery is 123-167 µg/g of dry tissue, which represents 93 percent of total nucleic acids from ancient bone. Other authors have reported the presence of ribosomal RNA (rRNA) in corn seeds (*47*), plants (*48*), marsupial wolf, 12S mithocondrial rDNA (*49*) and 18S and 28S rRNA in mummies (*26*).

In order to prove that the isolated material corresponds to nucleic acids, the following criteria were used: (1) UV absorbance spectrum. The absorbance spectrum of the DNA is shown in Figure 2, where the spectrum of contemporary DNA, extracted from human fibroblast grown in culture (BAZD) was compared to that of DNA purified from ancient bone. In bone tissue, a color shade ranging from yellow to a dark brown was observed that modified the typical absorbance spectrum of nucleic acids (Figure 2). This pattern is similar to the fulvic acids ultraviolet absorbance spectrum (*50*). To eliminate the pigment, samples were subjected to gel filtration chromatography (*19*), dialysis and the different treatments to obtain high molecular weight DNA as described in Materials and Methods.

The quality of the absorbance spectrum after gel column chromatography and different treatments improves and results are typical of the nucleic acids absorbance spectrum (Figure 2). Through gel filtration chromatography, approximately 25 percent of the genetic material is lost in the process. Dialysis and electroelution also eliminated the pigment from the sample, but the loss of genetic material is greater: up to 60 percent (results not shown). However, recovery with DE-81 filters, NA-45 membranes or by electroelution into troughs is between 65-85 percent (Figure 2). Nevertheless, all DNA recovered by any one of these methods shows a typical absorbance spectrum of nucleic acids (Figure 2). This pattern is different from the DNA from Egyptian mummies (16).

These pigments are not only present in bone remains but have also been reported in other ancient tissues (11, 12, 16, 21, 27, 34, 51). Pääbo (21) suggested that the pigments consist of Maillard products; however these could not be identified in the present study (results not shown). Furthermore, during 1% agarose electrophoresis, the pigments migrated faster than bromophenol blue. Other authors have suggested that this pigment could be derived from humic acids (37).

The origin of the variations among the pigments is unknown: They may be related to the kind of soil where the remains were buried, to length of time and the type of pit, as well as other physical and chemical factors. Although the pigments have not been chemically characterized, their presence is independent of the mechanism by which the remains were preserved (11, 16, 21, 27, 34, 51) and also interferes with the polymerase chain reaction (PCR) amplification (22, 34, 52, 53).

(2) Agarose gel electrophoresis. An alternative procedure to identify genetic material was agarose gel electrophoresis. The pattern of DNA obtained from prehispanic bone shows that the molecular weight of this genetic material ranges between 0.125-23.1 Kilobases (Kb) (results not shown). These values contrast with those obtained for genetic material from other human remains which have exhibited DNA molecular weights ranging from 0.120-12 Kb, in remains preserved under conditions of extreme anaerobiosis (19) or in other environments (19, 27-29, 33, 34, 37). Low molecular weight DNA seems to be common in human remains. The cause of the degradation of the DNA may be attributed to processes such as the effect of environmental factors like radioactivity, *post-mortem* autolytic processes, hydrolysis and oxidation of sugar and nitrogenous bases (18, 21, 27, 54) or instability of DNA glycosyl bonds under different solvent conditions as a function of temperature, pH, ionic strength and nucleic acid secondary structure and also non-enzymatic DNA methylation (54). The transformation of nitrogenous bases is also important, as exemplified in the alterations found in pyrimidines in mummified tissue (18).

However results from the present study show that high molecular weight DNA from ancient human remains is not exclusively related to the burial conditions found at the moment of unearthing. The burial undergoes many environmental modifications with time. In this sense importance must be given to such aspects as the archaeological context, location of the burial, type of dwelling, primary and secondary nature of burial, presence of offerings, type of soil, as well as the following factors: land displacement, temperature, filtration, humidity, uses given to the land, pH, oxide reduction potential, presence of organic and inorganic salts and the presence of microorganisms, fungi, plants and animals. It is therefore difficult to explain differences between nucleic acid recovery that belong to human remains which have been exposed to similar, if not equal, "environmental conditions" as exemplified in several prehispanic DNA samples (results not shown). Even so, high molecular weight DNA was isolated from human bone remains between 750-1050 years-old (*11, 12*).

The next step was to demonstrate that the high molecular weight DNA samples were of human origin. This was tested by Southern blot analysis with the human gene of elongation factor -1α (HEF-1α).

Figure 3 depicts Southern hybridization of high molecular DNA from 14 randomly chosen individuals from the archaeological sites (750-1050 years-old) of Malinche, Zapata II and Charnay (Museum Area) which were compared to present day human DNA from human cell-line grown in culture (BAZD). The DNA was digested with EcoRI and hybridized as explained in Materials and Methods. Results

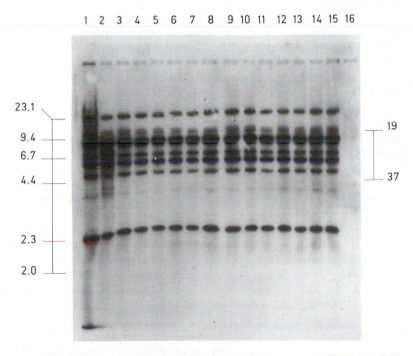

Figure 3. Southern blot analysis of elongation factor 1-α (HEF-1α) DNA. Contemporany DNA (Lane 1), prehispanic DNA (Lanes 2-15), λ phage DNA digested with HindIII (Lane 16). Molecular size markers are indicated in numbers of base pairs. Kilobases (Kb).

show that ten fragments invariably appear both prehispanic DNA and in contemporary DNA.. However, 19 Kb fragments are only present in prehispanic samples, 3.7 Kb fragments in only 6 of the prehispanic samples and 2 Kb in both.

Polymorphism of the human gene HEF-1α has been detected in Caucasian populations (55). The present study revealed EcoRI RFLPs of 3.7 Kb in 6 out of 14 studied individuals from the archaeological site of Tula, Hidalgo, which suggests a possible polymorphism of the gene HEF-1α in this prehispanic Mexican population. Thus, the possibility arises of studying polymorphisms and genetic variability in ancient Mexican populations, until now only described in living population (56, 57). These studies may be applied to determine kinship, migration patterns, relationships to present day ethnic groups, sex and genetic diseases of ancient Mexican populations. The systematic application of these studies will enrich the knowledge derived from the classical methodologies of physical anthropology.

Acknowledgments

Authors wish to thank Dr. Mario Castañeda for supplied HEF-1α probe, Ernesto Guerrero for help with preparation of histological sections and Blanca Paredes have generously provided bone samples.

Literature Cited

1. Genovés, S. *Introducción al diagnóstico de la edad y del sexo en restos óseos prehispánicos*. Instituto de Historia: Universidad Nacional Autónoma de México. México, México, D.F., 1962.
2. Genovés, S. *La proporcionalidad de los huesos largos y su relación con la estatura en restos mesoamericanos*; Cuadernos; Serie Antropológica, Núm. 19; Instituto de Investigaciones Históricas: Universidad Nacional Autónoma de México, México, D.F., 1962.
3. Jaen, M. T. ; López-Alonso, S. En *Antropología Física*. Época Prehispánica; Romero J., Ed.; Panorama Histórico Cultural, III; Instituto Nacional de Antropología e Historia / Secretaria de Educación Pública, México, D.F., 1974; pp 113-118.
4. Comas, J. *Manual de Antropología Física*. Instituto de Investigaciones Antropológicas: Universidad Nacional Autónoma de México, México, D. F., 1976.
5. Dávalos-Hurtado, E. In *Handbook of Middle American Indians: Physical Anthropology;* Stewart, T. D., Ed. ;University of Texas Press: Austin, 1970;p 9.
6. Jaen, M. T. ; Serrano, C. En *Antropología Física*: Epoca Prehispánica; Romero, J., Ed.; Panorama Histórico Cultural, III; Instituto Nacional de Antropología e Historia / Secretaria de Educación Pública, México, D.F., 1974; pp 153-168.
7. Serrano, C. ; Ramos, R. Ma. *Perfil bioantropológico de la población prehispánica de San Luis Potosí*. Serie Antropológica, Núm 40; Instituto de Investigaciones Antropológicas: Universidad Nacional Autónoma de México, México, D.F., 1984.

8. Toral, R. E. Tesis Profesional. Escuela Nacional de Ciencias Químicas, México, 1949.
9. Salazar-Mallén, M. *Gaceta Médica de México*, **1951**, *81*, 122- 128.
10. Romano, A..; Villalobos, R. F. ; Balcorta, L. A. *Boletín del Instituto Nacional de Antropología e Historia*, **1976**, *16*, Epoca III, p 45.
11. Vargas S. R. ; Sánchez, A. R. En *Estudios de Antropología Biológica;* López-Alonso S.; Ramos, R. Ma. Eds. ; Instituto de Investigaciones Antropológicas: Universidad Nacional Autónoma de México / Instituto Nacional de Antropología e Historia, México, D.F., 1990, Vol V; pp 219-242.
12. Vargas, S. R. Ph. D. Thesis,Universidad Nacional Autónoma de México, 1993.
13. Wang, G. H. ; Lou, C. C. *Sheu Wu Hua yu Sheng Wu Li Chin Chan,* **1991**,*99*, 70.
14. Hansen, H. E. ; Götler, H. *Am. J. Phys. Anthropo.*, **1983**, *61*, 447-452.
15. Pääbo, S. *Das Altertum*, **1984**, *30*, 213-218.
16. Pääbo, S. *J. Archeol. Sci.*, **1985**, *12*, 411-417.
17. Pääbo, S. *Nature*, **1985**, *314*, 644-645.
18. Pääbo, S. *Cold Spring Harbor Symp. Quant. Biol.*, **1986**, *51*, 441-446.
19. Doran, G. H. ; Dickel, D. N. ; Ballinger, W. E. Jr. ; Agee, O. F. ; Lapais, P. J.; Harswirth, W. W. *Nature,* **1986**, *323*, 803-806.
20. Pääbo, S. ; Glifford, J. A. ; Wilson, A. C. *Nucleic Acid Res.*, **1988**, *16*, 9775- 9887.
21. Pääbo, S. *Proc. Natl. Acad. Sci. U. S. A.*, **1989**, *86*, 6196-6200.
22. Pääbo, S. ; Higuchi, R. G. ; Wilson, A. C. *J. Biol. Chem.*, **1989**, *264*, 9707-9712.
23. Vargas, S. R. *Información Científica y Tecnológica*, **1989**, *11*, 19-21.
24. Schmill, N. ; Fritzler, D. A. *Clin. Invest. Med.*, **1989**. *12B*, 65.
25. Hagelberg, E. ; Sykes, B. ; Hedges, R. *Nature*, **1989**, *342*, 485.
26. Rogan, P. K. ; Salvo, J. J. In *Molecular Evolution;* Clegg, M. T.; O'Brien, S. J. Eds. ; Wiley: New York, N Y, 1990; pp 223-234.
27. Rogan, P. K. ; Salvo, J. J. *Year Book Phys. Anthropol.*; **1990**, *33*, 195-214.
28. Thuesen, Y. ; Engberg J. *J. Archaeol. Sci.*, **1990**, *17*, 679-689.
29. Haenni, D. ; Landet, V. ; Sakka, M. ; Begue, A. ; Stehelin, M. *Comptes Rendus Acad. Sci.* III, **1990**, *310*, 365-370.
30. Pedro, J. ; Chinay, E. *Am. J. Hum. Genet.*, **1991**, *49* (4 suppl), 115.
31. Hummel, S. ; Herrman, B. *Naturwissenschaften*, **1991**, *78*, 266-267.
32. Meijer, M. F. ; Perizonius, W. R. K. ; Geraedts, J. P. M. *Am. J. Hum. Genet.*, **1991**, 49 (suppl), 440.
33. Hagelberg, E. ; Bell, L. S. ; Allen, T. ; Boyde, A. ; Jones, S. J. ; Clegg, J. B. *Phil. Trans. R. Soc. Lond.* B, **1991**, *333*, 399-407.
34. Hagelberg, E. ; Clegg, J. B. *Proc. R. Soc. Lond.* B, **1991**, *244*, 45-50.
35. Meijer, M. F. ; Perizonius, W. R. K. ; Geraedts, J. P. M. *Biochem. Biophys. Res. Comm.*, **1992**, *183*, 367-374.
36. Foo, Y. ; Solo, W. L. ; Aufderheis, A. C. *Biotech.*, **1992**, *12*, 811-817.
37. Hagelberg, E. ; Clegg, J. B. *Proc. R. Soc. Lond.* B, **1993**, *252*, 163-170.
38. Hagelberg, E. ; Quevedo, S. ; Turbon, D. ; Clegg, J. B. *Nature*, **1994**, *369*, 25-26.
39. Handt, O. ; Richards, M. ; Trommsdorff, M. ; Kilger, C. ; Simanainen, J. ; Georgiev, O. ; Bauer, K. ; Stone, A. ; Hedges, R. ; Schaffner, W. ; Utermann, G. ; Sykes, B. ; Pääbo, S. *Science*, **1994**, *264*, 1775-1778.

40. Paredes, B. *Unidades Habitacionales en Tula, Hidalgo.* Serie Arqueológica. Instituto Nacional de Antropología e Historia, México, D. F. 1990.

41. Spirin, A. S. *Biokhimia,* **1958**, *23,* 656-662.

42. Schneider, W. C. In *Methods in Enzymology*; Colowick, S. P.; Kaplan, N. O., Eds. ; Academic Press: New York, 1957; Vol 3; pp 680-684.

43. Burton, K. In *Methods in Enzymology*; Grossman, L. ; Moldave, K., Eds. ; Academic Press: New York; 1968; Vol. 12, Part B; pp 163-165

44. Maniatis, T. E.; Frisch, E. F. ; Sambrook, J. *Molecular Cloning.* Cold Spring Harbor Laboratory: Cold Spring Harbor, New York, 1982.

45. Dreztzen, G. ; Bellard, M. ; Sassone-Corsi, P. ; Chambon, P. *Proc. Natl. Acad. Sci.* U. S. A., **1981**, *112,* 295-298.

46. Southern, E. *J. Mol. Biol.*, **1975**, *95,* 503-510.

47. Venanzi, F. ; Rollo, F. *Nature,* **1990**, *343,* 25-26.

48. Rogers, S. O. ; Bendich, A. *J. Plant. Biol. Mol.*, **1985**, *5,* 69-76.

49. Tuross, N. *Experientia,* **1994**, *50,* 530-535.

50. Thomas, R. H. ; Shaffner, W.; Wilson, A. C. ; Pääbo, S. *Nature*, **1989**, *340,* 465-467.

51. Johnson, B. H. ; Olson, C. B. ; Goodman, M. *Biochem. Physiol.* (B), **1985**, *81.* 1045-1051.

52. Akane, A. ; Shiono, H. ; Matsubara, K. ; Nakahori, Y. ; Seki, S. ; Nagafuchi, S. ; Yamada, M. ; Nagakome, Y. *Forensic Science International,* **1991**, *49,* 81-88.

53. Akane, A. ; Shiono, H. ; Matsubara, K. ; Nakamura, H. ; Hasegawa M. ; Kagawa, M. *J. Forensic Sci.,* **1993**, *38,* 691- 701.

54. Lindhal, T. *Nature*, **1993**, 362, 709-715.

55. Opdenakker, G. ; Cabeza-Avelaiz, Y. ; Fiten, P. ; Dijkmans, R. ; Van Dame, J. ; Volckaert, G. ; Billiau, A. ; Van Elsen, A. ; Van der Shueren, B. ; Van den Berghe, H. ; Cassiman, J. J. *J. Human Genet.*, **1987**, *75,* 339-344.

56. Lisker, R. *Estructura Genética de la Población Mexicana;* Salvat Mexicana Ediciones: México, D.F. 1981.

57. Helmuth, R. ; Fildes, N. ; Blake, E. ; Luce, M. C. ; Chinera, J. ; Madej, R. ; Gorodezky, C. ; Stoneking, M. ; Schmill, N. ; Klitz, W. ; Higuchi, R. ; Erlich, H. *Am. J. Hum. Genet.*, **1990**, *47,* 515-523.

RECEIVED December 6, 1995

Chapter 29

Radiocarbon Dating of Ancient Rock Paintings

W. A. Ilger[1], Marian Hyman[1], J. Southon[2], and Marvin W. Rowe[1,3]

**[1]Department of Chemistry, Texas A & M University,
College Station, TX 77843–3255**
**[2]Center for Accelerator Mass Spectrometry, Lawrence Livermore
National Laboratory, Livermore, CA 94551–9900**

We report here progress on our technique for ^{14}C dating of
pictographs. We use low-temperature oxygen plasmas coupled
with high-vacuum technology to selectively remove carbon-
containing material in the paints without contamination from
inorganic carbon from rock substrates or accretions. Pictograph
samples dated generally agree with ages expected on the basis
of archaeological inference. We also used the technique on
eight samples of known ^{14}C activity. In each case our results
agree with previously determined ages. Each of these determinations supports our conclusion that the technique has the
potential of producing accurate and reliable ages. Four new ^{14}C
dates were obtained on a quartered Pecos River style pictograph
sample (41VV75-37A-D). We used an idealized model to
estimate a lower limit for the age of the pictograph 41VV75-37.
The "age" itself should not be taken seriously as a meaningful
limit as the measured age indicates that background organic
material in the basal rocks and accretions can be a serious
problem.

Traditionally, the study of prehistoric rock art has been marginal to mainstream
American archaeological research. There are two primary reasons for this: (1)
no rigorous models had been constructed for interpreting the art, or for integrating existing interpretations of it into larger reconstructions of prehistory;
and (2) no means existed for directly dating the art. This situation began to
change during the past decade. Recent trends suggest, in fact, that rock art
research now may be at the theoretical and methodological forefront in
archaeological studies of hunter-gatherers, providing information not accessible using traditional excavation and settlement pattern data (*1*). The second
concern, the ability to relate rock art on cave and shelter walls to other
archaeological remains, is addressed partially in this paper.

[3]Corresponding author

0097–6156/96/0625–0401$12.00/0
© 1996 American Chemical Society

Tandem accelerator mass spectrometry (AMS) greatly reduced the amount of carbon necessary to obtain a ^{14}C date (2-4) and encouraged attempts to date rock art around the world. Van der Merwe *et al.* (5) were first to date a charcoal pigment from a pictograph; others soon followed, also dating charcoal pigments (6-13). As with any archaeological charcoal, caution must be exercised when interpreting charcoal-derived ^{14}C dates due to the "old wood problem"; but otherwise no special problems appear to attend dating of pictograph charcoal. However, inorganic pigments are used far more frequently for pictograph pigments than is charcoal. Iron- and manganese- oxides and hydroxides are the most common pigments and these cannot be dated directly. Nonetheless, if ancient painters added organic materials as binders or vehicles to their inorganic pigments, this organic carbon can be ^{14}C dated. In 1990, we introduced a plasma-chemical technique to date pictographs, even those that used inorganic pigments (14). We have tested our method by determining ^{14}C in samples of known activity: ^{14}C-free Albertite, ^{14}C-free IAEA, and Axel Heiberg wood. Our determinations on these ^{14}C-free samples showed that we were not introducing significant modern carbon with our technique. We have also studied previously dated charcoal and Third International Radiocarbon Intercomparison (TIRI) wood (Figure 1). Our dates agree with the previous ages on charcoal (Beta Analytic, Inc. and University of Texas Radiocarbon Laboratory) and TIRI wood. In all but one case, our measured dates agree within one standard deviation with those reported earlier. In the other case, the agreement is at the two standard deviation level. These measurements support the general validity of the technique.

Pictographs painted with inorganic pigments can therefore be dated with our technique. But to do so with accuracy and reliability requires that the following conditions be met: (1) organic matter was added initially to the paints as binder or vehicle and enough of the organic material has survived to yield a sufficiently large carbon sample; (2) organic carbon can be extracted without contamination from either atmospheric, or inorganic carbon often found either as limestone ($CaCO_3$ and $MgCO_3$) or in mineral accretions as carbonates and oxalates found both below and above pictographs; (3) the extraction technique used does not introduce significant mass fractionation that would influence a ^{14}C date; (4) carbon originally added to the paint does not exchange with other sources of carbon after paint was applied; and (5) basal rock of pictographs and associated mineral accretions do not contain enough organic material (contamination) to invalidate a date. Our previous work (14-20) demonstrates that condition (1) is met for most, but not all, pictographs we have studied. Conditions (2) and (3) are well met with the plasma-chemical technique introduced in our laboratory in 1990 (19, 20). Evaluating how well condition (4) is met is more difficult. Ten pictograph dates have been obtained which can be compared with the large age ranges inferred from archaeology. Archaeological inferences giving the expected age ranges for Pecos River, Red Linear and Red Monochrome styles in Texas have been discussed by us elsewhere (15). Measured ages generally fall in the ranges expected. Similarly, pictographs from Utah (All American Man) and Montana agree with archaeologically expected ages; they also were discussed elsewhere (13, 16). The general agreement of our dates with inferred ages based on archaeological information implies that there is not large scale exchange of pictograph binder or

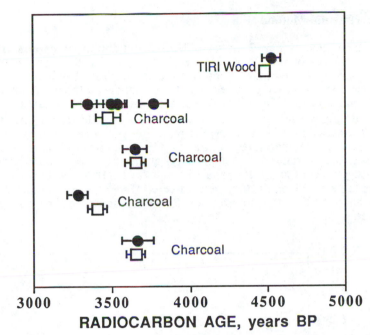

Figure 1. Comparison of our results with previously determined ages on TIRI wood and four charcoal samples. Symbols: ● our measurements; □ known age.

vehicle organic carbon with extraneous carbon from carbonates, oxalates, ground water or the atmosphere; thus, condition (4) is probably met, although further study is needed. We are currently conducting experiments to determine the source of the binder or vehicle used in the paint of two Pecos River style pictographs from Seminole Canyon, Texas site, 41VV75 (*21*). Using DNA phylogenetic analysis, our work shows that even-toed hoofed mammalian (Order Artiodactyla) organic matter was incorporated into the paints. Once the exact nature of the carbonaceous matter being dated is established, strict validity of condition (4) can be evaluated in detail.

Condition (5) is often violated in our experience. Contamination levels we have found range from insignificant to a level such that background unpainted samples yield carbon in amounts comparable to painted samples. In each of two approaches to correct for background contamination (*15*), reasonable ages resulted. Unfortunately, archaeologically inferred age ranges are too broad to provide stringent tests of corrections, or of the technique in general. In order to address reproducibility of the materials normally sampled for a pictograph date and reproducibility of AMS for pictograph samples, we subjected each section of a quartered Pecos River style pictograph sample to our plasma-chemical extraction technique for ^{14}C dating.

Experimental Method

Our technique is based on the selective extraction of carbon from any organic binder or vehicle used in the pictograph. For that purpose we chose O-plasmas (14). As many samples are painted on largely inorganic carbon-containing materials, e.g., limestone, it is essential that the plasma not convert these materials to CO_2. Therefore low temperatures are required. Temperature during plasma operation was determined as follows: a 1 torr O-plasma was maintained for 30 minutes; the radio frequency (RF) power was turned off; and the temperature recorded for 15 minutes. Figure 2 shows a temperature vs. time plot. Temperatures at time zero were found using linear regression. Thus at 50, 100, and 150 watts at 1 torr of O_2, the chamber was heated to 80, 125, and 150 °C, respectively. Under these conditions, Chaffee, Hyman & Rowe (19) reported no decomposition of either carbonates or oxalates.

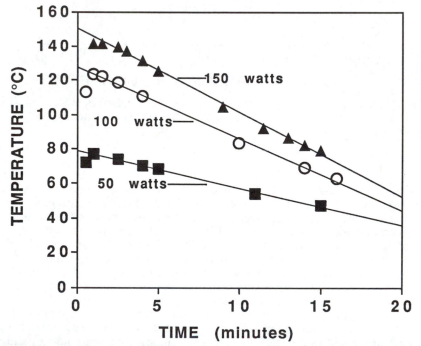

Figure 2. Extrapolation of time versus temperature for 50 watts (~80 °C), 100 watts (~125 °C), and 150 watts (~150 °C) to time zero.

We have built five systems since 1987; four are still in use. All are similar in principle, but differ in detail. The following is a generic description of the systems. The glass reaction chamber is ~10 cm in diameter and ~20 cm long with two ports. One port is for sample insertion, a second is connected to a glass manifold used to condense gases. CO_2 extracted from a sample by an O-plasma is frozen, along with H_2O, in a liquid nitrogen cooled tube. Isolation of CO_2 is accomplished by removing the liquid nitrogen and replacing it with an ethanol slurry at its melting point (-117.3 °C) to retain moisture, but release $CO_2(g)$. CO_2 is re frozen in another tube that is sealed off and sent to an AMS facility for ^{14}C dating.

Vacuum pumping is performed primarily by oil-free pumps, as even μg levels of contamination can affect ^{14}C results. An oil-based rotary pump is used occasionally to pump out two zeolite-filled sorption pumps through an oil trap. Sorption pumps are routinely used in series at liquid nitrogen temperature; one evacuates gas from atmospheric pressure (760 torr) to $\sim10^{-5}$ torr; the second to $\sim10^{-6}$ torr. When deemed necessary, the rough sorption pump is baked out to the rotary pump; the hard sorption pump is baked out onto the rough sorption pump that is cooled with liquid nitrogen. An ion pump maintains the chamber vacuum at $\sim10^{-7}$ torr. Power is supplied by one of several RF generators. Typical operating powers are <150 watts. Current is passed to the chamber through external Cu electrodes. These connect to the power supply through a matching network that is adjusted to simultaneously maximize delivered power and minimize reflected power. Ultra-high purity Ar and O_2 gases (99.999%) are used. Gas lines are purged by bleeding to atmosphere. A glass coil is attached to the gas inlet line, which is immersed in liquid nitrogen or an ethanol slurry during gas addition to freeze out condensable gas contaminants in Ar and O_2. All-metal, bakeable valves are used in locations exposed to plasma, as an O-plasma would attack any organic based gasket.

Gas lines are filled and evacuated three times after purging, and then evacuated to ~200 mtorr. Chamber pressure is adjusted to the following pressures for plasma generation: 0.5-1.0 torr for O_2 and 0.4 torr for Ar. Application of RF energy accelerates randomly occurring free electrons until they possess adequate energy to partially ionize the gas, thus generating additional free electrons. Under our conditions, probably less than 1% of the gas is ionized, with 10 to 20% of the O_2 present as excited atoms. The end result of this is a plasma with low bulk temperature, ≤150 °C, and excited particles, electrons, ions, atoms, and molecules, sufficiently energetic to break organic molecular bonds. In O-plasmas, virtually all organic materials are oxidized to CO_2 and water.

Atmospheric CO_2 adsorbed on surfaces inside the vacuum chamber would contaminate CO_2 extracted from a sample. We first minimize this by heating the empty system and pumping away the CO_2. Possible contamination of the empty sample chamber by either organic C or CO_2 is further reduced prior to introduction of a sample by running several O-plasmas and pumping away the products. The first is run at 100 watts and 1 torr O_2 for ≥1 hour. This removes most contamination from the chamber. The second O-plasma is run at 150 watts, and 0.5 torr of O_2 which increases plasma temperature by a factor of ~2 (*18*). These conditions are more vigorous than those employed during sample oxidation, and efficiently clean the chamber. O-plasmas are typically repeated until C mass is ~0.5 μg.

For sample insertion, we purge the line with O_2, rather than Ar, because the Ar response on the thermocouple pressure gauge maximizes at ~2 torr, even at atmospheric Ar pressure and above. O_2 is bled into the chamber until the pressure is slightly above atmospheric. The sample port flange is removed only then to prevent contamination from atmosphere being drawn into the chamber. Chamber pressure will be high enough to gently force the blank flange away from the chamber. The sample is then placed in the chamber, and if powdered, spread out. The Cu-gasket and flange are replaced and the O_2 is evacuated. Depending on sample size and condition, it takes from 30 minutes to several days to transfer vacuum pumping from rough sorption pump to hard sorption pump to ion pump.

After heat is applied overnight or longer, we use Ar plasmas (0.4 torr) to bombard sample and chamber surfaces with excited atoms and ions of chemically inert Ar to assist desorption of CO_2. Even a modest amount of atmospheric CO_2 could cause error in the final result. In most cases, this combination of heat and Ar plasmas clean sample and chamber surfaces to the extent that <1 μg C is generated from the final Ar plasma. Carbon masses are calculated using CO_2 pressures and approximate system volumes.

A vacuum integrity check is performed before sample oxidation to verify that no appreciable leakage is occurring. This involves cessation of chamber pumping, and monitoring pressure rise over time. Usual times utilized for a vacuum integrity check are 2 hours, twice as long as the usual 1 hour or less used for sample oxidation plasmas. Final gas pressure determines the maximum amount of contamination introduced into the chamber in 2 hours. Vacuum integrity checks are shown in Figure 3 for four samples of 41VV75-37. In the worst case for these samples, *if all* of the gas causing the temperature rise after an hour were CO_2, the mass of the contamination would still be ≤ 0.2 μg C, an insignificant amount, compared to typical backgrounds at AMS facilities at Lawrence Livermore National Laboratory (2 to 4 μg of C) (Southon, J., Center for Accelerator Mass Spectrometry, Lawrence Livermore National Laboratory, Livermore, CA, personal communication, 1993), and at the University of Arizona (2 to 3 μg of C) (Jull, A. J. T., NSF-Arizona AMS Facility, University of Arizona, Tucson, AZ, personal communication, 1993).

After completion of a vacuum integrity check, the sample is ready for oxidation. One torr of O_2 is added to the chamber with liquid nitrogen kept on the glass coil. The O-plasma is run for 1 hour at 100 watts. After plasma termination, O_2 is pumped from the chamber to ~10^{-7} torr. The chamber is then isolated from pumps. CO_2 is released from the cold finger by removal of liquid nitrogen; an ethanol slurry is applied to the cold finger to keep H_2O frozen. The P_{CO_2} is measured to calculate the weight of carbon. CO_2 is then re frozen in another cold finger, sealed off, and sent to an AMS facility. An effort is made to generate a second portion of CO_2, typically using 150 watts of RF power. If an adequate weight of C is generated, a second tube is sealed off and reserved for future studies. Standard plasma conditions used on pictograph samples are listed in Table I.

Whenever possible, we now routinely study a matrix background of unpainted stone removed from the wall near the pictograph. This is to evaluate the extent of contamination from organic C not derived from pictograph pigment and to permit correction for that component. Both sampling and treatment are as nearly identical as possible. Calculations allowing for the effect of contaminant carbon on a pictograph ^{14}C result were discussed in an earlier paper (*15*).

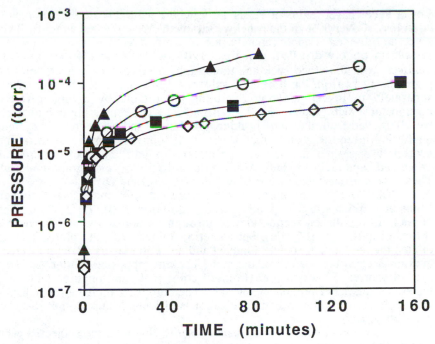

Figure 3. Vacuum integrity checks performed prior to oxidation of 41VV75-37A, B, C and D. These curve fits were drawn using third order polynomials. Symbols: ◇ 41VV75-37A; ○ 41VV75-37B; ▲ 41VV75-37C; ■ 41VV75-37D.

Table I. Plasma Types and Usual Conditions

Plasma Purpose	Gas	Pressure (torr)	Typical Power (watts)
Initial empty chamber cleaning	O_2	1.0	100
Subsequent empty chamber cleanings	O_2	0.5	150
Sample and chamber cleanings	Ar	0.4	25-40
Sample oxidation #1	O_2	1.0	100

Results and Discussion

Replicate Analyses of a Pecos River Style Pictograph, 41VV75-37. The pictograph we selected for this study was removed from site 41VV75 in the

Lower Pecos River region of Texas, a site where natural spallation has resulted in loss of over 50% of the panel we sampled. We collected an ~5 by 7 cm
piece and divided it approximately into quarters. After ultrasonicating the
sample in doubly distilled, deionized H_2O, we scraped off two opposite
quarters from the stone backing, attempting to remove only accretion and
pigment layers from the limestone substrate. We processed these quarters
separately to investigate reproducibility obtainable by dating different
portions of a pictograph (41VV75-37A and B). The remaining two quarters
were removed from the stone backing, combined, and homogenized by four
quartering it ten times on a small sheet of Al foil. This mixed pigment and
accretion was then divided approximately in half, each of which was
processed separately to study AMS reproducibility on pictograph material.
These two portions were identified as 41VV75-37C and D.

The color of the pigment with accretion mixture A was between
Munsell pinkish gray to pinkish white, notation 5YR 7.5/2 (23). The
pigment in a polished section viewed through a microscope was far redder
(Munsell dark red, 10R 3/6), but accretion mixed with the pigment layer
diluted the red hematite color. Pigment and accretion layers were removed as
quarter B from the surface of the stone backing opposite quarter A. This
second quarter (B) was redder (Munsell color pink, and notation 5YR 7/3)
than other quarters from this pictograph. The color of sample C was indistinguishable from D due to homogenization and were Munsell pinkish gray,
notation 5YR 7/2.

After completing plasma cleaning of the chamber, we subjected each
sample to the plasma extraction procedure. Table II sets forth details of plasmas used to process these four samples.

[14]C results returned from Lawrence Livermore National Laboratory
Center for Accelerator Mass Spectrometry (CAMS), as well as calibrated age
ranges for this series are listed in Table III. Samples 41VV75-37C and D
demonstrate AMS reproducibility of two homogenized opposite quarters of
the same pictograph. The ages are statistically indistinguishable. The average age of pictograph quarters C and D homogenized pigment is 3225 years
BP, while the average age of the first two pictograph quarters is 3265 years
BP. Thus, the average age of portions A and B is statistically indistinguishable from the average age of homogenized pigment and accretion mixture C
and D as well.

As previously mentioned, we ran opposite quarters A and B of the
Pecos River style pictograph separately without homogenization; their AMS
ages were different by 630 years. The colors of the two pigment and accretion powders, A and B, were also different from each other and from the C-D
color. The pigment with accretion mixture 41VV75-37B was redder than
41VV75-37A, indicating that B has a larger fraction of pigment, and thus
implying a higher proportion of organic binder or vehicle, than A. Portions
C-D were intermediate in color between A and B. Since B, the pigment-
enriched sample generated the oldest ^{14}C age, it follows that organic carbon
in the pictograph is older than that of contamination carbon in the accretion.
Mixing the pigment layer with overlying accretion generates a composite
[14]C date consisting of organic C from both sources. Differences in ages
between three results (A, B and C-D) are presumably due to a variation of the
proportion of organic carbon from pictograph and accretion layers in the
samples.

Table II. Plasmas Performed for Chamber Cleaning and Sample Processing of 41VV75-37A-D

Sample/ Plasma Gas	Plasma Purpose	Power (watts)	Final Duration (min)	Final Carbon (μg)
41VV75-37A				
O_2	chamber cleaning	100-150	65	0.08
Ar	sample/chamber cleaning	30	240	1.2
O_2	sample oxidation	100	62	715
O_2	sample oxidation	150	62	330
41VV75-37B				
O_2	chamber cleaning	100-150	110	0.3
Ar	sample/chamber cleaning	30	60	5
O_2	sample oxidation	100	62	870
O_2	sample oxidation	150	62	650
41VV75-37C				
O_2	chamber cleaning	100-150	60	0.2
Ar	sample/chamber cleaning	30	135	6
O_2	sample oxidation	100	62	730
O_2	sample oxidation	150	62	330
41VV75-37D				
O_2	chamber cleaning	100-150	75	0.9
Ar	sample/chamber cleaning	30	60	1.2
O_2	sample oxidation	100	62	960
O_2	sample oxidation	150	60	420

By utilizing differences in ages, we can limit ranges of possible ages of both pictograph and accretion. These limits are based on quarters A and B, as they exhibited the largest differences in age; inclusion of C-D would not affect the arguments. Two basic assumptions are used in these idealized model calculations. First, ages of both accretion and pictograph are assumed to be constant over the area sampled. Presumably this is true for the organic material that had been added to the paint, but it is far less likely to be true for the accretion, which may well vary in age from paint to paint on the pictograph. Second, ^{14}C activities reported from the AMS are assumed to be a weighted average of ^{14}C activities of accretion and pictograph organic materials. This assumption is probably valid by definition, since we are

Table III. Results and calibration of ^{14}C ages on Pecos River style pictograph at 41VV75-37A-D

41VV75-37 Quarter	CAMS Sample I. D.	[14]C Years (BP)[a]	[14]C Years (BP)[b]
A	CAMS-14087	2950 ± 60	3240 - 2960
B	CAMS-14088	3580 ± 60	3960 - 3730
C	CAMS-14089	3240 ± 60	3550 - 3360
D	CAMS-14090	3210 ± 60	3470 - 3360

[a]From LLNL
[b]Calibrated results were calculated by the program of Stuiver and Reimer (24). These results are 1s age ranges, as calculated using method A for each of the four experiments.

combining all non-paint organic matter into accretion. With these assumptions, we can write for each sample:

$$a_m = Xa_p + (1 - X)a_a \tag{1}$$

where a_m is ^{14}C activity of a mixture of pictograph pigment layer and accretion, a_p is ^{14}C activity of the pictograph itself, a_a is ^{14}C activity of accretion, and X is the fraction organic C in the pigment layer from the pigment and accretion mixture. [14]C activities of the first two quarters are 0.6926 (2950 years BP) and 0.6404 (3580 years BP), so that

$$0.6926 = X_A a_p + (1 - X_A)a_a \tag{2}$$

$$0.6404 = X_B a_p + (1 - X_B)a_a \tag{3}$$

where X_A represents the fraction of organic C in pigment from the pigment with accretion mixture for 41VV75-37A, and X_B represents the same fraction for 41VV75-37B. We now have two equations with four unknowns (a_p, a_a, X_A and X_B), so a unique solution is not possible. Addition of the other two quarters does not help as two additional unknowns are added with each equation. We can, however, use equations 2 and 3 to limit possible extreme values of X_A and X_B, that in turn can be used to constrain possible ^{14}C ages of both pictograph and accretion. We use maximum and minimum possible values for a_p and a_a. As maximum and minimum values for activity are 1 and 0, it follows that if a_a is assumed to be modern (1), and a_p is assumed to be ancient (0), we can calculate maximum limiting values for both X_A and X_B using equations 2 and 3. Conversely, if a_p is assumed to be ancient (0), and a_a is assumed to be modern (1), another limiting set of X_A and X_B values are generated. These possible X values are shown in Table IV.

**Table IV. Limiting Values of X, Based on Maximum and Minimum
Values of Activity for Pictograph and Accretion**

a_p	a_a	X_A	X_B
1	0	0.6926[a]	0.6404
0	1	0.3074	0.3596

[a]Although this is the maximum value of X_A based solely on activities, the additional constraint that $X_B > X_A$ restricts X_A further to <0.6404.

Based on our observation that the color of the pigment and accretion mixture for quarter B is redder than the mixture for quarter A, for any given pair of X_A and X_B values selected, we assume that $X_B > X_A$. There are three extreme cases to which we can apply X values in Table IV. First is the use of the maximum allowable value for X_B and the minimum value of X_A. The second uses the largest allowable value of X_B, in combination with the largest possible value of X_A (remembering that X_B must be greater than a given X_A). The third case uses minimum allowable values for X_B and X_A, again with $X_B > X_A$. Of these, only the first yields useful information; cases 2 and 3 will be discussed briefly below.

Thus, using case 1 values for X_A and X_B which differ to the greatest extent possible, e.g. 0.3074 for X_A and 0.6404 for X_B, we get simultaneous equations 4 and 5 below.

$$0.6926 = 0.3074a_p + (1 - 0.3074)a_a \qquad (4)$$

$$0.6404 = 0.6404a_p + (1 - 0.6404)a_a \qquad (5)$$

Solution of these equations yields activities of 0.7408 for a_a, and 0.5840 for a_p, which can then be converted to ages of 2410 years BP for the accretion and 4320 years BP for the pictograph. Thus the *maximum* age possible for the accretion is 2410 years BP while the *minimum* age for the pictograph is 4320 years BP. These ages should not be taken as meaningful limits. Rather they point out that the measured ages may be several hundreds of years from the "true", i.e., uncontaminated age. It is clear that a practicable means for dealing with the effect of contamination is needed.

Maximum X_A and minimum X_B are not selected, as that violates the condition that $X_B > X_A$. It is difficult to limit the *upper end* of the possible age range for the pictograph. Application of case 2, the largest allowable X_A and X_B combinations, and case 3, minimum allowable X_A and X_B values, do not lead to useful maximum and minimum limits on pictograph and accretion ages. Without knowing the accretion age, cases 2 and 3 yield possible pictograph ages of 10,000 years BP and above, which are excessively ancient. Further constraint of this range at the upper end is possible with additional assumptions. If accretion is assumed to deposit at a constant, uniform rate after the pictograph was painted, incorporating ambient carbon as it grows, then the age of the accretion would be one-half the time that has elapsed since the pictograph was painted. This would then place the pictograph age at 4820 years BP. Thus, with this simple, idealized model, our estimate for the age of pictograph 41VV75-37, though not well constrained at the older end, lies between 4320 and 4820 years BP. However, as this age range is based on questionable assumptions, it is probably not meaningful.

Taken at face value, X-values we arrived at with the idealized model indicate that a significant fraction of the organic C present in the sample is due to contamination in this pictograph. Percent contamination is represented by (1-X) x 100, 69.3% organic C contamination of 41VV75-37A, and 36% organic C contamination of 41VV75-37B for the limiting case we discussed. Because of these results, we will in the future change our procedure to remove as much accretion as possible prior to removal of the pigment layer. Both physical and chemical methods for reducing contamination will be investigated. Smaller amounts of accretion would not present such a significant contamination problem, and could be adequately corrected for. At 10-20% levels of contamination, reasonable corrections can be made with confidence and tolerably small uncertainty (15).

Evidence derived from the four samples of 41VV75-37 indicates that background contamination may introduce substantial uncertainty to the dating of Pecos River style pictographs. An improvement is possible if values of X_A and X_B, or even the ratio, X_B/X_A, could be obtained. If X_A and X_B were known, equations 2 and 3 could be solved uniquely for a_p and a_a, thus determining the ages of both the pictograph and accretion directly. This, however, rests on the ability to determine X_A and X_B, or the organic content of the pictograph layer/accretion ratio for each age determination. As we can think of no way to make those measurements, it does not appear to us to be a practicable solution to the contamination problem at this time.

Conclusions

We have found that the variability of the background contamination in a single Pecos River style pictograph sample on a few cm scale is appreciable, corresponding to a radiocarbon age difference of 630 years for adjacent samples. Even with utmost care in selecting unpainted background rock to correspond closely to the pictograph samples taken, uncertainties of several centuries may be unavoidable when the magnitude of the background is over 10-20% using the correction methods suggested earlier (15). The problem could be solved if (1) a means for determining the fraction (X_A and X_B) of contaminant-organic-carbon to pictograph-organic-carbon could be found and (2) by measuring two samples of the same pictograph and solving equations (2) and (3) simultaneously, which seems unlikely. Because we are unable to measure X_A and X_B directly, we will examine both physical and chemical means of removing the contamination from the sample as an alternative approach. This study underscores the need for extreme care in sample and background collection and their subsequent analyses, as well as a suitable means of dealing with contamination. Based on the findings from these quartered samples, previously reported ages for Pecos River style pictographs from our group (summarized in 14) may be better thought of as minimum ages; it is likely that contamination is younger than the paint.

We remain cautiously optimistic that pictographs painted with inorganic pigments can be dated directly using the plasma-chemical treatment followed by AMS radiocarbon determinations. Some situations, as the one described here for Pecos River style pictographs, 41VV75-37A-D, require more extreme measures than have been routine in our laboratory before adequately accurate and reliable ages can be obtained.

Acknowledgments

We are grateful to the following for partial support: Research Corporation Grant R-157; Donors of Petroleum Research Fund Grant ACS-PRF 20252-AC8 administered by the American Chemical Society; and Robert A. Welch Foundation Grant A-1235. Lawrence Livermore National Laboratory is funded by the U. S. Department of Energy under contract W-7405-Eng-48. Ruth Ann Armitage permitted us to include some of her data in Figure 1. Beth Shapiro assisted with the calculations associated with sample 41VV75-37. We are grateful to David Ing and Emmet Brotherton, Texas Parks and Wildlife Department, for permission to sample the pictographs at 41VV75.

Literature Cited

1. Whitley, D. S. and Loendorf, L. L. In *New Light on Old Art: Recent Advances in Hunter-Gatherer Rock Art*, Whitley, D. S. and Loendorf, L. L., Eds.; Monogram 36; UCLA Institute of Archaeology; University of California Press: Los Angeles, 1994; pp xi-xx.
2. Bennet, C. L.; Beukens, R. P.; Clover, M. R.; Gove, H. E.; Lievert, R. B.; Litherland, A. E.; Purser, K. H.; Sondheim, W. E. *Science* 1977, *198*, 508-509.
3. Nelson, D. E.; Kortling, R. G.; Stott, W. R. *Science* 1977, *198*, 507-508.
4. Muller, R. A.. *Science* 1977, *196*, 489-494.
5. Van der Merwe, N. J.; Sealy, J.; Yates, R. *South African Journal of Science* 1987, *83*, 56-57.
6. McDonald, J.; Officer, K.; Donahue, D.; Head, J.; Ford, B. *Rock Art Research* 1990, *7*, 83-92.
7. Valladas, H.; Cachier, H.; Arnold, M. *Rock Art Research* 1990, *7*, 18-19.
8. Farrell, M.; Burton, J. *North American Archaeologist* 1992, *13*, 219-247.
9. Geib, P.; Fairley, H. *Journal of Field Archaeology* 1992, *19*, 155-168.
10. Valladas, H.; et al. *Nature* 1992, *357*, 68-70.
11. Clottes, J.; Courtin, J.;Valladas, H. *Rock Art Research* 1992, *9*, 122-123.
12. David, B. *Rock Art Research* 1992, *9*, 139-141.
13. Chaffee, S.; Hyman, M.; Rowe, M. W.; Coulam, N.; Schroedl, A.; Hogue, K. *American Antiquity* 1994, *59*, 769-781.
14. Russ, J.; Hyman, M.; Shafer, H. J.; Rowe, M. W. *Nature* 1990, *348*, 710-711.
15. Ilger, W.; Hyman, M.; Rowe, M. W. *Radiocarbon* 1995, in press.
16. Chaffee, S. D.; Loendorf, L. L.; Hyman, M.; Rowe, M. W. *Plains Anthropologist* 1994, *39*, 195-201.
17. Russ, J.; Hyman, M.; Shafer, H. J.; Rowe, M. W. *Plasma Chemistry & Plasma Processing* 1991, *11*, 515-527.
18. Russ, J.; Hyman, M.; Rowe, M. W. *Radiocarbon* 1992, *34*, 867-872.
19. Chaffee, S.; Hyman, M.; Rowe, M. W. In *Time and Space: Dating and Spatial Considerations in Rock Art Research*, Steinbring, J.; Watchman, A.; Faulstich, P.; Taçon, P., Eds.; Occasional Publication No. 8, Australian Rock Art Research Association: Caulfield, Australia, 1993, pp 67-73.

20. Chaffee, S.; Hyman, M.; Rowe, M. W. In *New Light on Old Art: Recent Advances in Hunter-Gather Rock Art Research*; Whitley, D. S. and Loendorf, L. L., Eds.; Monograph 36; UCLA Institute of Archaeology; University of California Press: Los Angeles, 1994; pp 9-12.   ı
21. Reese, R. L.; Derr, J.; Hyman, M.; Rowe, M. W.; Davis, S. *J. Archaeological Science*, **1995**, in press.
22. Reese, R. L.; Mawk, E. J.; Derr, J.; Hyman, M.; Rowe, M. W.; Davis, S. This volume **1994**, Chap. Z, xxx-yyy.
23. *Munsell Soil Color Charts* **1975**, MacBeth Division of Kollmorgen Corporation, Baltimore, MD.
24. Stuiver, M.; P. J. Reimer *Radiocarbon* **1993**, *35*, 215-230.

RECEIVED August 28, 1995

Chapter 30

Radiocarbon Calibration and the Peopling of North America

C. M. Batt and A. M. Pollard

Department of Archaeological Sciences, University of Bradford, Bradford BD7 1DP, United Kingdom

The debate about the date of the first humans to enter North America has raged for over 100 years - in particular, the quest to find the earliest datable to Pre-Clovis site has brought up a number of contenders with sound radiocarbon stratigraphies. The chronological relationship between Clovis culture and the demise of the megafauna has also been vigorously debated. This paper will demonstrate that these questions may not be answerable by radiocarbon dating alone, in the light of the recently available programs extending calibration back to the Late Glacial. This could, of course, explain why the problem has never been satisfactorily resolved.

The study of the earliest settlers of the Americas, where they came from, when they arrived and how they spread through the continent has been among the most controversial areas of North American archaeology for much of the last century. Despite an increasing number of sites investigated and the use of many and diverse strands of evidence, the question of when the first people came to North America remains largely unanswered (1). Much of the debate has polarized into two positions, which agree that humans first entered North America from Siberia, but disagree on the subsequent events and timings. The supporters of the "short chronology" argue for entry into North America at the end of the last glaciation, between 14,000 and 12,000 y BP, followed by a rapid expansion into a vast, empty continent. Those advocating the "long chronology" argue that colonization took place before 15,000 y BP, possibly as early as 35,500 y BP, using as evidence a number of sites with radiocarbon dates which appear to indicate earlier human habitation. The short chronologists reject these apparently early sites on the grounds of poor dating or lack of conclusive evidence of human occupation linked to the dates. They interpret the apparent explosion of human activity starting between 11,500 y BP and 11,000 y BP, characterized by stone tools known as Clovis fluted projectile points, as evidence of

0097–6156/96/0625–0415$12.00/0

the spread of a vigorous population of efficient large mammal hunters, termed the "Clovis folk". The suggestion is that these people spread from North to South America in a few thousand years, leaving as evidence the characteristic stone tool assemblage of the Clovis horizon throughout American prehistory (2). This model is viewed favorably by those paleontologists seeking to explain the sudden demise of American megafauna (mammoths, camelids, sloths, etc.) at around the same time on the grounds of overkill by the Clovis folk. Those advocating the long chronology dispute this model of rapid human expansion and megafaunal extinction, emphasizing instead the handful of sites scattered throughout North and South America which yield evidence of pre-Clovis occupation, termed "the ones that will not go away" by one of the leading proponents (3).

Many different strands of evidence have been examined in an attempt to resolve this issue (1) including linguistics (4), physical anthropology (5), genetics (6-8), Pleistocene geology (9) and ecological evidence (10). However, despite the accumulation of such ancillary evidence, this is often circumstantial and the problem remains essentially an archaeological one, whose ultimate resolution depends upon archaeological evidence (3). In 1969, Haynes (11) suggested that the minimum requirements to establish the age of purportedly early sites are as follows:

"The primary requirement is a human skeleton, or an assemblage of artefacts that are clearly the work of man. Next this evidence must lie *in situ* within undisturbed geological deposits in order to clearly demonstrate the primary association of artefacts with stratigraphy. Lastly, the minimum age of the site must be demonstrable by primary association with fossils of known age or with material suitable for reliable isotopic age dating" (11: 714).

The sites proposed as pre-Clovis have been subject to intense scrutiny on all of these fronts; the objective of this paper is to examine one aspect of a very complex debate, namely whether radiocarbon dating can provide the requisite minimum age which is so crucial to the debate.

The General Use of Radiocarbon in the Debate

In her perceptive critique of the evidence for pre-Clovis occupations, Dincauze (12) draws attention to the tendency for archaeologists to depend uncritically on physicochemical dating techniques at the expense of more contextual data and interpretations. She particularly focuses on difficulties which arise in the use of new techniques, the archaeologists' frequent inability to evaluate particular results critically and on an incomplete communication between dating laboratories and the archaeologist. Many of the possible difficulties in the use of radiocarbon dates have been evaluated and discussed extensively, for example skepticism over the association of the dated material with the anthropogenic material (13) and the problems of contamination by older carbon (14). The advent of accelerator mass spectrometry (AMS) dating (15-17) has facilitated the dating of rare human remains directly, thus removing some of the uncertainties of contextual association of dating less valuable material such as charcoal, and has allowed improved chemical purification processes, thus reducing the chances of error due to contamination. However, two areas of difficulty within radiocarbon dating have received scant attention in relation to the dating of the first Americans, and yet may have serious implications for the fidelity of the dates upon which the debate is constructed. These two areas are the recent studies demonstrating the limitations of radiocarbon dating arising from diagenetic considerations and the necessity of calibrating radiocarbon ages in order to obtain calendar dates. Throughout this paper the following accepted convention is used: BP means "years before present" where "present" is defined as cal AD 1950; cal AD, cal BC and cal BP are used to refer to calibrated dates and "y BP" can be taken to be uncalibrated years before AD 1950.

Bone Chemistry Issues

The most direct way of settling the question of the antiquity of human arrival in the Americas is to date the human bones themselves. Although radiocarbon dating - the most obviously applicable method - has been available since the 1950s, the large sample requirement for the dating of bone (between 100 and 500 g) precluded its use until small sample techniques became widely available in the 1980s, chiefly as a result of the development of AMS for radiocarbon dating (*15*).

In the interim, a highly controversial series of dates were published by Bada and co-workers (*18*) using the measurement of the degree of racemization of one or more of the amino acids in bone collagen. This method put the arrival of native North Americans in California at somewhere around 70,000 y BP, based on the accumulated percentage of D-aspartic acid in the bone collagen. (Although organic chemistry uses the Cahn-Ingold-Prelog convention for the identification of enantiomorphs, the geochemical literature has retained the D- and L-notation, and for simplicity we repeat this here). This view was hotly contested, not least because the morphology of these supposedly mid-Pleistocene skeletons was apparently modern, unlike any skeletal material of a similar date from elsewhere in the world (*12*). The debate was settled by the subsequent re-measurement of the same amino acid extracts used in the racemization dating in Oxford using the AMS facility, confirming that the material was no older than Holocene, less than 11000 radiocarbon years BP (*19*). As such, they certainly could no longer be used to support the "long" chronology. The discrepancy, rather unfortunately, was traced back to a radiocarbon date of 17,150 y BP on a bone which was used to calibrate the amino acid racemization procedure. Unlike radiocarbon, which is based on the radioactive decay of a nucleus, racemization is a chemical process, and, as such, its rate is controlled by external factors such as temperature, pH and possibly the presence of ions capable of catalyzing the reaction. In order to overcome these uncertainties, it has become common to use an independently dated sample from a similar geochemical environment to calibrate the rate of racemization for that particular environment. Unfortunately, the sample chosen for this study was erroneously dated, and this has marred the reputation of amino acid racemization in archaeological chemistry ever since. Re-calculation of the racemization dates on the Californian paleoindian skeletons using a correctly dated calibration sample gives dates which are broadly in agreement with AMS and other dates (*20*), but the damage to the reputation of amino acid work had been done.

Although this controversy was seemingly settled by the publication of the definitive series of Holocene AMS radiocarbon dates, it has recently been resurrected as a result of developments in sample preparation chemistry of bone for radiocarbon dating. It has been known since the end of the 1980's that increasing purification of the bone sample, from acid insoluble collagen through to highly purified single amino acids (especially proline and hydroxyproline) can result in variations in radiocarbon date, usually, the more specific the chemical nature of the sample, the older age estimate produced (21). The explanation for this was simple, in that it was assumed, not unreasonably, that the purer the sample, the more reliable the radiocarbon date should be. The ultimate sample should of course be a pure single amino acid, especially proline or hydroxyproline, which are much more common in mammalian collagen than in other proteins, and hence should be less prone to contamination from other sources.

Study of the purity of bone samples for radiocarbon dating by Stafford and co-workers (17, 22, 23) has produced a rather unsettling conclusion. They have been working on rigorous purification procedures of mammalian collagen from a number of North American sites, and they have observed a sharp difference in the dating results obtained on samples which they have labeled as "non-collagenous" as opposed to "collagenous" samples. They have used the surviving weight percentage of nitrogen in a sample as a measure of the level of organic preservation in the bone. The pattern of dates obtained from increasingly pure samples from collagenous bones conforms to that expected, with increasingly old dates being obtained as the sample is purified, and a consistent date being given from a range of different amino acids in the sample. The pattern from non-collagenous samples is different, and much less coherent - the dates tend to get younger on purification, and the individual amino acids give less consistent dates. The definition of collagenous and non-collagenous is slightly disturbing, with (mammoth and human) bones having around 0.7-0.8% N qualifying as collagenous, whilst non-collagenous bone had less than 0.1% N. These figures should be compared with the figure of more than 4% N in fresh bone. Clearly even the collagenous bone has suffered a considerable degree of nitrogen loss.

The conclusion drawn from this work by Stafford is that non-collagenous bones cannot be relied upon to give a precise date, but that accurate radiocarbon dates can be obtained from bones if they are carefully selected and chemically characterized (17). Even if the Californian paleoindian skeletons had been mid-Pleistocene, their poor state of collagen preservation would have made it likely that they would have yielded a Holocene date (22). This somewhat alarming conclusion, although not jeopardizing the radiocarbon picture of the pre-Holocene colonization of North America completely, does counsel a great deal of caution when dealing with bone in such poor condition. One possible avenue for further research is to focus not on the collagen fraction of the organic component of bone, but on the other material present (at around 10% of the collagen level in fresh bone) - the so-called non-collagenous proteins. There is some evidence that these may survive better than collagen in some circumstances, and might therefore be useful for dating non-collagenous bone. Several groups around the world are assessing the relative survival rate of proteins such as osteocalcin and albumin compared to collagen in a range of environments with a view to providing more reliable radiocarbon dates on degraded bone (24). A helpful summary of the problems encountered in the radiocarbon dating of bone from early American material is provided by Taylor (*25*).

The Calibration Issues

One of the basic assumptions of radiocarbon dating when it first developed was that the atmospheric ratio of carbon-14 to carbon-12 has been constant. However, it has become apparent that this assumption is only approximately true. In the recent past the ratio has been affected by the combustion of fossil fuels (*26*) and nuclear weapons testing. Over a longer timescale it has been shown that concentration variations may also be caused by changes in the geomagnetic field, sunspot activity and by changes in the amount of carbon in the exchange reservoir, for example by melting of glaciers or release of carbon from the deep ocean (*27*). Hence the age in conventional radiocarbon years given by laboratory measurements is not the same as the age in calendar years and a calibration is necessary in order to convert from one to the other. Such calibration is carried out using curves constructed using radiocarbon determinations from samples whose age is known from other methods, most commonly wood samples dated by dendrochronology.

The publication in 1986 of internationally agreed calibration curves covering the period cal AD 1950 to 5210 cal BC (*28-30*) provided a means to correct for the known variations in natural radiocarbon flux. Since then the calibration has been extended back in time by the study of progressively older trees, for example the German oak and pine sequence which extends back to 9400 cal BC (*31*). Speaking in very general terms, prior to around 500 cal BC radiocarbon ages are consistently too recent whereas from 500 cal BC to cal AD 1300 radiocarbon ages tend to overestimate the calendar age (*32*). As the dendrochronological record cannot extend into the last glaciation, in order to determine the longer term trend, recent studies have compared radiocarbon determinations with uranium/thorium dates (*33, 34*). The most recent comparison of U-Th and radiocarbon ages using corals (*33, 35, 36*) has suggested that the period during which radiocarbon gives a substantial underestimate of calendar age obtained from U-Th studies extends back to at least 30,000 years ago and the discrepancy increases to as much as 3000-3500 years at around 15,000 radiocarbon years BP.

Unfortunately, calibration of radiocarbon dates is not a straightforward procedure leading to a simple pushing back of dates by a fixed amount; there are two other factors which need to be considered. Firstly, there is ambiguity in the interpretation of results from the calibration curve as it is not a simple linear function. Hence, the span in the calendar date corresponding to the error limit span of the radiocarbon age may be considerably greater and in some cases there may be several possible calendar date spans corresponding to a single radiocarbon determination. Secondly, the calibration curve itself will have an error limit band and so this will inevitably widen the calendar date span. This means that, in practice, apparently precise uncalibrated dates may result in calendar dates with much greater error ranges (*37*). It must be emphasized that calibration of radiocarbon dates is not merely an optional extra but is fundamental to any use or interpretation of the information. The error bounds associated with a radiocarbon age or a calendar date are an integral part of the information available; the range cannot be neglected at the expense of the central value. It is also accepted practice to use the 95.4% confidence levels, rather than the 68.3% level commonly used in this debate. At the lower confidence range there is nearly a one in three chance of the true result lying outside the range. In order to avoid confusion over which values are being used there is an accepted convention for stating dates (38), as outlined above, but this has not always been adhered to in this debate.

The Specific Sites and Dates in Question

To examine the implications of calibration on the debate surrounding the peopling of North America it is necessary to examine the dates in question from both the Clovis and pre-Clovis sites and the way in which they are used in the context of this debate. Adovasio (3) identifies four sites that he maintains meet the criteria outlined above for establishing the age of early sites. We are not qualified to, and do not attempt to, evaluate the archaeological evidence from the sites or the fidelity of the radiocarbon ages themselves, but rather to examine how the radiocarbon determinations are used.

The four sites advocated by Adovasio (3) are the Bluefish Cave Complex in the Yukon, the Nenana Valley Site Complex in Alaska, Monte Verde in south-central Chile and the Meadowcroft Rockshelter and related Cross Creek Sites in Pennsylvania. Of these sites, Monte Verde and Meadowcroft Rockshelter were situated to the south of the ice sheets that separated Alaska and northwest Canada from the rest of the Americas in the relevant period, and are therefore significant in the pre-Clovis - Clovis debate. These two sites apparently have considerable evidence for human occupation which has been interpreted as being before 11,500 y BP, the so-called "Clovis horizon", on the basis of the uncalibrated radiocarbon dates.

Monte Verde in Chile may represent the oldest evidence of humans in South America (3) and has been under investigation since 1976 by Dillehay and colleagues (39). There is a suggestion of an early phase of occupation, dated to the 34th millennium BP using carbonized wood, consisting of three possible cultural features and 26 stone items. However, uncertainty over the extent and character of this early occupation and its association with the carbonized wood dated (2) have led to some skepticism. The later evidence is interpreted as including wooden structures and hearths; with wood, bone and stone artifacts (40). This evidence is dated with over a dozen conventional and AMS radiocarbon dates to between 12,000 and 13,000 y BP (41). The calibration of radiocarbon dates from Monte Verde is currently in progress (Batt, C. M.; Pollard, A. M., in preparation).

Meadowcroft Rockshelter is probably the most appropriate of the purportedly early sites on which to examine the use of radiocarbon dating in the debate, meeting, as it does, the criteria given above for the acceptance of early sites, with a well-defined stratigraphy, artifacts of indisputable human manufacture and 52 radiocarbon dates, many with clear stratigraphic associations with artifacts and ecofacts (3, 42). The site itself is a deeply stratified, south-facing, sandstone rockshelter located 47 km southwest of Pittsburgh, Pennsylvania. The shelter is situated within an unglaciated portion of the Allegheny Plateau on the north bank of Cross Creek, a minor west-flowing tributary of the Ohio River. The site and its environs have been extensively studied by Adovasio and associates since 1973 (*43*) and the findings on, and interpretations of, the site have been the subject of controversy and debate from the first publication of the radiocarbon sequence (*44*).

The site comprises 11 attritionally and/or colluvially emplaced strata and appears to be the longest aboriginal occupational sequence available from the New World (*3*), with the upper strata spanning the entire Holocene and the lower strata extending back into the late Pleistocene. There have been bone, wood and fiber artifacts from the site, as well as a distinctive lithic assemblage. Although there are only about 700 lithic artifacts, they appear to be characteristic of secondary and tertiary core reduction, biface thinning and refurbishing of finished implements, and include a small lanceolate bifacial projectile point, called the Miller Lanceolate point (*3*). Such a lithic assemblage has been said to be not at variance with possible Siberian prototypes (*45*). The faunal remains are more difficult to characterize as, although 115,166 identifiable bones and bone fragments have been recovered from the rock shelter (*3*), only 11 of these come from the middle and lower strata and there is no evidence of predation of large game animals which are now extinct. The remains also include assemblages of snail and freshwater mussel shells, charcoal, nutshells, seeds and fruits of many plant species, and two fragments of human bone (*43*).

While the main strength of Meadowcroft is the number of strands of archaeological information that are available, the cornerstone of much of the discussion surrounding the site rests on the large number of radiocarbon dates and their purported consistency; "...to our knowledge, no other New World locality and few, if any, Old World localities have so many internally consistent radiocarbon assays." (*3*). Thirty nine dates cover 175 y BP to 12,800 y BP (*3*), but most of the discussion of the radiocarbon dates in the literature has concentrated on the reliability of the 13 older dates covering 12,800 y BP to 31,000 y BP, although the five oldest of these, between 21,000 y BP and 31,000 y BP, are not associated with cultural material. The discussion of these dates has centered on the possibility of particulate or non-particulate contamination of the earlier strata (*14, 46*) or post-depositional disturbance by animal or human activity (*13*). However, the fact that the radiocarbon determinations are in stratigraphic order mitigates against these two suggestions and Adovasio and colleagues have argued strongly that the radiocarbon dates are entirely consistent with what little is known of the flora and fauna at the time and that contamination or bioturbation is very unlikely (*47, 48*). Adovasio discounts particulate contamination on the basis that there is no obvious source for such contamination and it would have been detectable in the optical and scanning electron microscope examination of the specimens. Non-particulate contamination is discounted again on the grounds of there being no obvious sources and no mechanism for inclusion (*48*). The lack of contamination has been supported by similar AMS results being obtained from both the solid and the soluble fraction from a sample from one of the early, pre-cultural layers (*49*) and by amino acid racemization determinations (Hare, 1988 reported in *48*).

Table I. List of Radiocarbon Dates Associated with Clovis Sites (from *54*)

Agate Basin, pre-Folsom	charcoal	$11,840 \pm 130$[a]	I-10899
	charcoal	$11,700 \pm 95$	SI-3731
	charcoal	$11,450 \pm 110$	SI-3734
	average of 3[b]	$11,650 \pm 60$	
U.P. Mammoth site	tusk organics	$11,280 \pm 350$	I-449
Clovis type site	carbonized plants	$11,630 \pm 350$	A-491
	carbonized plants	$11,170 \pm 110$	A-481
	carbonized plants	$11,040 \pm 240$	A-490
	average of 3	$11,170 \pm 100$	
Domebo site	wood	$11,045 \pm 650$	SM-695
	bone organics	$11,220 \pm 500$	SI-172
	average of 2	$11,150 \pm 400$	
Dent site	bone organics	$11,200 \pm 500$	I-622
Murray Springs site	charcoal	$11,190 \pm 180$	SMU-18
	charcoal	$11,150 \pm 450$	A-805
	charcoal	$11,080 \pm 180$	Tx-1413
	charcoal	$10,930 \pm 170$	Tx-1462
	charcoal	$10,890 \pm 180$	SMU-27
	charcoal	$10,840 \pm 70$	SMU-41
	charcoal	$10,840 \pm 140$	SMU-42
	charcoal	$10,710 \pm 160$	Tx-1459
	average of 8	$10,900 \pm 50$	
Lehner site	charcoal (F_1)	$10,770 \pm 140$	SMU-168
	charcoal (F_1)	$10,940 \pm 100$	A-378
	charcoal (F_1)	$11,080 \pm 200$	SMU-181
	charcoal (F_2/D,Z)	$10,950 \pm 110$	SMU-194
	charcoal (F_2/D,Z)	$10,860 \pm 280$	SMU-164
	charcoal (F_2/D,Z)	$10,950 \pm 90$	SMU-290
	charcoal (F_2/D,Z)	$11,080 \pm 230$	SMU-196
	charcoal (F_2/F_0)	$10,620 \pm 300$	SMU-347
	charcoal (F_2/F_0)	$10,700 \pm 150$	SMU-297
	charcoal (F_2/F_0)	$10,710 \pm 90$	SMU-340
	charcoal (F_2/F_0)	$11,170 \pm 200$	SMU-264
	charcoal (F_2/F_0)	$11,470 \pm 110$	SMU-308
	average of 12	$10,930 \pm 40$	

SOURCE: Adapted from ref. 54
[a]Errors in radiocarbon ages are given as one standard deviation.
[b]Dates combined using procedure outlined in ref. 55

Calibration of Clovis and Pre-Clovis Dates

The concentration of attention on the fidelity of the radiocarbon dates themselves, and particularly the early dates, has perhaps drawn attention away from the difficulties associated with the way the radiocarbon dates have been used in more general terms. Much of the debate surrounding the date of the Clovis phenomenon in the Americas has been carried out without consideration being given to the effect of calibration on the radiocarbon age estimates; an exception is Whitley and Dorn (*50*) who have calibrated the dates and used 95% confidence limits, but in our view, have not fully explored the implications of calibration. Meltzer (*51*) has drawn attention to the possible effects of calibration, including the use of the coral calibration, but quantitative investigation has not been carried out. The general lack of calibration is understandable, since a calibration curve for the time period prior to 10,000 radiocarbon years BP has only been available since 1993 (*36*), based on the U/Th and ^{14}C dating of coral prior to 10,000 y BP (*33*). It has generally been assumed that the effect of calibration in this debate, if anything, will simply be to push the calibrated dates slightly further back in calendar years, without affecting any of the conclusions (*25*). This has not been the experience elsewhere, where it has been found that calibration - especially the need to use 95% confidence intervals rather than central estimates with a potentially ignorable error range - has seriously affected the type of inferences which may be made from radiocarbon evidence (*37*). We believe that the time has now come to review the radiocarbon evidence for pre-Clovis occupation in the Americas in the light of the preliminary calibration data now available through programs such as OxCal (*52*).

It is generally agreed by both the supporters of the "long chronology" and of the "short chronology" that the Clovis horizon is very sharply defined in radiocarbon years. To quote from Haynes, "there is now ample evidence that reliably well-dated Clovis sites fall into a very narrow period from 11,200 to 10,900 y BP" (*53*:446). This statement is based on 30 AMS radiocarbon dates from seven Clovis-age sites first published and analyzed by Haynes *et al.* (*54*), which therefore seems a reasonable starting point for our reconsideration of these conclusions. The dates are reproduced in Table I. There is no doubt that these dates cluster remarkably around the age range in radiocarbon years given by Haynes. Some of these sites have multiple dates (especially Lehner and Murray Springs), which have been combined using the procedure outlined by Long and Rippeteau (*55*) to give single estimates of the radiocarbon age of each site. Long and Rippeteau clearly set out some of the problems associated with averaging radiocarbon dates - there is a hierarchy of cases with increasing dangers which may result from combination. If two or more dates are derived from samples which are effectively drawn from the same parent sample (e.g., two charcoal samples from the same branch) then it is clearly appropriate to average the dates. Current practice would be to combine them before calibration. Less clear is the case of several samples from the same stratigraphic unit, in which case combination might be complicated by the duration of existence of the stratigraphic unit. Combining dates from several stratigraphic units to give the "date" of the site, or from several sites to give the date of a cultural horizon, clearly increases the risk of blurring the resulting date. Statistical tests are available to suggest whether dates are

Table II. Calibration of Clovis Age Sites (54) using OxCal.

Site	n	Average ^{14}C date[a] from ref. 54	Combined ^{14}C age[a] (OxCal R_COMB)	Calibrated 95% date range[b] (cal BP) (Figure 1)	Summed date range[b] (cal BP) (OxCal SUM) (Figure 2)
Agate Basin	3	11650±60	11651±62	13838-13383	14083-13201
UP Mammoth	1	(11280±350)		(14046-12456)	(14046-12456)
Clovis type site	3	11170±100	11183±96	13330-12878	14192-12526
Domebo	2	11150±400	11154±396	14081-12065	14394-11076
Dent	1	(11200±500)		(14428-11586)	(14428-11586)
Murray Springs	8	10900±50	10885±48	12944-12665	13542-12292
Lehner	12	10930±40	10947±37[c]	12988-12741	13612-12215
All Clovis sites	30		11067±25[c]	13090-12867	13990-12142

[a]Errors in radiocarbon ages are given as one standard deviation.
[b]Date ranges are given as two standard deviations
[c]signifies that χ^2 test fails
Figures in brackets indicate that a single radiocarbon date has been used.

being combined inappropriately (see below). Despite these problems, it seems not unreasonable, in view of the narrow range of dates, to combine the dates for the Clovis age sites, as has been done by Haynes and others.

Using the program OxCal (version 2.01; *52*) it is possible to calibrate these dates, and combine them in a number of ways. The results of these procedures are summarized in Table II. Using the R_COMB facility, which combines individual dates from the same radiocarbon event before calibrating, the resulting combined uncalibrated dates for each site are extremely close to the uncalibrated dates published by Haynes *et al.* (*54*), as would be expected (columns 3 and 4 in Table II). Subsequent calibration pushes these dates back in time, and the use of the 95% confidence intervals gives calendrical dates which fall between 12,665 and 13,838 cal BP for the sites with multiple dates (see Figure 1). Essentially the picture of remarkably tight dating for sites such as Lehner (12,988-12,741 cal BP) and Murray Springs (12,944-12,665 cal BP) is unchanged, although sites such as Dent (14,428-11,586 cal BP) and UP Mammoth site (14,046-12,456 cal BP) give a wider age estimate, since they are based on single radiocarbon age estimates. It should be noted, however, that this procedure for Lehner results in an error being flagged - the χ^2 test used to check the homogeneity of the dates being combined is failed at the 5% level, suggesting that the procedure is inappropriate. Cautiously, however, it is possible to go further, and combine all 30 radiocarbon estimates to give a calibrated age estimate for the Clovis event as evidenced by these data (again, the χ^2 test is failed during this process). The resulting age range is 13,090-12,867 cal BP - a spread of 223 calendar years, not dissimilar from the age range of 300 radiocarbon years accepted by Haynes and others as representative of Clovis.

It is doubtful if this process is valid. The fact that the χ^2 test is violated suggests that it is not, apart from philosophical considerations as to whether Clovis can be seen as a single radiocarbon event. An alternative procedure is offered by OxCal, called SUM, which is used to combine radiocarbon ages for multiple events. According to Bronk Ramsey (*52*), "Combining probability distributions by summing is usually difficult to justify statistically but it will generate a probability distribution which is a best estimate for the chronological distribution of the items dated. The effect of this form of combination is to average the distributions and not to decrease the error margins as with other forms of combination." Applying this procedure to the Clovis-age dates results in the calibrated age ranges given in column 6 of Table II, and shown in Figure 2. The effect, as expected, is to considerably widen the age ranges for all the sites with multiple dates. Lehner, for example, is now predicted to lie between 13,612 and 12,215 cal BP - a range of nearly 1400 calendar years, as opposed to 250 using R_COMB. If all 30 dates are now combined using SUM, the duration of the Clovis horizon now becomes 13,990-12,142 cal BP - a range of nearly 1850 calendar years.

To those with experience of using calibrated radiocarbon dates in periods closer to the present, the prediction of a range of 1850 calendar years for Clovis appears intuitively much more likely than one of 223 years based on the dates published by Haynes *et al.* (*54*), and is probably closer to the truth. It is clearly unsatisfactory that the choice of method can have such a major effect on the outcome, but experience suggests that the wider age range is more likely to be correct. It goes

Calibrated Clovis Dates Combined using R_COMB

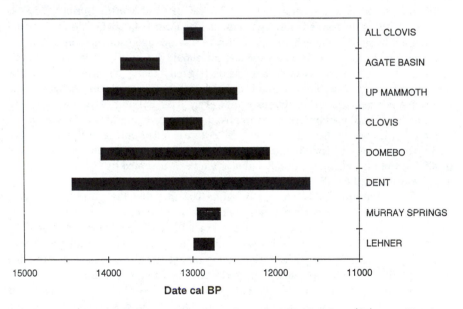

Figure 1. The published radiocarbon dates for Clovis sites (54), combined using the R_COMB facility and calibrated using OxCal. Ranges represent 2 standard deviations, that is 95.4% confidence.

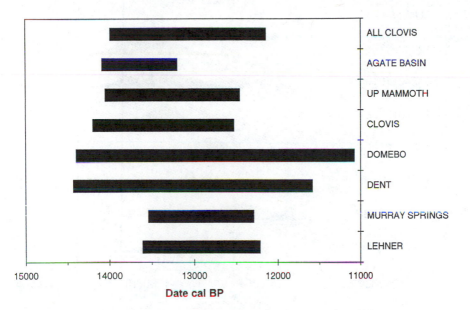

Figure 2. The published radiocarbon dates for Clovis sites (*54*), calibrated using OxCal and then combined using the SUM facility. Ranges represent 2 standard deviations, that is 95.4% confidence.

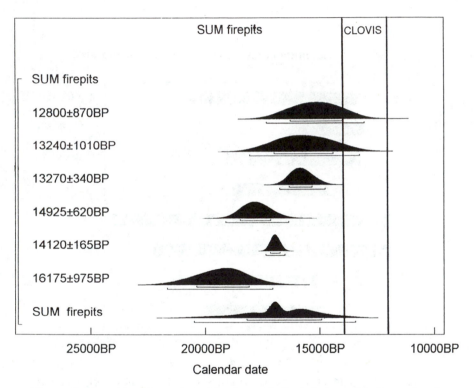

Figure 3. The oldest dates from Meadowcroft Rockshelter with firm cultural associations (57), calibrated using OxCal and combined using the SUM facility. Dates are presented in the order originally published, that is in order of radiocarbon date within the stratigraphic context. Ranges represent 2 standard deviations.

without saying that this is only an illustrative conclusion. Radiocarbon dates published for Clovis sites since 1984 (e.g. *56*), as well as revisions to the dates for the sites considered here in the light of previous discussions regarding sample preparation, need to be treated in the same way to give the best estimate of the timespan of Clovis. It is also true to say that the calibration curve prior to the end of the dendrochronological calibration may itself be subject to revision and, if it were ever possible to measure them, would certainly show the "wiggles" which have a further complicating effect on the calibrated dates. Nevertheless, this preliminary study suggests that, after calibration, Clovis should be regarded as lasting from approximately 14,000 to 12,150 cal years BP.

In order to see, at least on a preliminary basis, what effect (if any) this might have on the pre-Clovis controversy, we have taken the radiocarbon sequence for the Meadowcroft Rockshelter, and calibrated it using OxCal. The Meadowcroft radiocarbon chronology has been challenged and discussed from several angles (*48*, *57*), but has not yet to our knowledge been published and considered in a calibrated form. Unfortunately, the calibration programs presently available only extend as far back as 18,400 radiocarbon years BP, approximately 22,000 calendar years BP (*36*), but fortunately this allows us to calibrate the oldest dates at Meadowcroft with firm cultural associations (*57*). These consist of six dates associated with "stratum IIa (F-46 Lower)" described as "charcoal from fire pits, lowest one-third of stratum." The result of calibrating these dates and combining them using the SUM facility (see above) are shown in Figure 3, with the revised estimate for the Clovis episode superimposed. We see immediately that two of the six dates said to represent pre-Clovis occupation at Meadowcroft (SI-2489, 12800 ± 870 y BP and SI-2065, 13240 ± 1010 y BP) now overlap significantly with the dates for Clovis derived from other sites. The remaining four are still clearly pre-Clovis, but if all six dates are combined, a different picture emerges. Combination using the R_COMB procedure in OxCal gives a χ^2 test failure warning (suggesting that the dates do not represent the same radiocarbon event), but ignoring this results in a calibrated age range of 17,966 - 17269 cal BP - clearly before our revised estimate of the onset of Clovis at ca.14,000 cal BP. As discussed above, however, we believe there are serious grounds for doubting the validity of this procedure. Using the alternative SUM routine, we obtain a combined age range of 20,499 - 13,457 cal BP. This now brings the younger end of the range into overlap with the revised Clovis age band, and calls into question the assumption that these dates give clear evidence of pre-Clovis occupation at Meadowcroft. It is, of course, possible to take the alternative view, and say that these combined calibrated dates show that it is possible that humans were at Meadowcroft as early as 20,500 calendar years BP. All we are saying is that with our current state of knowledge the radiocarbon dates themselves cannot distinguish between these two hypotheses.

Clearly, this cannot be the last word on the pre-Clovis evidence from Meadowcroft. We have had to rely on the published table of dates (*57*), and we have accepted the evidence at face value. We have not attempted to comment on the two earlier dates with "possible cultural associations" (SI-2062, 19,100 ± 810; SI-2060, 19600 ± 2,400). These cannot be calibrated yet, but would almost certainly be well before Clovis, if the possible cultural associations were to be established. On the basis of the published radiocarbon evidence alone, however, we suggest that the best estimate for the context described as clearly pre-Clovis in radiocarbon years should be regarded as falling between 20,500 and 13,500 cal BP, with a consequent possible overlap with the earliest calibrated dates for Clovis from other published sites. In legal terms, this would amount to a judgment of "Case not Proven".

Summary

Dating evidence, particularly that provided by radiocarbon, has been regarded as one of the strongest lines of evidence for or against the existence of pre-Clovis colonization in America south of Alaska. This preliminary review has suggested that radiocarbon alone may not be capable of resolving the questions asked. Recent work has demonstrated that the most direct dating evidence - radiocarbon dating of the bones of themselves - may be misleading if the collagen fraction has suffered extensive degradation. Increased understanding of the degradation pathways, and improved purification procedures, may in time remove this problem. Alternatively, a shift of emphasis from the extraction and dating of collagen to other proteins in bone may offer a way forward.

Turning to the dates themselves (both on bone and other materials, principally charcoal), we find that the entire debate, with some justification, has been carried out to date largely without conversion of the radiocarbon timescale into calendar years. A preliminary (but as yet not internationally agreed) calibration for the relevant period back to approximately 22,000 cal BP is now available. Calibration using the program OxCal has suggested that the Clovis-age sites published (54) are best thought of as covering the range approximately 14,000 to 12,150 cal BP, rather than the much narrower range of 11,200 to 10,900 radiocarbon years BP generally used to date Clovis. Applying this to the calibrated radiocarbon dates from Meadowcroft Rockshelter reduces the number of individual dates with clear cultural associations which are pre-Clovis from six to four. If these six dates for the putative pre-Clovis context are combined in the most realistic manner, we suggest that this yields a calibrated range of 20,500 - 13,500 cal BP, which allows the interpretation that the context overlaps with other Clovis-age sites.

In the light of this, we suggest that the time is just about ripe (pending consolidation and international acceptance of the pre-Holocene calibration curve) for a radical overhaul of the evidence for the dating of the Clovis horizon using methods similar to those outlined here, and particularly in relation to the dating of those sites argued to be evidence for pre-Clovis occupation in North and South America. Another aspect in urgent need of review is the relationship between the arrival of Clovis and the last dates for the demise of the various species of megafauna. Powerful new techniques, based on Bayes theorem combined with calibration of radiocarbon dates, are now available to study time lapses between events in the archaeological record (58), and the application of these to good quality radiocarbon dates may go a considerable way towards resolving topics which have been debated in the literature for many years. It is possible, as may be the case at Meadowcroft, that we must accept that the radiocarbon evidence alone simply cannot resolve the issue.

Acknowledgments

Helpful discussion and assistance during the production of this paper was provided by Dr. Randolph Donahue, Department of Archaeological Sciences, University of Bradford; Janet Ambers of the Department of Scientific Research at the British Museum; Dr. Christopher Bronk Ramsey of the Research Laboratory for Archaeology and the History of Art, University of Oxford and Dr. Suzanne Young, Archaeometry Laboratory, Peabody Museum, Harvard. The views expressed herein are, however, the sole responsibility of the authors.

Literature Cited

1. Meltzer, D. J. *American Antiquity* **1989,** *54*, 471-490.
2. Fagan, B. M. *Ancient North America. The Archaeology of a Continent.* Thames and Hudson Ltd: London, 1991.
3. Adovasio, J. M. In *From Kostenki to Clovis: Upper Palaeolithic-Palaeo-Indian Adaptations;* Soffer, O.; Praslov, N. D., Eds.; Plenum Press, New York, 1993; pp 199-218.
4. Greenburg, J. H. *Language in the Americas*; Stanford University Press, Stanford, 1987.
5. Turner, C. G. *Acta Anthropogenetica* **1984,** *8*, 23-78.
6. Suarez, B. K.; Crouse, J.; O'Rourke, D. *American Journal of Physical Anthropology* **1985,** *67*, 217-232.
7. Szathmary, E. J. E. In *Out of Asia: Peopling the Americas and the Pacific;* Kirk, R.; Szathmary E. Eds.; The Journal of Pacific History, Canberra, 1985; pp 79-104.
8. Williams, R.; Steinberg, A.; Gershowitz, H.; Bennett, P.; Knowler, W.; Pettitt, D.; Butler, W.; Baird, R.; Dowdarea, L.; Burch, T.; Morse, H.; Smith C. *American Journal of Physical Anthropology* **1985,** *66*, 1-19.
9. Wendorf, F. *The American Naturalist* **1966,** *100*, 253-270.
10. Schweger, C.; Matthews, J.; Hopkins, D.; Young, S. In *Palaeoecology of Beringia*; Hopkins, D.; Matthews, J.; Schweger, C.; Young, S., Eds.; Academic Press, New York, 1982; pp 425-444.
11. Haynes, C. V. *Science* **1969,** *166*, 709-715.
12. Dincauze, D. F. *Advances in World Archaeology* **1984,** *3*, 275-323.
13. Kelly, R. L. *Quaternary Research* **1987,** *27*, 332-334.
14. Haynes, C. V. *American Antiquity* **1980,** *45*, 582-587.
15. Gowlett, J. A. J.; Hedges, R. E. M., Eds.; *Archaeological Results from Accelerator Dating*; Oxford University Committee for Archaeology Monograph 11: Oxford, 1986.
16. Taylor, R. E. In *Method and Theory for Investigating the Peopling of the Americas*; Bonnichsen, R.; Steele, D. G., Eds; Oregon State University, Oregon, 1994; pp 27-44.
17. Stafford, T. W. In *Method and Theory for Investigating the Peopling of the Americas*; Bonnichsen, R.; Steele, D. G., Eds; Oregon State University, Oregon, 1994; pp 45-55.

18. Bada, J. L.; Schroeder, R. A.; Carter, G. F. *Science* **1974,** *184,* 191-193.
19. Taylor, R. E.; Payen, L. A.; Prior, C. A.; Slota, P. J.; Gillespie, R.; Gowlett, J. A. J.; Hedges, R. E. M.; Jull, A. J. T.; Zabel, T. H.; Donahue, D. J.; Stafford, T. W.; Berger, R. *American Antiquity* **1985,** *50,* 136-140.
20. Bada, J. L. *American Antiquity* **1985,** *50,* 645-647.
21. Aitken, M. J. *Science-based Dating in Archaeology;* Longmans, London, 1990; p 88.
22. Stafford, T. W.; Hare, P. E.; Currie, L.; Jull, A. J. T.; Donahue, D. J. *Journal of Archaeological Science* **1991,** *18,* 35-72.
23. Stafford, T. W.; Hare, P. E.; Currie, L.; Jull, A. J. T.; Donahue, D. *Quaternary Research* **1990,** *34,* 111-120.
24. Ajie, H. O.; Kaplan, I. R.; Hauschka, P. V.; Kirner, D.; Slota, P. J.; Taylor, R. E. *Radiocarbon* **1992,** *34,* 296-305.
25. Taylor, R. E. In *The First Americans: Search and Research*; Dillehay, T. D.; Meltzer, D. J., Eds.; CRC Press, Boca Raton, Florida, 1991; pp 77-111.
26. Suess, H. E. *Science* **1955,** *122*, 415-417.
27. Stuiver, M.; Braziunas, T. F. *Radiocarbon* **1993,** *35*, 137-190.
28. Pearson, G. W.; Stuiver, M. *Radiocarbon* **1986,** *28*, 839-862.
29. Pearson, G. W.; Pilcher, J. R.; Baillie, M. G. L.; Corbett, D. M.; Qua, F. *Radiocarbon* **1986,** *30*, 911-934.
30. Stuiver, M.; Pearson, G. W. *Radiocarbon* **1986,** *28*, 805-838.
31. Kromer, B.; Becker, B. *Radiocarbon* **1993,** *35*, 125-136.
32. Pearson, G. W.; *Antiquity* **1987,** *61*, 98-103.
33. Bard, E.; Arnold, M.; Fairbanks, R.G.; Hamelin, B. *Radiocarbon* **1993,** *35,* 191-199.
34. Edwards, R. L.; Beck, J. W.; Burr, G. S.; Donahue, D. J.; Chappell, J. M. A.; Bloom, A. L.; Druffel, E. R. M.; Taylor, F. W. *Science* **1993,** *260*, 962-968.
35. Bard, E.; Hamelin, B.; Fairbanks, R. G.; Zindler, A. *Nature* **1990,** *345*, 405-410.
36. Stuiver, M.; Reimer, P. J. *Radiocarbon* **1993,** *35*, 215-230.
37. Bowman, S. *Antiquity* **1994,** *261,* 838-843.
38. Accepted at the 1985 International Radiocarbon Conference at Trondheim, unpublished.
39. Dillehay, T. D. *Monte Verde: A Late Pleistocene Settlement in Chile;* Smithsonian Institution Press Blue Ridge Summit, Pennsylvania, 1989; Volume 1.
40. Dillehay, T. D. *Scientific American* **1984,** *251*, 106-117.
41. Dillehay, T. D.; Pino, M.; Davis, E. M.; Valastro, S.; Varela, A. G.; Casamiquela, R. *J. Field Archaeology* **1982,** *9*, 547-550.
42. Morlan, R. E. In *Americans Before Columbus: Ice-Age Origins*; Carlisle, R. C., Ed.; Department of Archaeology, University of Pittsburgh, 1988; 31-43.
43. Adovasio, J. M.; Carlisle, R. C. *Scientific American* **1984,** *250*, 104-108.
44. Adovasio, J. M.; Gunn, J.; Donahue, J.; Stuckenrath, R. *Pennsylvania Archaeologist* **1975,** *45*, 1-30.
45. Yi, S.; Clark, G. *Current Anthropology* **1985,** *26*, 1-20.
46. Tankersley, K. B.; Munson, C. A. *American Antiquity* **1992,** *57*, 321-326.

47. Adovasio, J. M.; Gunn, J. D.; Donahue, J.; Stukenrath, R.; Guilday, J. E.; Volman, K. *American Antiquity* **1980,** *45*, 588-595.
48. Adovasio, J. M.; Donahue, J.; Stuckenrath, R. *American Antiquity* **1990,** *55,* 348-354.
49. Gillespie, R.; Gowlett, J. A. J.; Hall, E. T.; Hedges, R. E. M.; Perry, C. *Archaeometry* **1985,** *27*, 237-246.
50. Whitley, D. S.; Dorn, R. I. *American Antiquity* **1993,** *58*, 626-647.
51. Meltzer, D. J. *Annual Review of Anthropology,* in press.
52. Bronk Ramsey, C. *OxCal. Radiocarbon Calibration and Stratigraphic Analysis Program*; Research Laboratory for Archaeology, Oxford University, Oxford, 1994.
53. Haynes, C. V. *Quaternary Research* **1991,** *35,* 438-450.
54. Haynes, C. V.; Donahue, D. J.; Jull, A. J. T.; Zabel, T. H. *Archaeology of Eastern North America* **1984,** *12*, 184-191.
55. Long, A.; Rippeteau, B. *American Antiquity* **1974,** *39*, 205-215.
56. Haynes, C. V. In *Radiocarbon after Four Decades;* Taylor, R. E.; Long, A.; Kra, R. S., Eds.; Springer-Verlag, New York, 1992; pp 355-374.
57. Adovasio, J. M.; Boldurian, A. T.; Carlisle, R. C. In *America Before Columbus: Ice Age Origins*; Carlisle, R. C. Ed.; Ethnology Monographs 12, Department of Anthropology, University of Pittsburgh, Pittsburgh, Pennsylvania, 1988; pp 45-61.
58. Buck, C. E.; Kenworthy, J. B.; Litton, C. D.; Smith, A. F. M. *Antiquity* **1991,** *65,* 808-821.

RECEIVED December 23, 1995

Chapter 31

Accelerator Mass Spectrometry Radiocarbon Measurement of Submilligram Samples

D. L. Kirner[1], J. Southon[2], P. E. Hare[3], and R. E. Taylor[1,4]

[1]Radiocarbon Laboratory, Department of Anthropology, University of California, Riverside, CA 92521
[2]Center for Accelerator Mass Spectrometry, Lawrence Livermore National Laboratory, Livermore, CA 94551–9900
[3]Geophysical Laboratory, Carnegie Institution of Washington, 5251 Broad Branch Road, N.W., Washington, DC 20015
[4]Institute of Geophysics and Planetary Physics, University of California, Riverside, CA 92521

Previous studies have determined that the total system ^{14}C background values in catalytically- reduced graphitic carbon samples are inversely proportional to their weights. We further examine this relationship down to 40 micrograms using both assumed ^{14}C "dead" background sample and a contemporary standard, Australian National University (ANU) sucrose. Our observations are consistent with those previously reported with respect to the inverse relationship between sample weight and ^{14}C activity. These observations support the view that a constant addition of modern carbon contamination during the graphitization process explains the observed background ^{14}C activity in graphitic carbon samples.

Accelerator mass spectrometry (AMS) radiocarbon dating combines the technology of ^{14}C dating with mass spectrometry and particle acceleration to accomplish high energy mass spectrometry to measure ^{14}C on an ion-by-ion basis *(1)*. At the outset, researchers anticipated three major advantages to AMS technology for ^{14}C applications. The first was a reduction in required sample size from gram to milligram amounts of carbon. The second benefit was a significant reduction in counting time from weeks to virtually minutes. The third anticipated advantage was an expected extension in the ^{14}C timescale from approximately 40/50,000 to 100,000 years *(2-4)*. The anticipation that AMS could extend the temporal range of ^{14}C dating was based, in part, on the consideration that cosmic radiation, a major source of background in conventional decay counting, is essentially eliminated in AMS technology *(5)*.

The previous decade of research has made AMS measurements on one milligram samples of graphitic carbon a routine operation. Subsequent studies that

0097–6156/96/0625–0434$12.00/0

have focused on submilligram samples have reported several problems. One of these issues is a significant increase in blank levels with reductions in sample weights below about 200 to 300 micrograms of graphitic carbon. Vogel *et al. (6)* suggest that this phenomenon is the result of the introduction of a constant amount of modern carbon during the preparation of the catalytically-condensed graphitic carbon, a step that is required in most of the AMS systems currently in operation.

We will focus our discussion on experiments that examine background levels and measured ^{14}C values in a contemporary standard as a function of decreasing sample size below one milligram.

Background Levels for Submilligram Samples

Various approaches have been used by AMS researchers to reduce background levels *(5-10)*. Currently, in routine operations, standard background values on one milligram samples are reported as ranging from about 0.1 to 1.0% modern. In ^{14}C years, this translated to ages of approximately 35,000 to 55,000 before present (BP).

Contamination Sources. Potential sources of contamination resulting in these background levels have been reported by several laboratories. At present, most AMS systems convert samples to CO_2 by combustion for non-carbonate organics and acidification for carbonate samples. This is followed by the production of catalytically-condensed graphitic carbon *(5-10)*, which is introduced into the AMS spectrometer ion source for measurement. Five major sources of background contamination can be identified: *in situ* contamination, the pretreatment chemistry, combustion or acidification, graphitization, and the AMS spectrometer *(5, 11)*.

Sample pretreatment requires separation of the datable material from contamination. This separation is accomplished by physical and chemical means. Incomplete separation of *in situ* contamination of the sample is one source of background contamination. Contamination may also occur during the pretreatment process due to handling, or from ^{14}C contamination in chemical reagents or adsorbed on glassware. Similarly, ^{14}C contamination from reagents or reaction vessel walls may be introduced during the production of CO_2 by combustion or hydrolysis, or during the subsequent graphitization step *(6)*.

Instrument background occurs when ^{14}C is detected in a sample that should be free of [14]C. This may be the result of a detector anomaly in which a ^{14}C pulse registers in the detector circuitry when no ^{14}C is actually present. Additionally, it may result from ^{14}C contamination of the AMS spectrometer ion source or beam line. This occurs when ^{14}C that derives from a spectrometer component reaches the detector.

AMS laboratories reporting experiments designed to quantify their background levels note a hierarchy of values beginning with instrument background. Some facilities define this value exclusively as reflecting activity generated in the beam line with the ion source closed off from the remainder of the tandem system. For example, at the Nuclear Physics Laboratory, University of Washington, Seattle, this has been measured at ≥90,000 BP (0.001% modern), i. e., no counts detected in 30 minutes. An equivalent result was obtained with their ion source open to the beam transport system containing an empty aluminum target holder *(8)*. The University of Toronto group *(5)* reports an apparent age on an empty aluminum target holder in their system at 85,000 BP (0.0025% modern). In experiments carried on with a FN tandem using a modified GIC Model 846 ion source at the University of California Lawrence Livermore National Laboratory (LLNL) AMS Laboratory (CAMS), we determined that the instrument background with the ion source closed off was ≥104,000 BP, i. e., no counts detected in 20.5 minutes. With the ion source containing an empty aluminum target holder open to the beam transport system, a ^{14}C count rate equivalent to 74,000 BP (0.009% modern) was measured *(11)*.

More typically, an overall instrument background in a tandem system is measured using some type of geologic graphite. As far as we are aware, the oldest reported ^{14}C value on geologic graphite is 69,030±1700 BP for a sample powdered and encapsulated under argon. Geologic graphite exposed to ambient air exhibited somewhat higher values, approximately 58,000 to 65,000 BP *(8)*. Typical values reported over the last decade by other AMS laboratories for the ^{14}C age of geologic graphite range from 50,000 to 65,000 BP *(5, 7, 9)*. The average (N=2) ^{14}C value obtained at LLNL CAMS laboratory on geological graphite used by the University of California, Riverside (UCR) ^{14}C laboratory is 64,460±3200 BP. The average (N=7) ^{14}C value on graphite powder used by the LLNL CAMS laboratory is 57,900±1500 BP.

Background Mass Dependence. In contrast, graphitic carbon that is produced from combusted CO_2 consistently registers higher background values, which increase in relationship to decreasing sample weights in the microgram range

During the last decade several laboratories have published studies of their efforts to reduce background levels during their production of graphitic carbon. To date, the most complete published study of the relationship between sample weight and background values was undertaken by Vogel *et al.* *(6)*. Using anthracite coal, they demonstrated that for samples below 500 micrograms the ^{14}C background values increased as a function of decreasing sample weight. The best fit of the ^{14}C activity to the sample weight relationship was interpreted to indicate that a constant amount of contamination, equivalent to 2.2±1.1μg of modern carbon, was added to each sample during the combustion or graphitization step. This suggests that as the sample decreased in size, the constant amount of contamination being added resulted in a progressive net decrease in the ^{14}C age of the sample.

The sources of contamination that were evaluated by Vogel *et al.* *(6)* for the combustion step were adsorbed CO_2 or CO on the walls of the Vycor^{TM} tubing and residual traces of carbon in the CuO used as the oxygen source. In the graphitization step they examined the possibility of memory effects in the vacuum system and traces of carbon in the Fe catalyst. They also considered the adsorption of CO_2 by the graphitic carbon and small amounts of contamination that may have been picked up during storage and handling. Their conclusions were that approximately 60%-70% of the contamination occurred as a result of the release of adsorbed CO_2 from the Vycor^{TM} combustion tube at the high temperatures used during combustion.

We have examined the relationship between sample size and background activity in submilligram samples prepared from wood recovered from reported Pliocene sediments *(11)*. The pretreatment regimen consisted of an acid wash with 2 M HCl followed by a wash in 1 M NaOH, distilled H_2O and a final treatment of 2 M HCl. The very low background values achieved on one milligram samples suggested that any measurable ^{14}C contamination had been removed. Graphitization was carried out at the UCR Radiocarbon Laboratory using methods based on those described by Vogel *et al. (6)*. The mean ^{14}C age of the 1000 microgram samples is 52,140±439 BP (N=19). The lowest ^{14}C value obtained on a 1000 microgram sample is 56,150±540 BP (0.09% modern). Figure 1 illustrates the relationship between the graphitized carbon weight and the ^{14}C activity (% modern) from 10 to 1000 micrograms. The mean for the two smallest samples at 10 micrograms is 20,370±1410 BP (0.9% modern). Accepting the Vogel *et al.* view *(6)* that the most probable interpretation of the data is a constant addition of modern carbon contamination, Figure 2 represents the best fit of the data characterizing the constant addition of the equivalent of 1.0± 0.4 micrograms of modern carbon.

Contemporary Standards for Submilligram Samples

Most AMS researchers who are pursuing microgram radiocarbon capability concentrate on the preparation and analyses of background material. This allows the researcher to evaluate the sources and amounts of modern carbon contamination in the various preparative steps. Only a small number of studies have examined the relationship using standards such as Oxalic acid or ANU sucrose. Such data is necessary to be able to infer meaningful ^{14}C age estimates for submilligram samples.

A recent study conducted by NIST in cooperation with the National Ocean Science AMS facility utilized materials of known ^{14}C abundance that spanned the entire range of concentrations from *supra* modern to effectively zero blanks *(12)*. Sample materials included Oxalic acid I and II, Carrara marble and graphite. The researchers reported that in modern carbon targets below 500 micrograms there was a significant mass dependence. In addition, they noted a significant processing blank variability that they suggest is the ultimate limiting factor in submilligram sample measurements.

Using ANU sucrose, we have examined the relationship between sample weight and ^{14}C activity for samples from 40 to 1000 micrograms. The accepted values for ANU sucrose is 150.85±0.8% modern *(14)*. Figure 2 is a plot of the relationship between the graphitized carbon weight and the ^{14}C activity for the ANU sucrose measurements. In each case, the ^{14}C values have been corrected for background using an algorithm proposed by Donahue *et al. (15)*. Above 100 micrograms, the ^{14}C value is within 1.5% of the expected activity. Below 100 micrograms, there is a reduction in the ^{14}C values. In order to account for this reduction in ^{14}C activity, several explanations were considered. The line in Figure

Background Activity vs. Sample Size

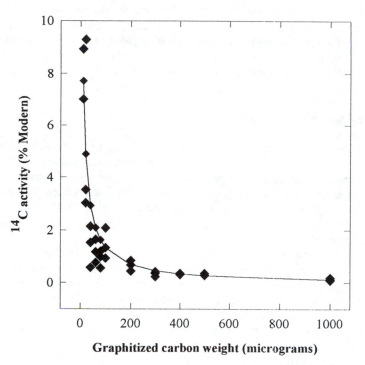

Graphitized carbon weight (micrograms)

Figure 1. Relationship of sample weight (micrograms) and ^{14}C activity (% modern) of presumed ^{14}C-free wood samples: 10 to 1000 micrograms.

Background Activity vs. Sample Size

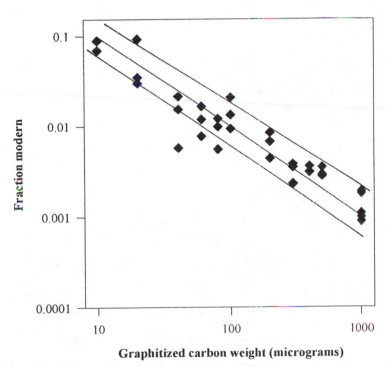

Figure 2. Relationship of sample weight (micrograms) and ^{14}C activity (fraction modern) of presumed ^{14}C-free wood samples: 10 to 1000 micrograms on a common logarithmic scale. The lines represent the fit of the data and the addition of 1.0±0.4 micrograms of modern carbon.

ANU Activity vs. Sample Size

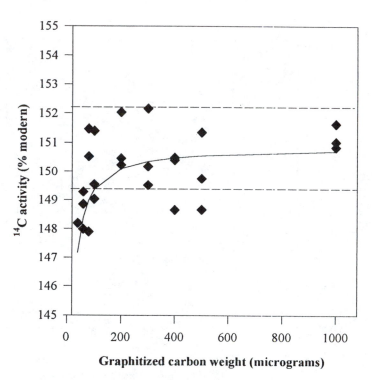

Figure 3. Relationship of sample weight and ^{14}C activity of ANU sucrose: 40 to 1000 micrograms. The solid line represents a hypothetical relationship between sample weight and ^{14}C activity of ANU sucrose with the addition of one microgram of "dead" carbon. The dashed line represents ±1% of 150.8% modern.. The ^{14}C activity was corrected using mass-matched backgrounds produced with a cobalt catalyst.

3 represents one scenario - the hypothetical relationship between sample weight and ^{14}C activity that would result from the constant addition of one microgram of dead (^{14}C-free) carbon contamination. The dead carbon contamination would cause a reduction in the ^{14}C values for ANU sucrose at the submilligram level while having little effect on the background measurements. Potential sources for this contamination may be dead carbon from pump oil in the backing pump in the graphitization apparatus that has been ^{14}C dated previously, and appears to be ^{14}C-free (Prior, C., University of California, Riverside, personal communication, 1995.) or from pump oil or vacuum grease in the AMS system.

We emphasize that contamination by dead carbon is not the only possible explanation for these results. An alternative explanation is size-dependent isotopic fractionation. Fractionation refers to the selective enrichment or depletion of one isotope at the expense of another. The graphitization reaction may fractionate strongly if it does not go to completion *(16)* and we suspect that this may be a particular problem for submilligram samples. In addition, changes in the carbon-to-catalyst ratios have been shown to cause fractionation *(16, 17)*. Samples with different carbon-to-catalyst ratios may have different thermal properties and thus may fractionate differentially while being sputtered in the ion source. Samples with lower carbon to catalyst ratios generally produce lower beam currents. There may be beam-dependent isotope fractionation in the transmission of particle beams through the spectrometer. Further investigation is needed in this area.

Conclusions

This investigation was undertaken to continue the examination of the relationship between ^{14}C activity and sample weight for catalytically-reduced graphitic carbon in the submilligram range. Our data confirm the previous conclusions of several investigators that in general samples >500 micrograms show little or no measurable mass dependency. However, below 500 micrograms both the background and ANU values reflect the increasing effects of microcontamination or fractionation as a function of sample size. In the case of background contamination, as was previously noted, the interpretation of the data suggests a constant addition of modern carbon on the order of 1.0 ± 0.4 micrograms. The ANU sucrose data can be interpreted in terms of this same background, plus dead carbon contamination or fractionation as previously discussed.

Finally, the overall outlook for those pursuing the microgram radiocarbon capability is encouraging. The problem of mass dependence can be compensated for by using mass matching techniques as suggested by Klinedinst *et al. (12)*, where submilligram samples are measured against similarly sized standards. However, the key to success is to be able to produce accurate, precise estimates of the size dependent backgrounds for submilligram samples. This is the primary limiting factor for successful submilligram sample measurement.

Acknowledgments

The studies reported in this paper were supported, in large part, by a grant from the University of California/Lawrence Livermore National Laboratory (UC/LLNL) with additional support by the Gabrielle O. Vierra Memorial Fund, the Dean of the College of Humanities and Social Science and the Intramural Research Fund, University of California, Riverside. Comments of Richard Burky, Lijun Wan, and Christine Prior on an earlier draft of this paper are very much appreciated. This is contribution 95/05 of the Institute of Geophysics and Planetary Physics, University of California, Riverside.

Literature Cited

1. Gove, H. E. In *Radiocarbon After Four Decades: An Interdisciplinary Perspective;* Taylor, R. E., Long, A., Kra, R. S., Eds.; Springer-Verlag: New York, New York, 1992; pp 214-229.
2. Muller, R. A. *Science* **1977**, *32*, 489-494.
3. Bennett, C. L.; Beukens, R. P.; Clover, M. R.; Gove, H. E.; Liebert, R. B.; Litherland, A. E.; Purser, K. K.; Sondheim, W. E. *Science* **1977**, *108*, 508-509.
4. Nelson, D. E.; Kortelling, R. G.; Scott, W.R. *Science* **1977**, *198*, 507-508.
5. Beukens, R. P. In *Radiocarbon After Four Decades: An Interdisciplinary Perspective*; Taylor, R. E., Long, A., Kra, R. S., Eds.; Springer-Verlag: New York, New York; pp 230-239.
6. Vogel, J. S.; Nelson, D. F.; Southon, J. R. *Radiocarbon* **1987**, *29*, 323-333.
7. Arnold, M.; Bard, E.; Maurice, P.; Duplessy, J. C. *Nuclear Instruments and Methods in Physics Research* **1987**, *B29*, 120-123.
8. Schmidt, F. H.; Balsey, D. R.; Leach, D. D. *Nuclear Instruments and Methods in Physics Research* **1984**, *B29*, 97-99.
9. Gillespie, R.; Hedges, R. E. M. *Nuclear Instruments and Methods in Physics Research* **1984**, *B5*, 294-296.
10. Gurfinkel, D. M. *Radiocarbon* **1987**, *29*, 335-346.
11. Kirner, D.; Taylor, R. E.; Southon, J. R. *Radiocarbon*, in press.
12. Klinedinst, D. B.; McNichol, A. P.; Currie, L. A.; Schneider, R. J.; Klouda, G. A.; von Reden, K. F.; Verkouteren, R. M.; Jones, G. A. *Nuclear Instruments and Methods in Physics Research* **1994**, *B92*, 166-171.
13. Vogel, J. S.; Nelson, D. E.; Southon, J. R. *Radiocarbon* **1989**, *31*, 145-149.
14. Currie, L. A.; Polach, H. *Radiocarbon* **1980**, *22*, 933-935.
15. Donahue, D. J.; Linick, T. W.; Jull, A. J. T. *Radiocarbon* **1990**, *32*, 135-142.
16. Vogel, J. S.; Southon, J. R.; Nelson, D. E. *Nuclear Instruments and Methods in Physics Research* **1984**, *B5*, 289-293.
17. Arnold, M.; Bard, E.; Maurice, P.; Valladas, H.; Duplessy, J. C. *Radiocarbon* **1989**, *31*, 284-291.

RECEIVED October 9, 1995

INDEXES

Author Index

Affiliation Index

Subject Index

A

Bestsellers from ACS Books

The ACS Style Guide: A Manual for Authors and Editors
Edited by Janet S. Dodd
264 pp; clothbound ISBN 0–8412–0917–0; paperback ISBN 0–8412–0943–X

Understanding Chemical Patents: A Guide for the Inventor
By John T. Maynard and Howard M. Peters
184 pp; clothbound ISBN 0–8412–1997–4; paperback ISBN 0–8412–1998–2

Chemical Activities (student and teacher editions)
By Christie L. Borgford and Lee R. Summerlin
330 pp; spiralbound ISBN 0–8412–1417–4; teacher ed. ISBN 0–8412–1416–6

Chemical Demonstrations: A Sourcebook for Teachers,
Volumes 1 and 2, Second Edition
Volume 1 by Lee R. Summerlin and James L. Ealy, Jr.;
Vol. 1, 198 pp; spiralbound ISBN 0–8412–1481–6;
Volume 2 by Lee R. Summerlin, Christie L. Borgford, and Julie B. Ealy
Vol. 2, 234 pp; spiralbound ISBN 0–8412–1535–9

Chemistry and Crime: From Sherlock Holmes to Today's Courtroom
Edited by Samuel M. Gerber
135 pp; clothbound ISBN 0–8412–0784–4; paperback ISBN 0–8412–0785–2

Writing the Laboratory Notebook
By Howard M. Kanare
145 pp; clothbound ISBN 0–8412–0906–5; paperback ISBN 0–8412–0933–2

Developing a Chemical Hygiene Plan
By Jay A. Young, Warren K. Kingsley, and George H. Wahl, Jr.
paperback ISBN 0–8412–1876–5

Introduction to Microwave Sample Preparation: Theory and Practice
Edited by H. M. Kingston and Lois B. Jassie
263 pp; clothbound ISBN 0–8412–1450–6

Principles of Environmental Sampling
Edited by Lawrence H. Keith
ACS Professional Reference Book; 458 pp;
clothbound ISBN 0–8412–1173–6; paperback ISBN 0–8412–1437–9

Biotechnology and Materials Science: Chemistry for the Future
Edited by Mary L. Good (Jacqueline K. Barton, Associate Editor)
135 pp; clothbound ISBN 0–8412–1472–7; paperback ISBN 0–8412–1473–5

For further information and a free catalog of ACS books, contact:
American Chemical Society
Customer Service & Sales
1155 16th Street, NW, Washington, DC 20036
Telephone 800–227–5558